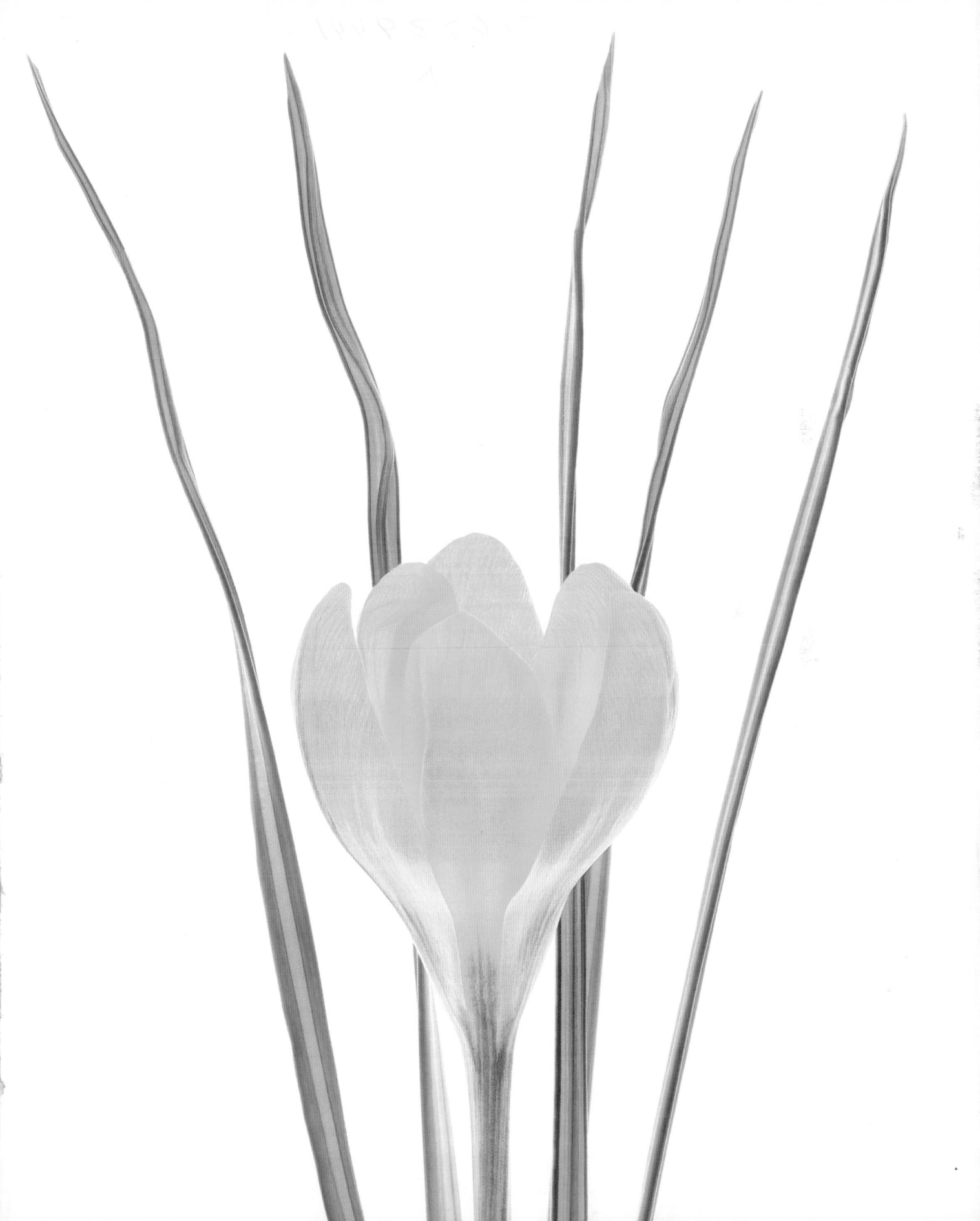

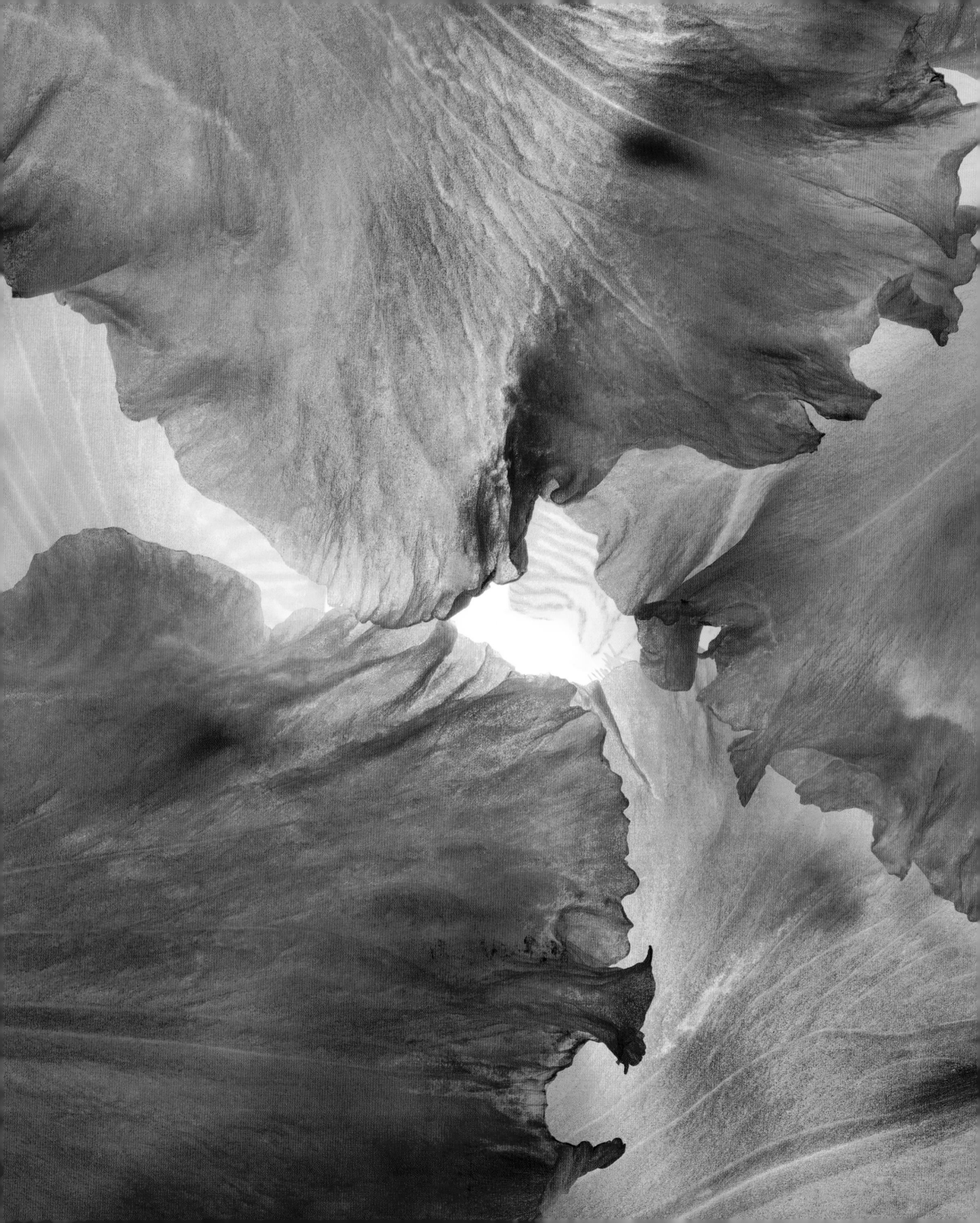

SEEING
FLOWERS

Discover the Hidden Life of Flowers

PHOTOGRAPHY BY **ROBERT LLEWELLYN**
TEXT BY **TERI DUNN CHACE**

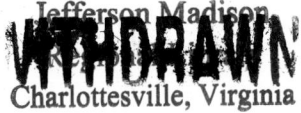
Timber Press
Portland | London

306239441
R

OPPOSITE A flamboyant poppy flower (Papaver orientale) seduces pollinators with bright colors and contrasting dark marks that direct attention toward the bountiful pollen in the center.

HALF TITLE This yellow crocus is a member of the iris family.

FRONTISPIECE Sultry, orange bearded iris—a beloved plant that comes in many, many hues.

Text copyright © 2013 by Teri Dunn Chace. Photographs copyright © 2013 by Robert Llewellyn.
All rights reserved.

Published in 2013 by Timber Press, Inc.

The Haseltine Building	2 The Quadrant
133 S.W. Second Avenue, Suite 450	135 Salusbury Road
Portland, Oregon 97204	London NW6 6RJ
timberpress.com	timberpress.co.uk

Printed in China
Designed by Breanna Goodrow

Library of Congress Cataloging-in-Publication Data

Dunn Chace, Teri.
  Seeing flowers: discover the hidden life of flowers/Teri Dunn Chace; photographs by Robert Llewellyn.–1st ed.
       p. cm.
  Includes bibliographical references and index.
  ISBN 978-1-60469-422-2
 1. Flowers–Identification. 2. Angiosperms–Identification. 3. Flowers–Pictorial works. 4. Angiosperms–Pictorial works. I. Llewellyn, Robert J. II. Title.
  SB407.D87 2013
  581--dc23
                                        2012050581

A catalog record for this book is also available from the British Library.

Dedicated to my sister in exploring and marveling at nature,
Nina Sandlin, and to the gardeners who nurtured
the flowers in this book–they did so with great love, respect,
and wonder for the life unfolding.

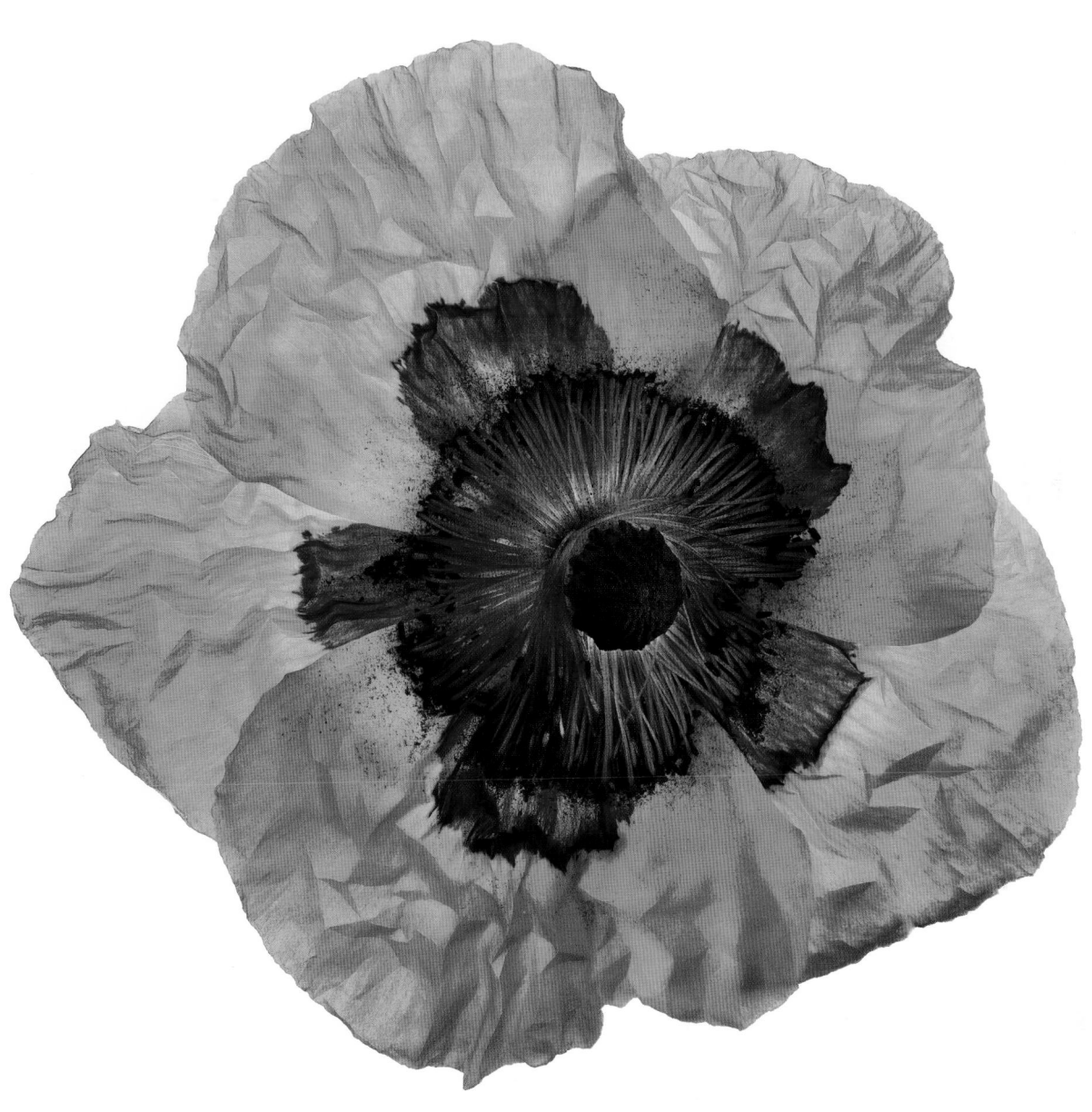

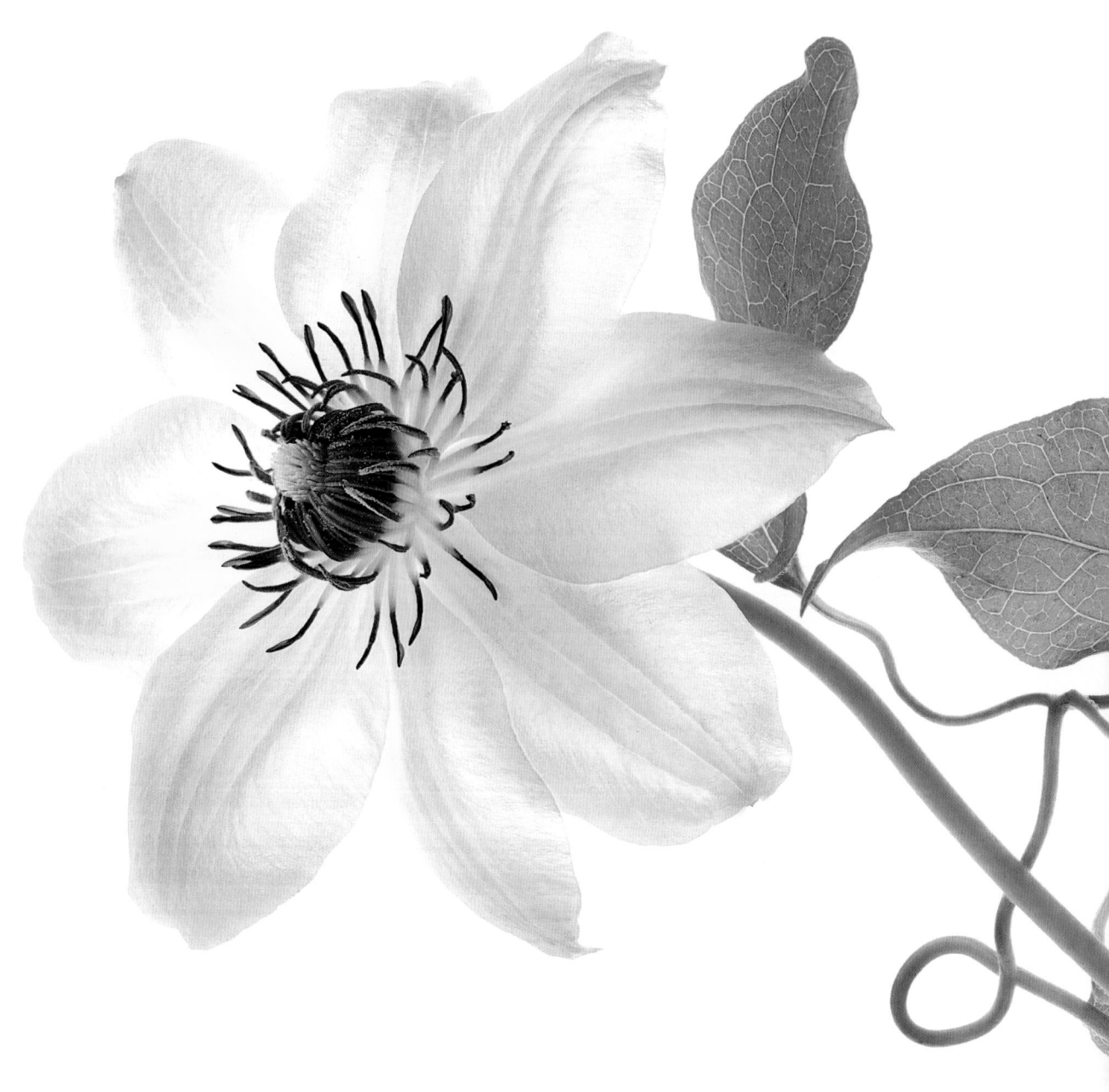

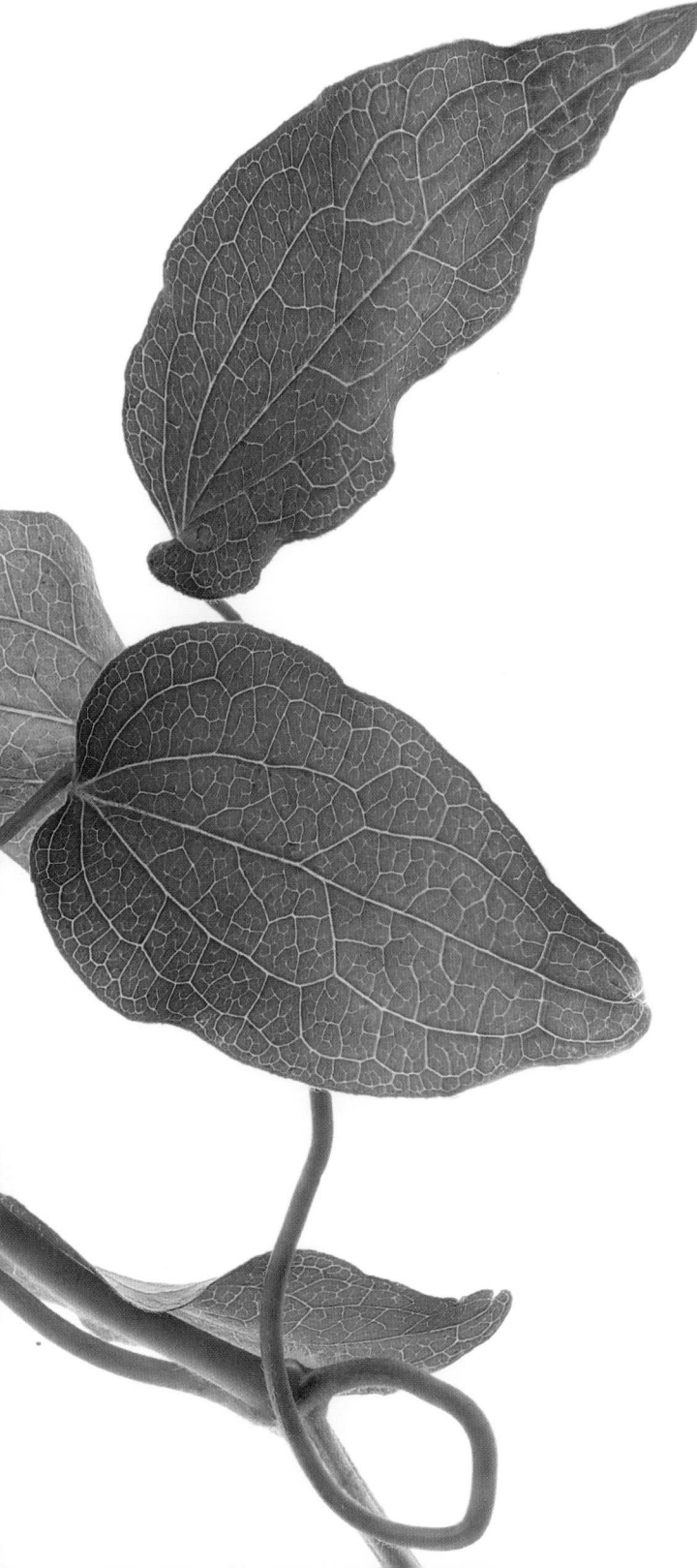

CONTENTS

Ravishingly beautiful *Clematis* is in the buttercup family.

Blue false indigo (*Baptisia australis*).

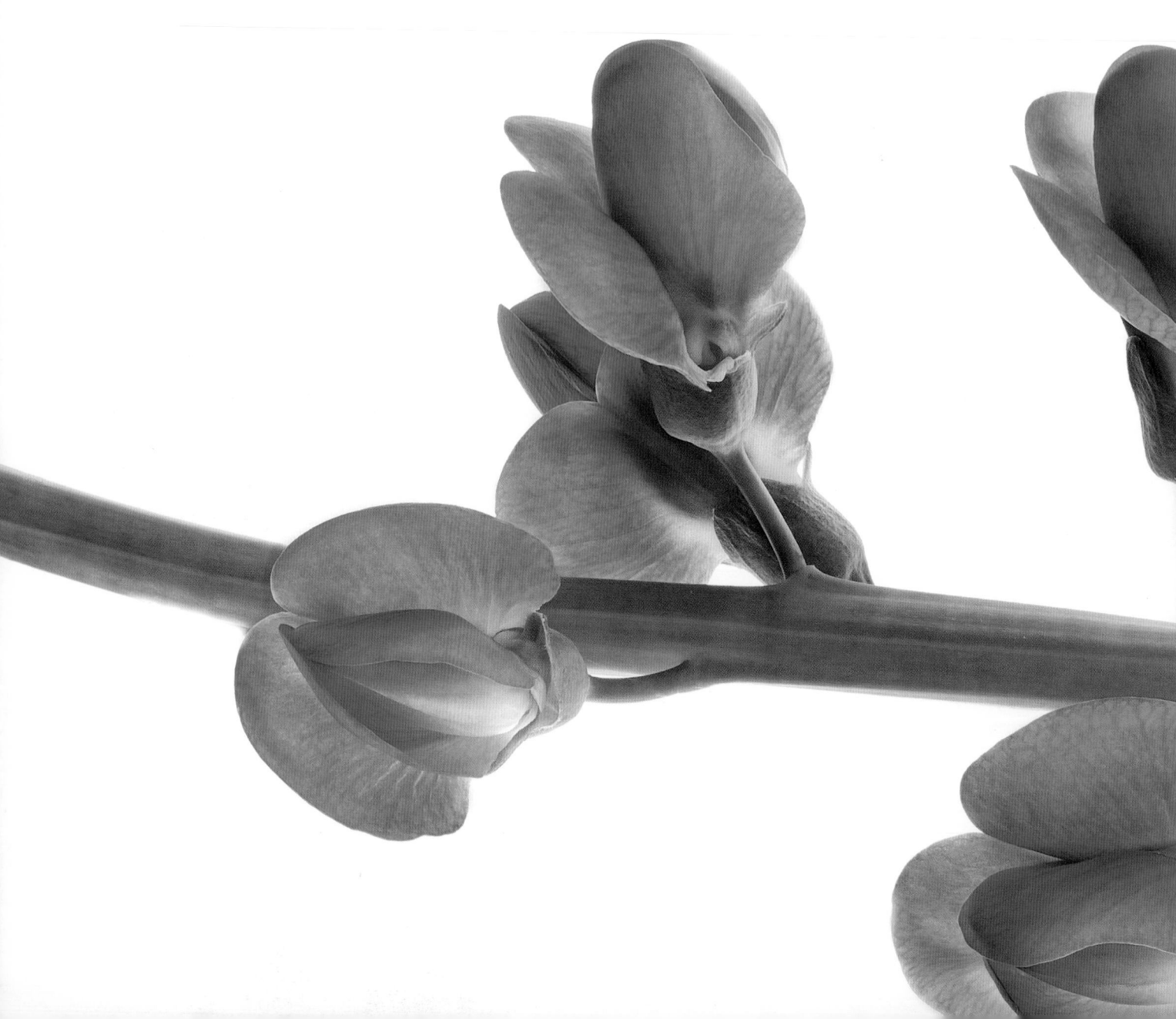

ACKNOWLEDGMENTS

Robert thanks the many dedicated gardeners who contributed their flowers to this project. Carolyn Achenbach was the most generous with her flowers and her calls to tell him what was blooming. Other contributing flower people are Fran Boninti, Candy Crosby, Anne Vanderwaker, Karen Lilleleat, Kathy Thomas, and Caroline McLean. Also thanks to gardeners Nick Nichols and Reba Peck, Mike and Karen O'brien, and Jan Glennie-Smith. Garden writer Nancy Ross Hugo gave both advice and her garden's flowers. The Lewis Ginter Botanical Garden in Richmond, VA, generously donated some of the flowers.

Peter Hatch, Director of Gardens and Grounds at Monticello, offered great advice and recommendations. Special thanks goes to Jon Golden for his computer and technology skills. Also for Timber Press editor Tom Fischer, and his never-ending list of flowers to search for (you won't ever have a bad day looking for flowers).

He would like to acknowledge his wife, Bobbi, for her support in this project and her willingness to let him pull off the road into a ditch and run off into a field to retrieve a specimen of viper's bugloss (and all of its thousand thorns).

Teri would like to give special thanks to Tom Fischer, Bob Llewellyn, Mike Dempsey, and Sarah Rutledge Gorman.

She also wishes to thank the following people for their help, encouragement, and/or inspiration while this project was under way: Alan Chace, Tristan Dunn, Wes Dunn, Fran and John Soulé, MaryJane and Jude Blau, Wayne Ferrell, Steve Frowine, Lynette Gaslin, Phyllis and Paul Chace, Christine Shahin, Mike and Maxwell Ryan, Anne Tomei, Ozzie at the post office, Jonathan Raban, Dave Alvin, and Lucinda Williams.

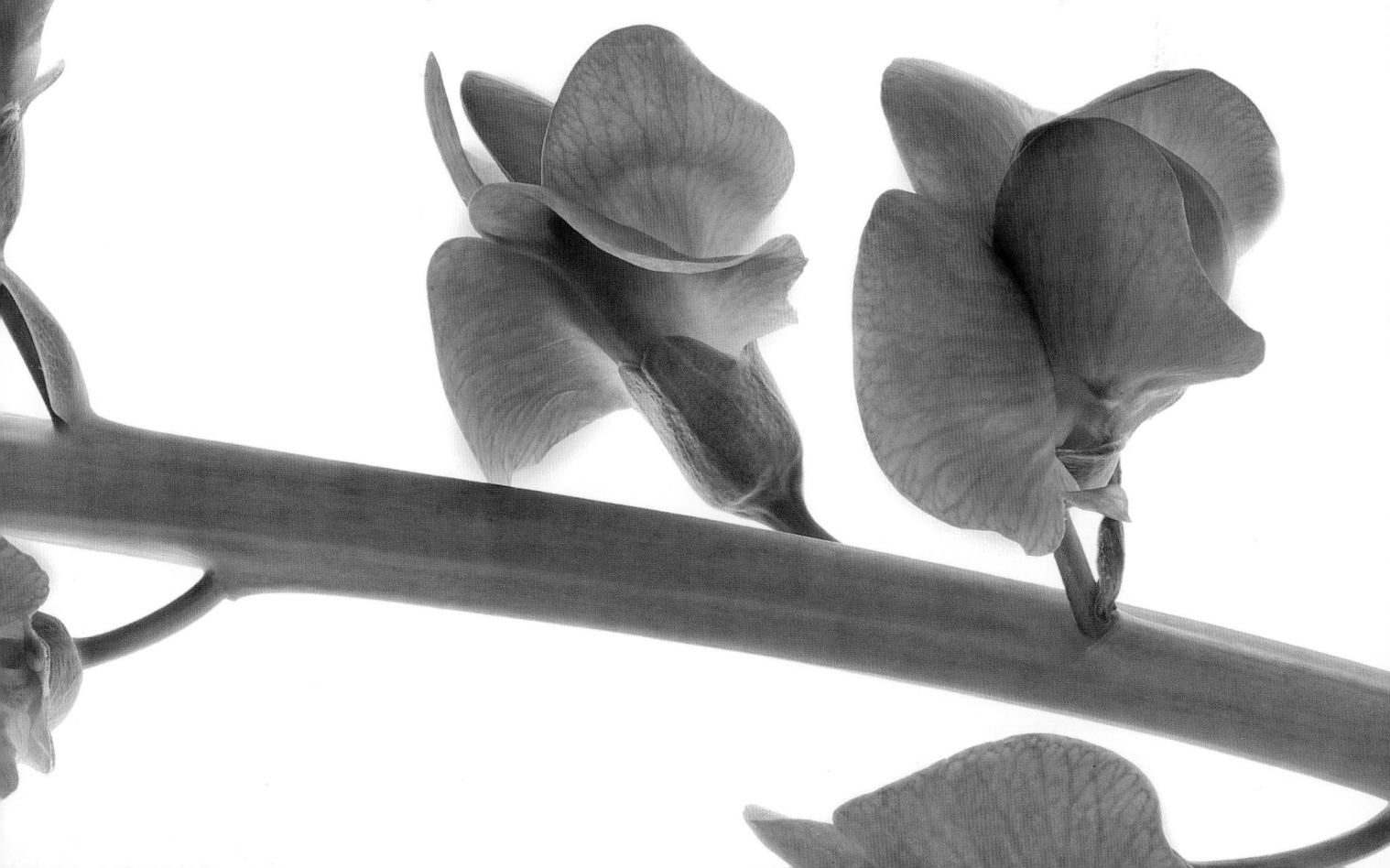

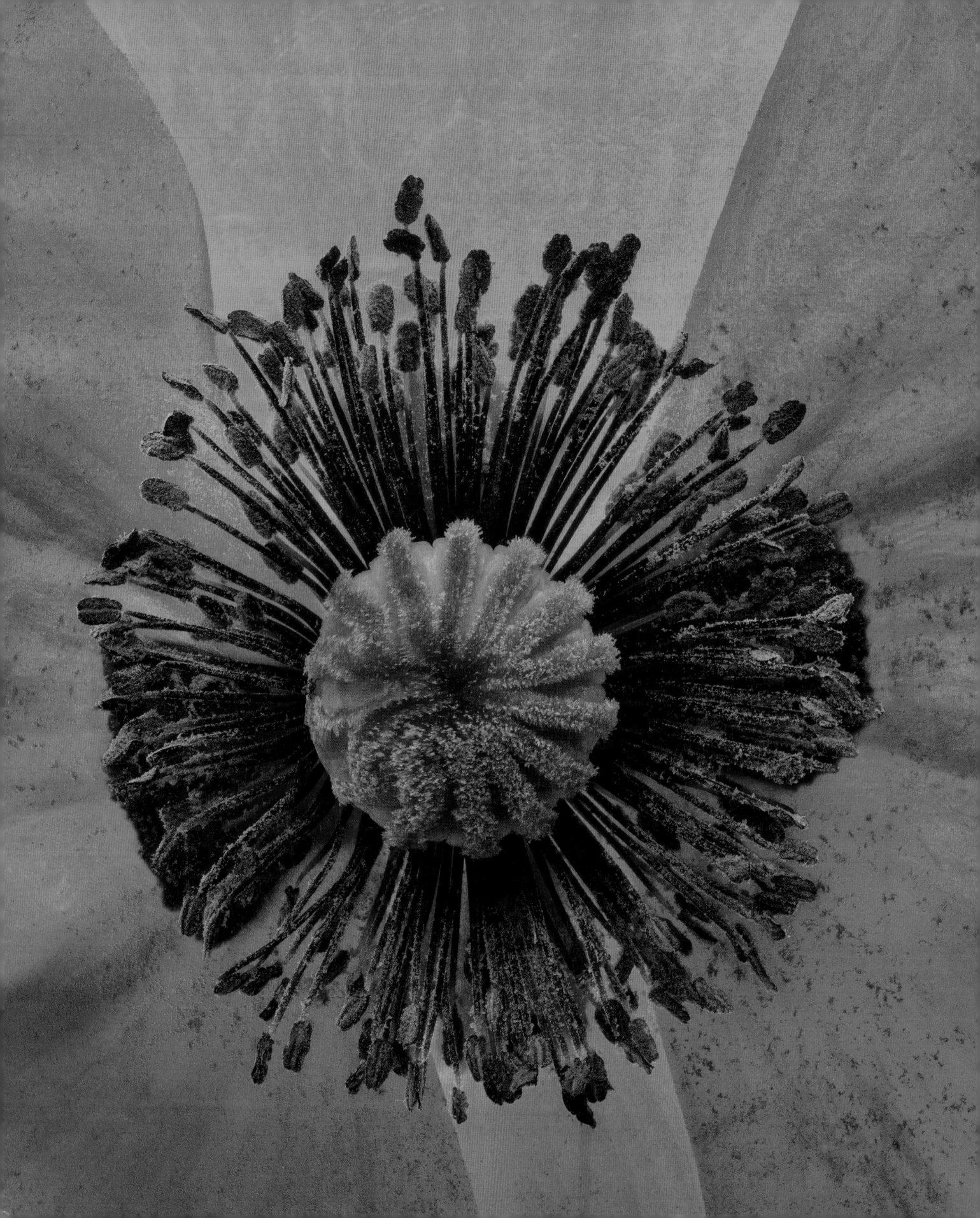

"Behold these infinite relations, so like, so unlike; many, yet one."

Ralph Waldo Emerson, address to the senior class of Harvard Divinity College, 1838

All my life, I have been looking at, prodding, poking, sniffing, and plucking at plants. These are habits born of long hours outdoors. I remember, as a very small girl in suburban southern California, squinting at and then tugging on a passionflower vine coming over our fence from the neighbor's yard. It was so mysterious, so complicated, and yet so symmetrical! I got pollen all over my fingers as I dismembered flower after flower, marveling. And I vividly recall the heady scent of orange blossoms in the nearby orchards. To this day, that fragrance is a Proustian trigger that returns me to my childhood, where I am tucked under the dappled shade of orange trees, spying on the bees browsing the sweet white-petaled flowers while the other kids in our game of hide-and-seek shouted in the distance. They should have known to look for me among flowers.

Later, transplanted to the East Coast, I knelt in cool woodlands to admire the small and pretty spring wildflowers, rue anemones, clintonias, and mayapples. In a small bed off the porch I planted and fussed over perennials: black-eyed Susans, lavender, campanulas, and various irises. When I installed my first vegetable garden, I kept vigilant watch as tomato flowers turned to red fruits and as plump white blossoms on twining vines segued to delicious sugar snap peas. I noticed how spicy-scented beach roses became spangled with stout orange hips in autumn. I kept an orchid on my windowsill at work and cheered when it actually bloomed. Through all these travels and observations, I accumulated knowledge about the ways of plants and their flowers.

I studied botany informally via mentors, as well as in college classrooms. And I was fortunate to work as an editor at Horticulture Magazine during its Boston glory days, when the issues were monthly and thick with fascinating information and ideas. Everywhere I went or lived, I habitually gardened and acquainted myself with the plants around me, both wild and cultivated.

My path is not necessarily unique. If the world of plants holds sway in your affections, you probably have comparable stories. Perhaps over a lifetime you too have accumulated impressions and information about flowers and plants.

But it was not until I beheld Bob Llewellyn's gorgeous, meticulous photographs of flowers that I truly understood how little I understood. This is not a case of "the veil of the familiar" clouding my perceptions. His photographs are different, unconventional. They actually comprise many small images—eight to forty-five—shot at a different point of focus, then stitched together

A central "stigmatic disk" is unique to poppies. Eventually it becomes a distinctive seedpod.

using software developed for work with microscopes. The results are astoundingly detailed, intimate images of the plants we thought we knew.

To take you on a journey through these remarkable views of flowers and flowering plants, this book is organized into a broad sampling of twenty-eight plant families. This is not arbitrary, but rather a way into a world both strange and familiar. Taxonomic botanists, all the way back to the innovative, diligent Carl Linnaeus, have long observed similarities between plants and grouped like ones together, primarily using the features of the flowers. Revisions, additions, and reshufflings have inevitably occurred—and continue to occur.

In recent years, the advent of genetic research—referencing the DNA sequences of plants—has revolutionized and, arguably, greatly simplified the classification work. You may never have heard of the APG, or Angiosperm Phylogeny Group, but it and its subsequent updates (as of this writing, we're on APG III) have made significant progress. Some of the rulings verify what the naked eye and/or microscopes have shown all along, while others are surprising. Some long-established plant families have suffered splits, while others have been submerged, making the work at times controversial. And yet it reflects the onward march of botanical knowledge.

The aim of *Seeing Flowers* is not to get mired down in the finer points of classification. Rather, it is to allow plant lovers, or anyone driven by curiosity about plants, to follow Bob's remarkable photographs and witness this world in fresh ways. Common to most flowers are features such as petals and the sepals that support them, plus the sexual organs that offer pollen to pollinators and eventually develop into seeds and fruit. I hope you will discover, as I did when I looked through these photographs, a sense to it all. As varied, weird, wonderful, sexy, and graceful as flowers are, ultimately they have always been the plant world's supremely resourceful way of staying alive.

In each chapter I acquaint you with a plant family by exploring its distinguishing characteristics and providing examples, some of which are also presented in Bob's amazing photographs. As I researched and wrote these chapters, I found fascinating relations, intricacies, exceptions, tales, and tidbits. I share them in the hope that they will expand your appreciation, as they did for me.

Once upon a time, as the late naturalist Loren Eiseley reminds us in his lyrical essay "How Flowers Changed the World," there were no flowering plants, or angiosperms. The earliest plants arose near water, which aided fertilization and thus reproduction. Later plants (gymnosperms), such as primitive conifers and spore-bearing ferns, depended on wind. The advent of true flowers and the seeds they produced was, as Eiseley declares, "a profound innovation."

Angiosperm means "encased seed," an astonishing item that grows in the heart of a flower. Here an embryonic plant is developed, and soon it falls, floats away, ejects, is eaten, or grabs onto fur or clothing, and thus survives and expands its realm. But before this can occur, a flower must be pollinated. Pollination is the ambitious goal of all flowers. Bear this in mind as you view and learn more about the enticements, trickery, shortcuts, and quirks they use. There is plenty of variety, but also a common and practical purpose.

Angiosperms have indeed changed the world. They have populated it explosively, greatly outnumbering the more primitive gymnosperm plants. They feed and thus support the planet's insects, birds, and animals. And they also they fill our world with incomparable beauty, especially with their flowers.

I now invite you—with Bob's glorious, innovative photographs to contemplate—to see flowers, in all their diversity, for what they are: wonders.

Common yarrow (*Achillea millefolium*) sprig.

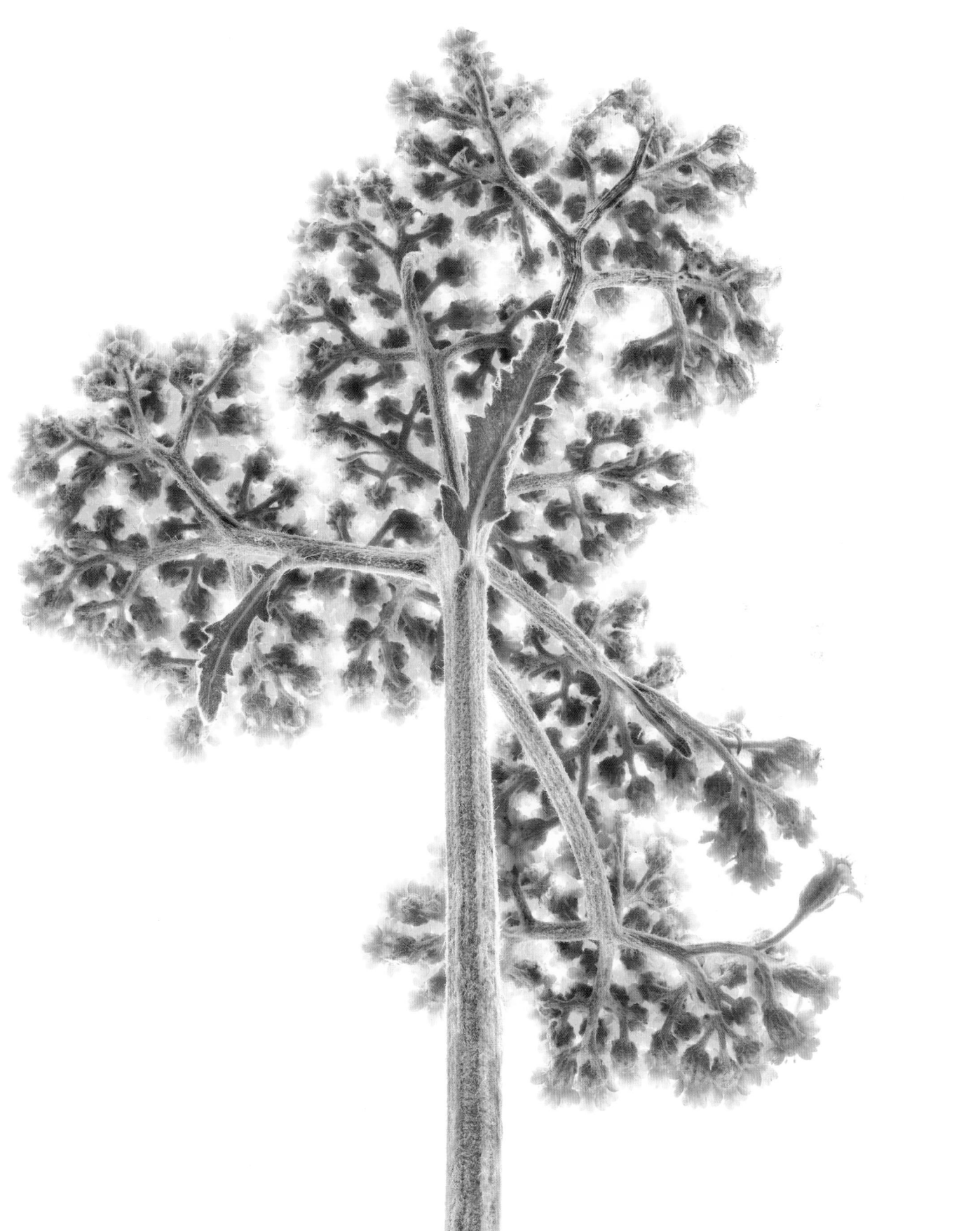

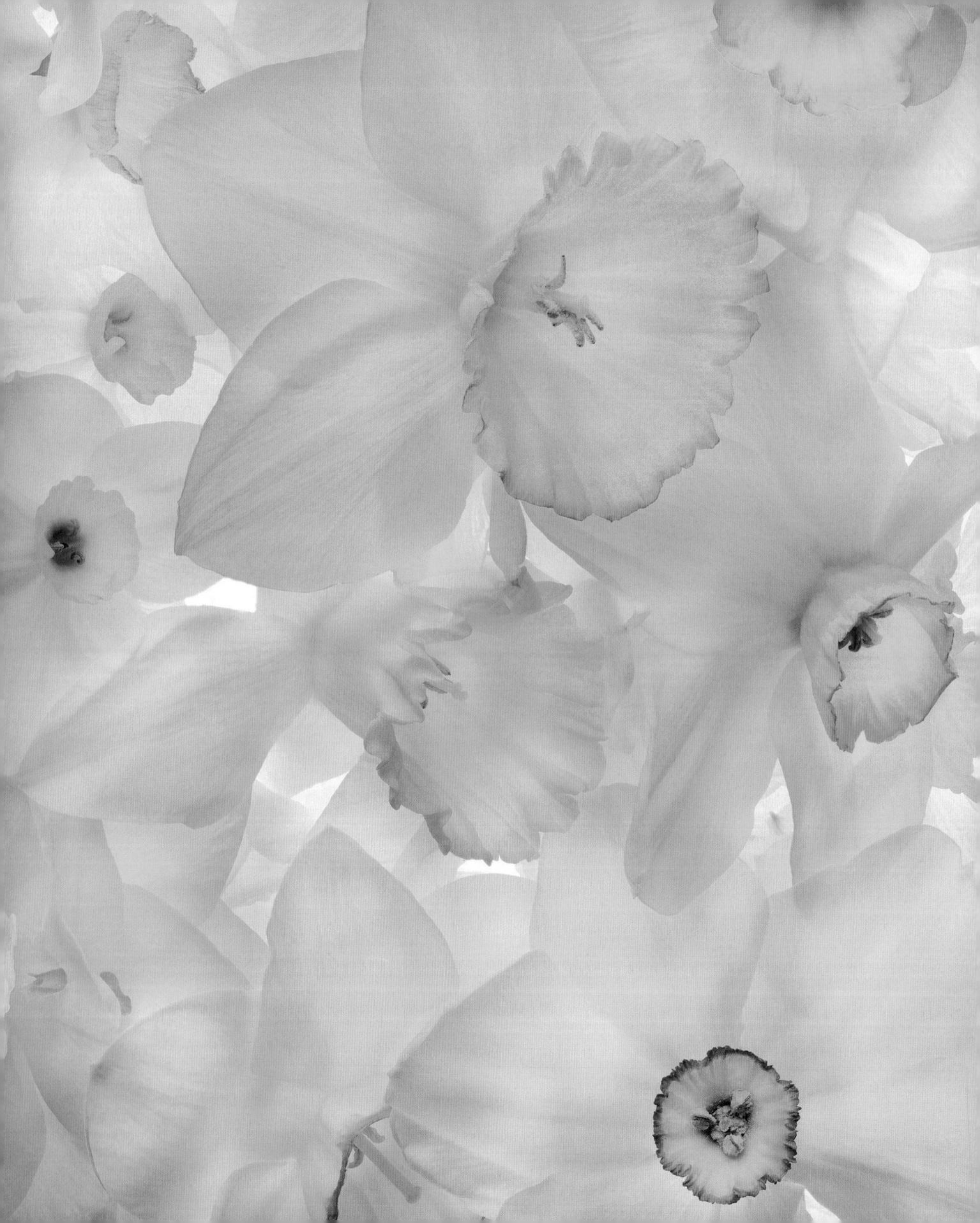

THE AMARYLLIS FAMILY

Amaryllidaceae

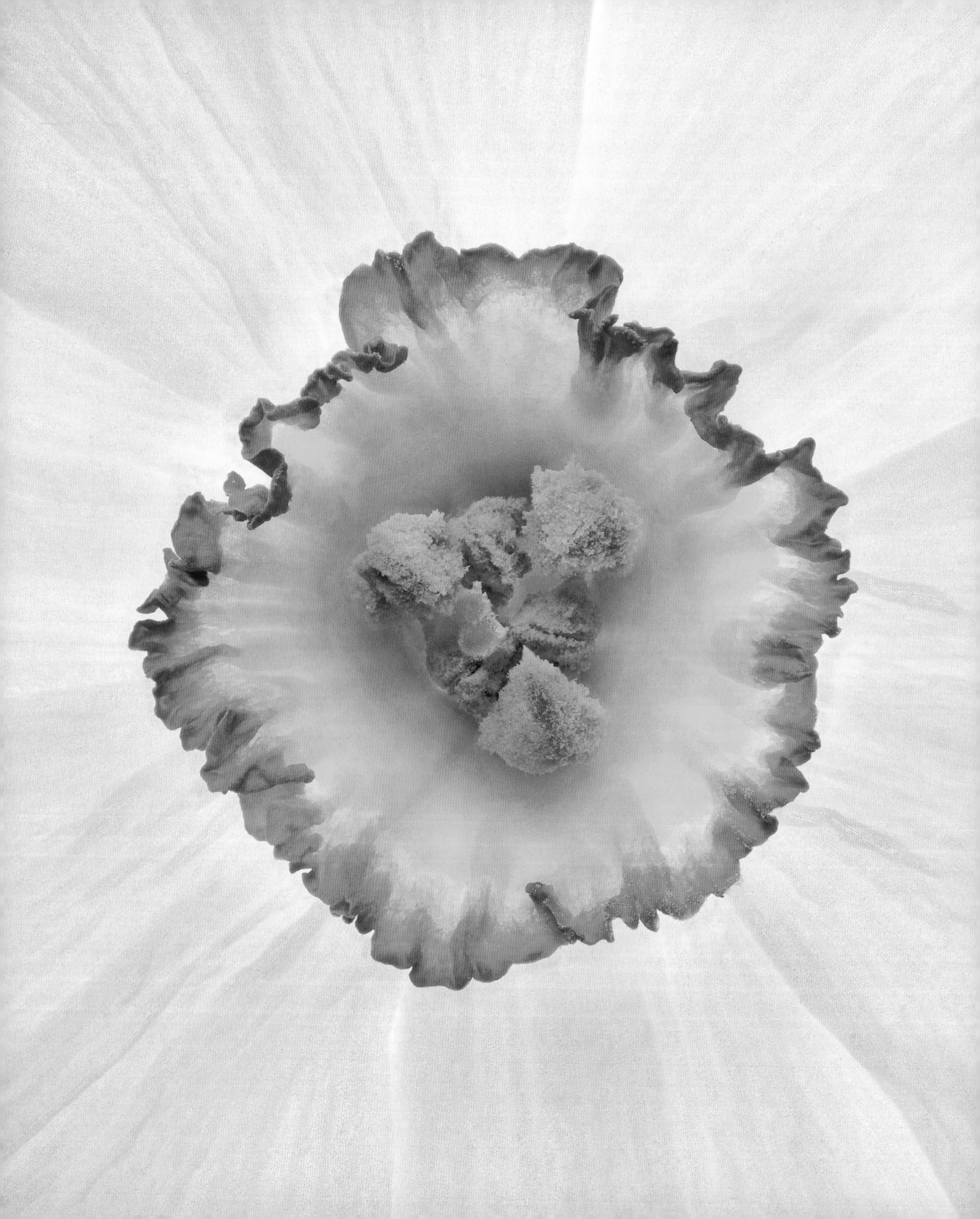

"AMARYLLIS I'm having a party on Saturday. I'd like it if you could come.

MRS. PAROO Well, Amaryllis asked you to her party. Are you going or not?

WINTHROP PAROO No.

MRS. PAROO No what?

WINTHROP PAROO No, thank you.

MRS. PAROO No thank you, who? You know the little girl's name.

AMARYLLIS I bet he won't say it.

MRS. PAROO "No, thank you, WHO," Winthrop?

WINTHROP PAROO No, thank you, AMARYLLITH!

AMARYLLIS Amaryllith! Amaryllith!"

MEREDITH WILLSON, *The Music Man*, 1957

LEFT AND PAGE 14 Lovely, assertive daffodils and jonquils may differ somewhat in form and colors, but are always recognizable as *Narcissus*. Botanists have identified a dozen distinct divisions within this varied genus, according to the relative size of the corona (center cup) and perianth (petals), as well as bloom size and whether blooms are in clusters or held singly.

Because it is such a popular, widely available holiday gift, many people have had occasion to admire the common amaryllis (*Hippeastrum*, family Amaryllidaceae). It also helps that the plant is brought to spectacular bloom inside in the off-season months, when the outdoor garden is generally quiet, perhaps slumbering under a blanket of snow. A fancy red or satiny white trumpet brightening up the indoors is a joy to behold and, perchance, to study.

The Northern Hemisphere winter is, of course, the Southern Hemisphere summer, which is why these natives of South America are programmed to bloom so readily then. As you might expect, they appreciate the warmth, protection from drafts, and regular deliveries of water that fuel their holiday-season show. Hybrid bulbs are big business, especially the bounty of Dutch imports. The bulbs generally range from 8 to 12 inches in circumference. Larger ones tend to produce more flower stalks and thus more flowers, so they are a bit more expensive. While *Hippeastrum* bulbs are probably the most substantial in this plant family, they are also typical in terms of shape: roundish. And they are tunicate, which refers to a papery coating over a fleshy interior. No gnarly rhizomes or wiry little tubers in Amaryllidaceae.

Once potted in a well-drained mix with at least half of that big bulb exposed, amaryllis care is a cinch. "Foolproof—just add water!" exclaim the catalogs and packaging boxes. For the most part, with quality bulbs free of fungal disease, this claim is completely correct, and

BELOW The cup in the center of a daffodil is called a corona. Some are short, like a shallow bowl, while others are longer, more like a trumpet. Some have frilly edges, and some are rimmed with a contrasting color. Some coronas are split up to half their length to create a flatter, showier flower.

RIGHT Common snowdrops, *Galanthus nivalis*, are among the tiniest and earliest flowers of spring. Their mildly poisonous bulbs and foliage make them tempting to neither deer nor rodents and keep them pest free.

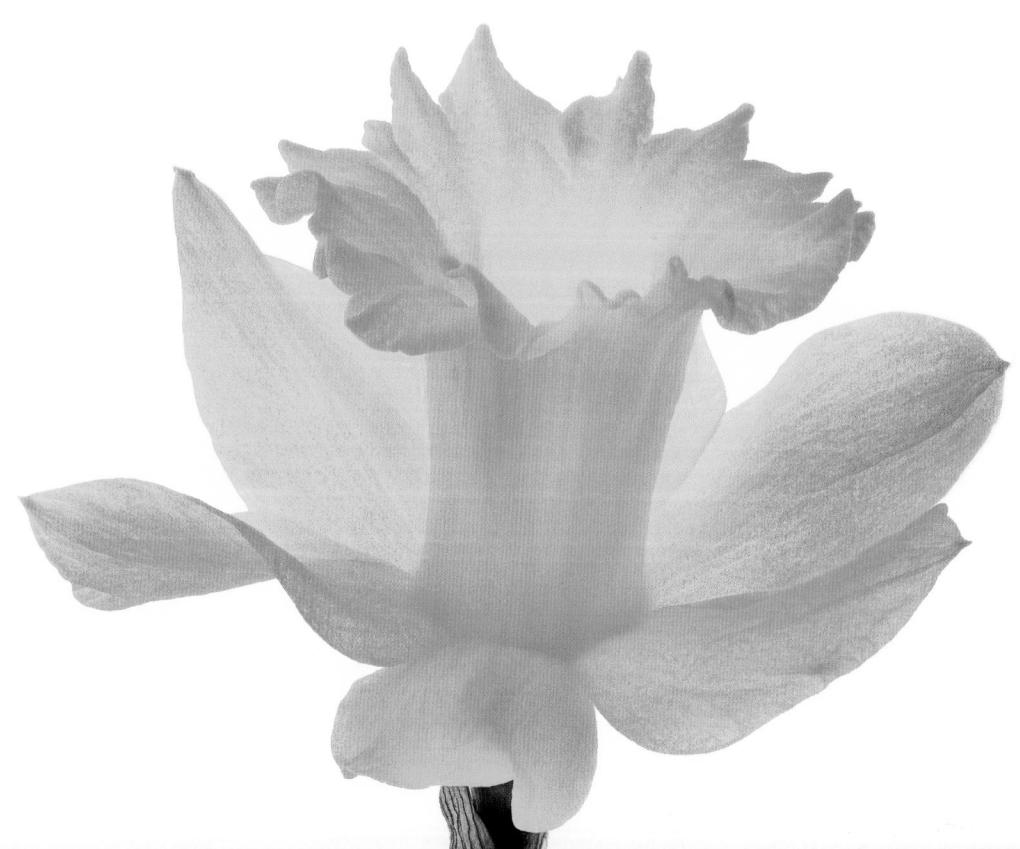

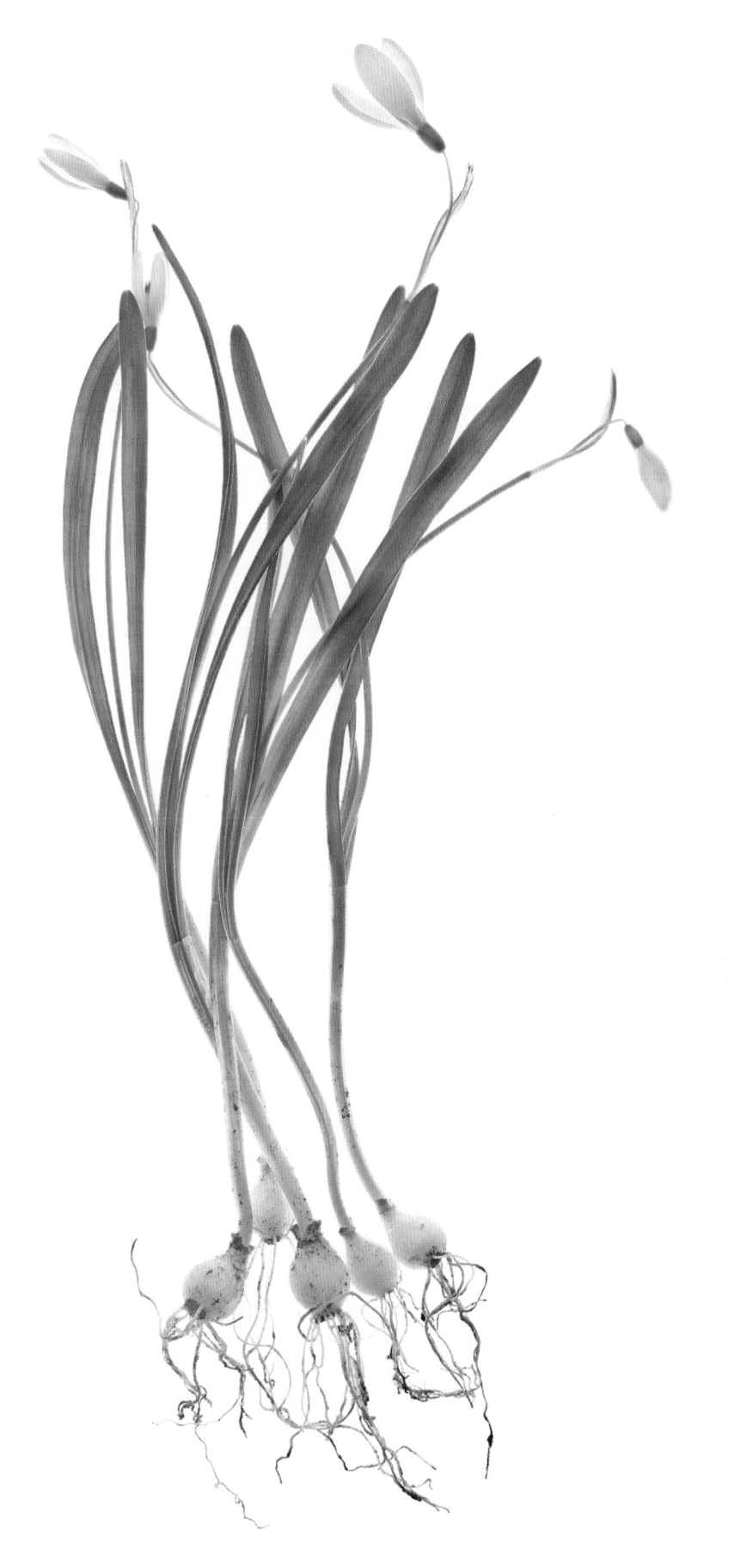

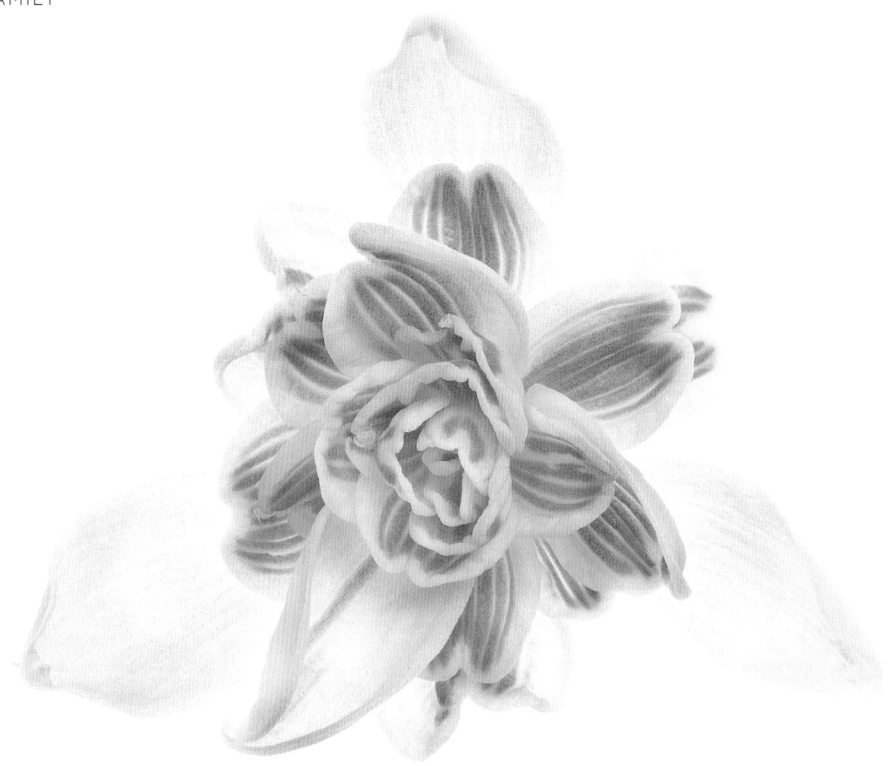

bringing these plants into bloom is easy. You might not even have to bother with staking, because plant breeders have encouraged shorter, stouter stems, although some cultivars end up looking rather top-heavy or unnatural.

Nonetheless, the blooms always dazzle. Six large, beautiful, slightly overlapping petals, in colors ranging from red to pink to white, are sometimes lightly brushed, dusted, or rimmed with a contrasting color. Or blooms are striped right down the middle, as with the enduringly popular 'Christmas Star', which is vivid red with a white starburst and a green eye. Doubles, of course, sport extra petals, in multiples of six, for a plush look. This may be impressive, but it is not to everyone's taste. Another variation is miniatures, whose many smaller flowers emerge from smaller bulbs. 'Green

Goddess' produces graceful clusters of pristine white blooms with light green centers that are in scale with the entire plant.

The interior, or eye, of these blooms is so classically constructed that it makes a fine lesson for junior botanists. You can easily spot the pistil, which consists of the style (with pollen tube inside). It is attached to the ovaries deep in the center of the blossom. At the tip, you will find the small, three-legged, puffy white stigma. This structure extends out furthest from the bloom's center. It is flanked by several stamen filaments, which curl slightly back inward and are topped with tiny yellow, spongy-looking anthers that harbor the pollen.

Another favorite indoor-forced winter bulb is also in this family: paperwhites, *Narcissus tazetta*, which also

ABOVE A petite treasure, double snowdrop, *Galanthus nivalis* 'Flore Pleno', has been in cultivation since 1731. The petal-packed blossoms stand out better than those of the species, especially when plants form a colony.

RIGHT Little as they are, sweet pendants of wee *Galanthus ikariae* give away their kinship with amaryllis with six petal-like tepals.

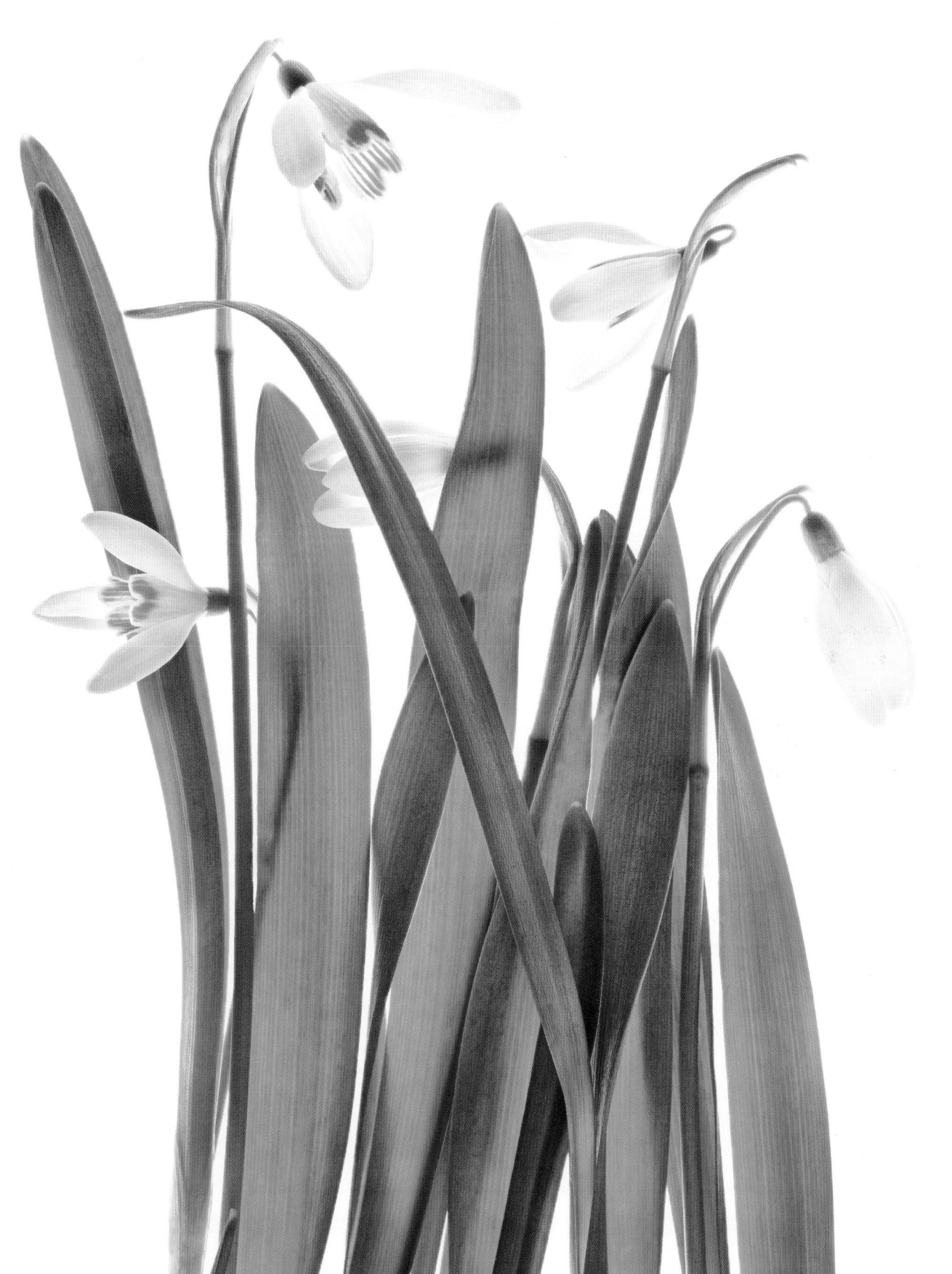

The exotic-looking red spider lily, *Lycoris radiata*, blooms in late summer to early fall. Notice the six prominent stamens and the six distinctive, recurved, petal-like tepals; parts of six are characteristic of this plant family.

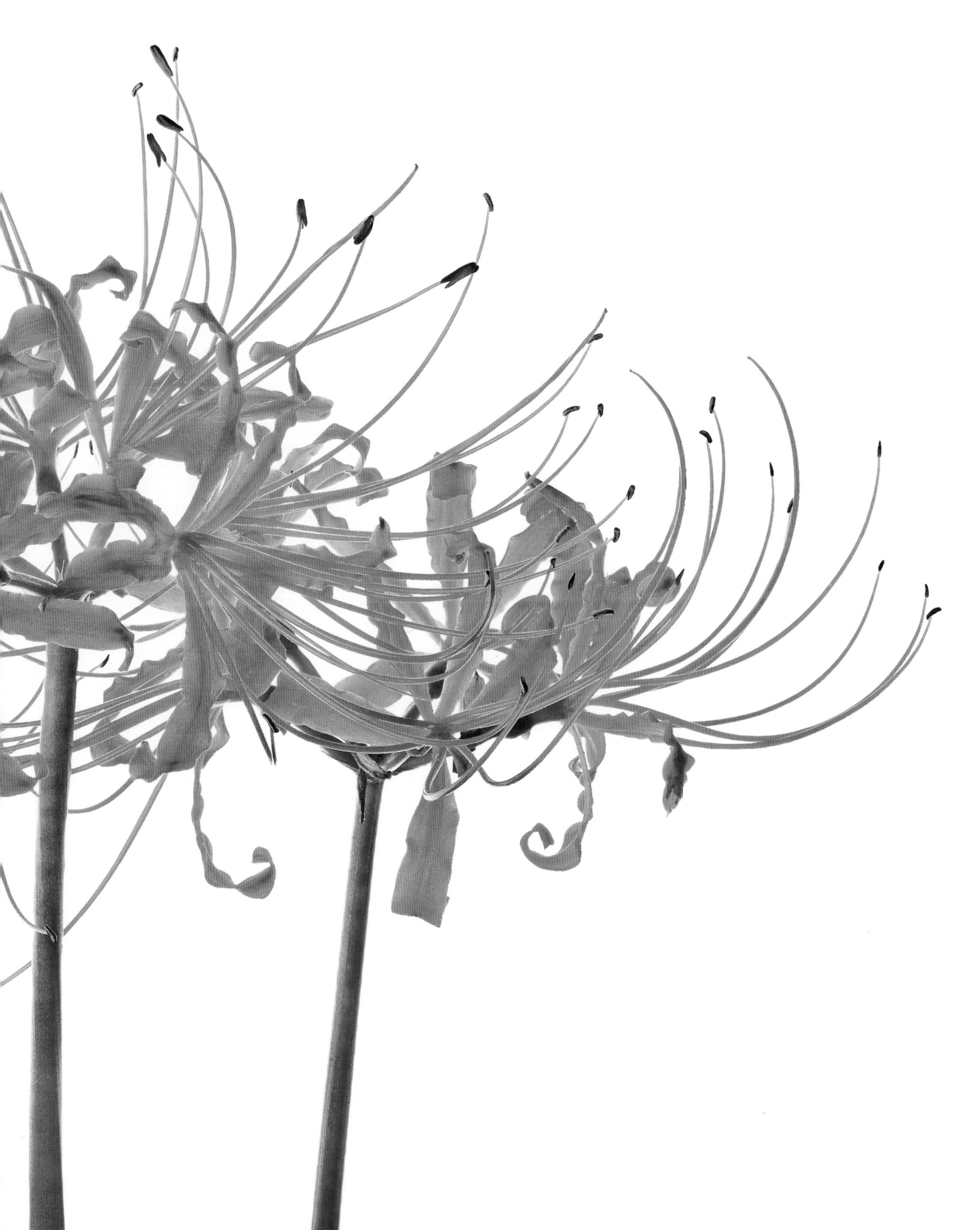

has a papery covering over a fleshy bulb. Paperwhites too can sit atop or partially buried in soil mix or pebbles, and prosper with the super simple "just add water" regimen.

Soon, as spring approaches, the outdoor parade of daffodils and jonquils commences. They are a complicated group that horticulturists have split into twelve divisions. Some have big, singly displayed flowers, some bloom in clusters, some are fragrant, and all sport cups or trumpets (coronas) and saucers (perianth segments). Their color range is traditionally white, orange, and/or yellow, with some varieties veering toward orange-red, orange, or pink. They give away their Amaryllidaceae membership with those papery tunicate bulbs, as well as by their six overlapping petals (six-part perianths) and by blooming on leafless stems.

This family also extends to the less familiar genus, the true *Amaryllis*, which hails from South Africa. Plant breeders have also tinkered with these, with similar good results. Those known as naked ladies or resurrection lilies, *A. belladonna*, are often seen growing outdoors in southern California and other similarly mild climates, where they become perennial. These have the trumpet- or funnel-shaped flowers you'd expect, nominally similar to their cousins. The summer blooms are usually pink, sometimes a soft pastel, and sometimes a much bolder, richer hue. They may be fragrant, with a sweet, cloying scent that repels both marauding deer and nibbling rodents (all plant parts, including the bulbs, range from mildly toxic to quite poisonous). These are carried on tall, bare, gray-green to purplish stalks between 2 and 3 feet high. Leaves do not coincide with the blooms, but occur separately in late fall or early spring. The softball-size bulbs, like some of their above-mentioned relations, may partially rear up out of the ground.

Enthusiastic collectors, especially those gardening in mild climates or possessing a greenhouse, love to tinker with flowers in this family. With care and patience, it is possible to make homemade crosses, as the pollen is readily accessible (in nature, bees pollinate the flowers).

Successful crosses lead to swollen green pods that ripen a few flat black seeds, which you may then sow and grow. It will be a couple of years, however, before they reach blooming size and you can assess the results. According to Internet message boards, amateur amaryllis breeders have succeeded in creating variations both handsome and strange, and the color red is dominant.

An intergeneric cross between family members *Amaryllis belladonna* and *Crinum* (swamp lily or spider lily) has also been successful. Dubbed *Crinodonna*, or ×*Amarcrinum*, it is available from bulb specialists, and hailed for its robust nature and lovely pink flowers. No doubt still more experiments within this family can be tried and evaluated.

One more member of Amaryllidaceae bears mentioning and praising: the glorious clivia. Usually the trumpet-shaped flowers are brilliant orange, although there is a yellow variety, *Clivia miniata* 'Aurea' or 'Citrina'. Blooms are carried in clusters atop leafless stems, as is seen in so many of their kin. Individual flowers have a short tube at the base and six petal-like tepals. Clivia is of South African origin, so it may be grown outdoors in mild climates. Elsewhere, it is prized as a houseplant or greenhouse plant.

With the six-part flowers and strappy leaves, the similarity between this family and the lily family is hard to miss. There are many reasons why they are not grouped together. For starters, the lily family is already vast and grouped into numerous divisions, so botanists who have split off similar yet different plants are undoubtedly justified. Furthermore, the unique physical characteristics and behavior of Amaryllidaceae argue persuasively for its own family. Gardeners in temperate areas are fortunate to be able to enjoy these dramatic beauties so far from their origins.

Deliciously scented paperwhites show off both the beauty and the durability of the genus *Narcissus*, lasting days and even weeks if kept in a cool room.

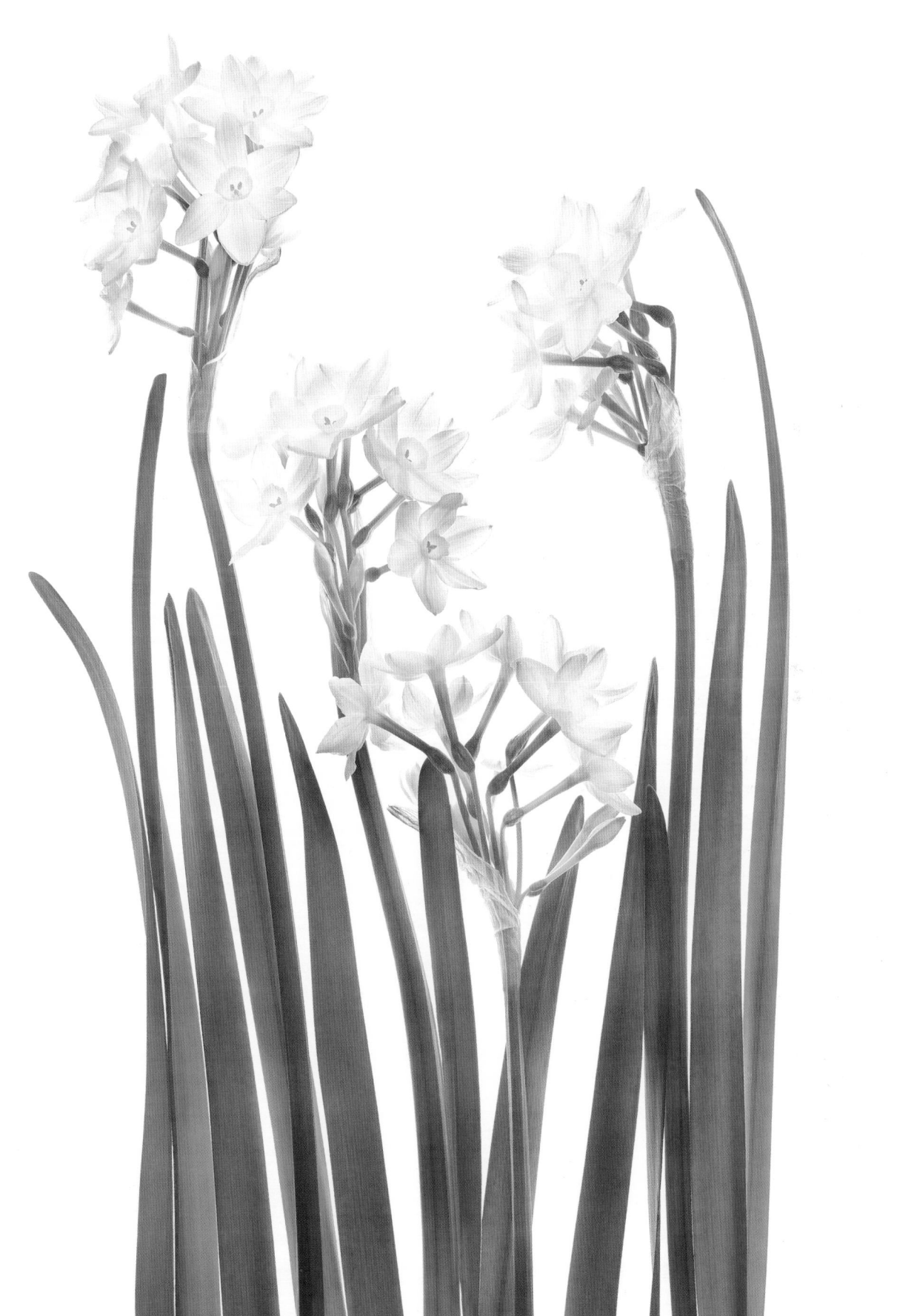

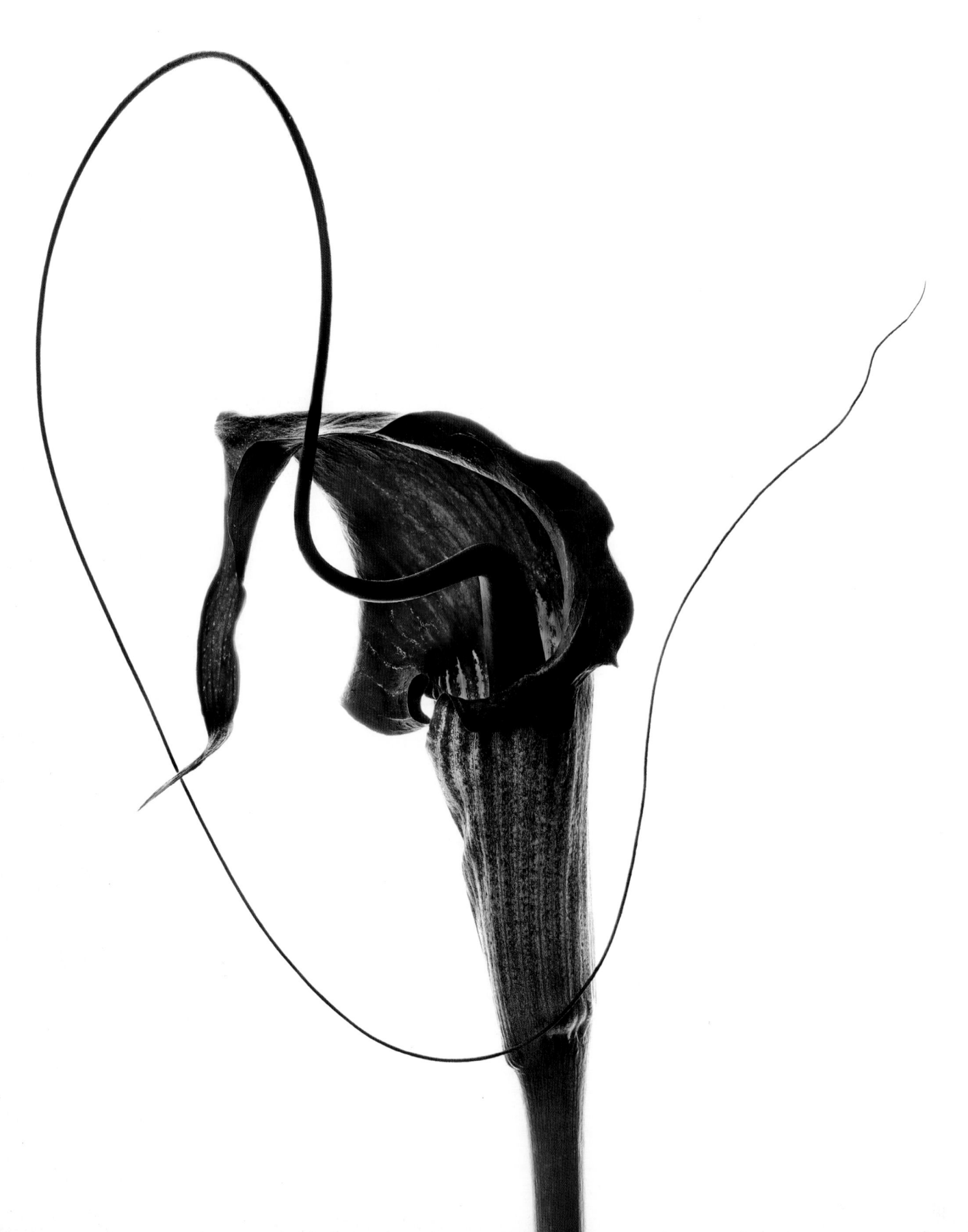

THE ARUM FAMILY

Araceae

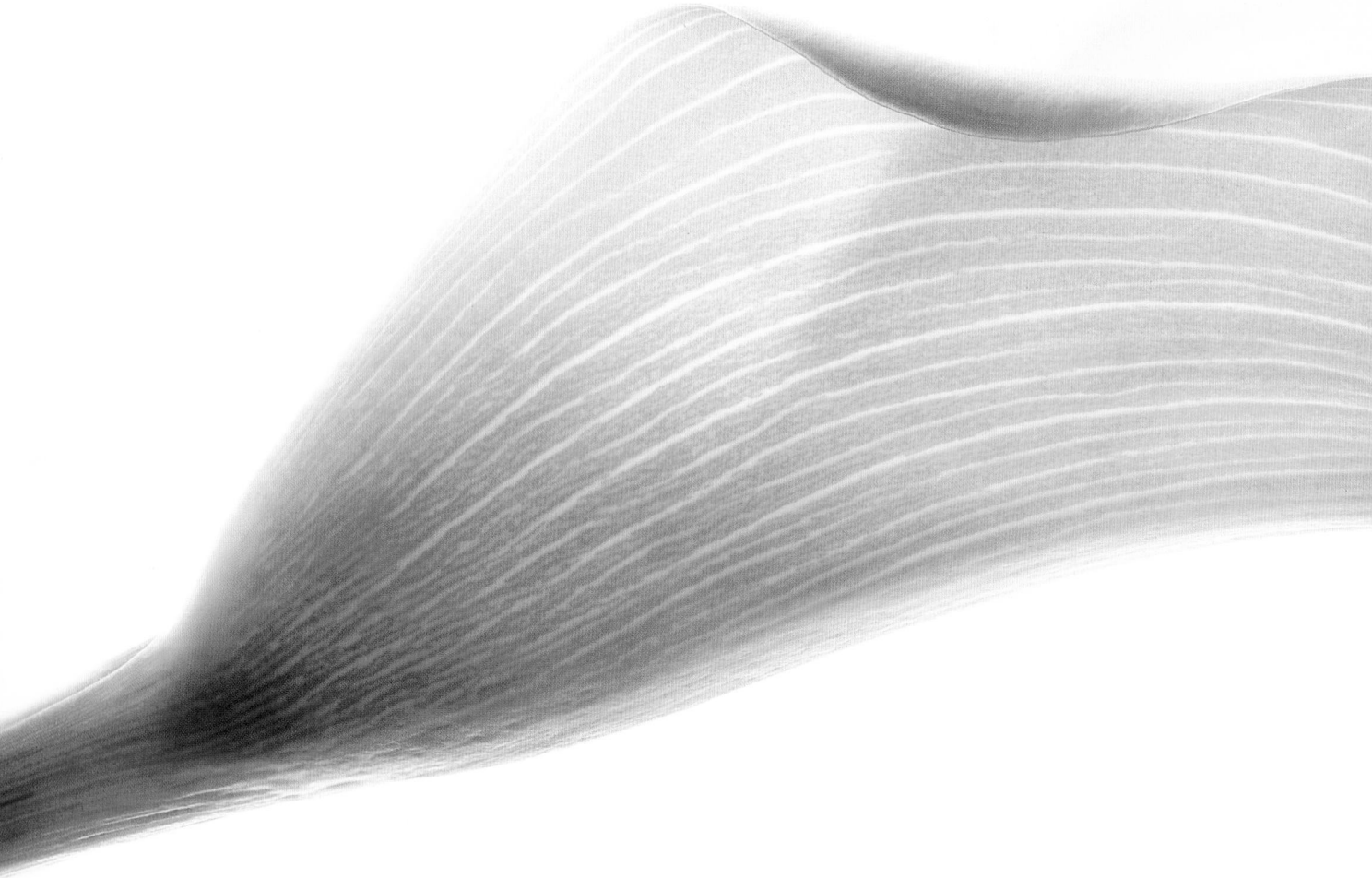

Calla lily, *Zantedeschia aethiopica*, flowers are perhaps the most familiar and appealing in the arum family. The elegant, colorful outer petal is actually a spathe, a modified leaf or leaflike bract.

PAGE 26 Spooky yet beautiful, the Japanese *Arisaema thunbergii* subsp. *urashima* goes by the racy common name dominatrix Jack-in-the-pulpit. The dusky-hued, whiplike tail is actually an elongated spadix, and has been known to extend up to 18 inches.

"O, mickle is the powerful grace that lies
  In herbs, plants, stones, and their true qualities;
  For nought so vile that on the earth doth live
  But to the earth some special good doth give!"

WILLIAM SHAKESPEARE, *Romeo and Juliet*, ca. 1594

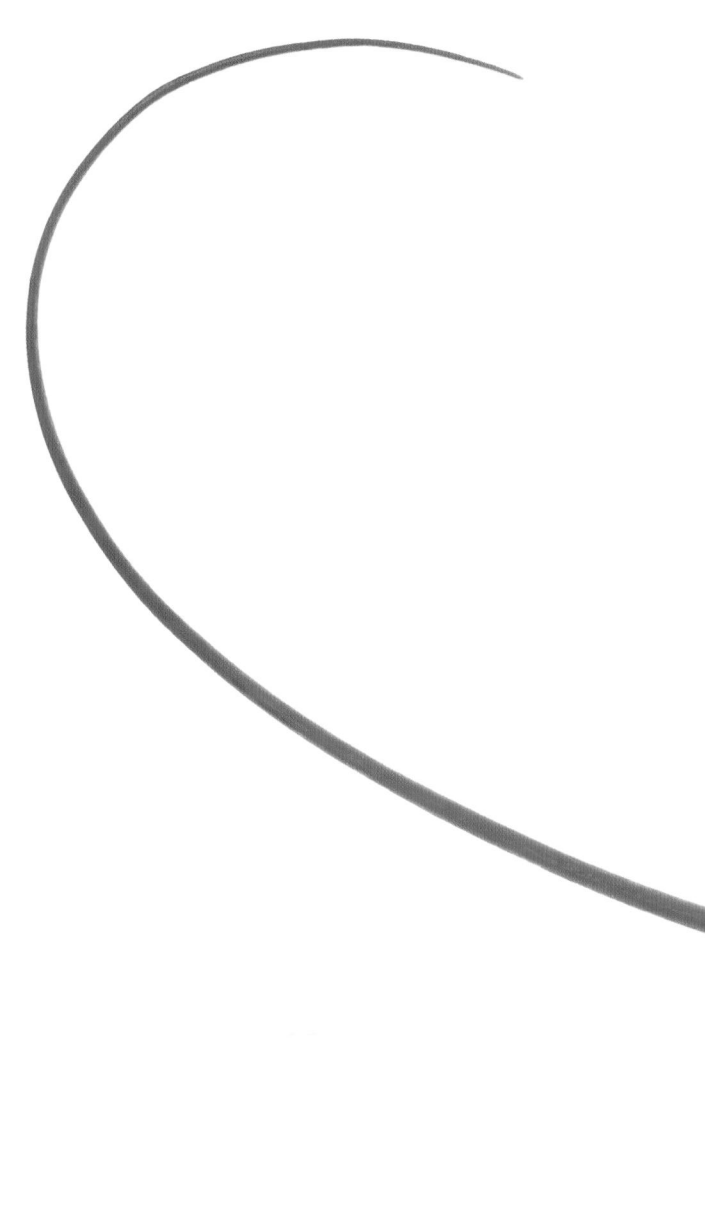

Of all the strange and wonderful families in the plant world, Araceae may be among the most eccentric. While a relatively small group (more than 100 genera), it occurs worldwide, with members large and small, terrestrial and aquatic. All are arguably beautiful in foliage and flower. And yet the plants are poisonous through and through, thanks to the presence of calcium oxalate crystals. Contact with some of these plants causes a rash, and eating improperly prepared foods from others allows the crystals to lodge in the mouth and throat tissues, causing, at minimum, swelling and an excruciating burning sensation.

Araceae, or aroids, include springtime woodland stalwarts familiar in temperate woodlands: skunk cabbage (*Symplocarpus foetidus*), Jack-in-the-pulpit (*Arisaema triphyllum*), and lords and ladies (*Arum maculatum*). The family also includes exotic, big-leaved bruisers like elephant's ear (*Alocasia*) and taro (*Colocasia*), which are native to tropical Asia. A few have been popular in cultivation for a long time, such as caladiums (*Caladium* species and cultivars), easygoing foliage plants that form carpets in shady areas in warm-climate gardens and do well as houseplants elsewhere. And let's not forget calla lilies, originally from South and East Africa (*Zantedeschia* species and cultivars), so beloved by florists and their customers. It sounds like a wildly divergent group, but it is not. The plants share many common qualities.

Attractive foliage is often a hallmark of plants that prosper in shady areas, and these are no exception. Leaves in Araceae are almost always glossy and substantial, and often marked or veined, which highlights their form and beauty. They all love moisture, and their broad

An eccentric-looking relative of the native woodland Jack-in-the-pulpit, miniature green dragon, *Pinellia cordata*, attains only about 6 inches in height. Its light green spathes shelter a slender spadix that becomes a long, twirling tail by late summer.

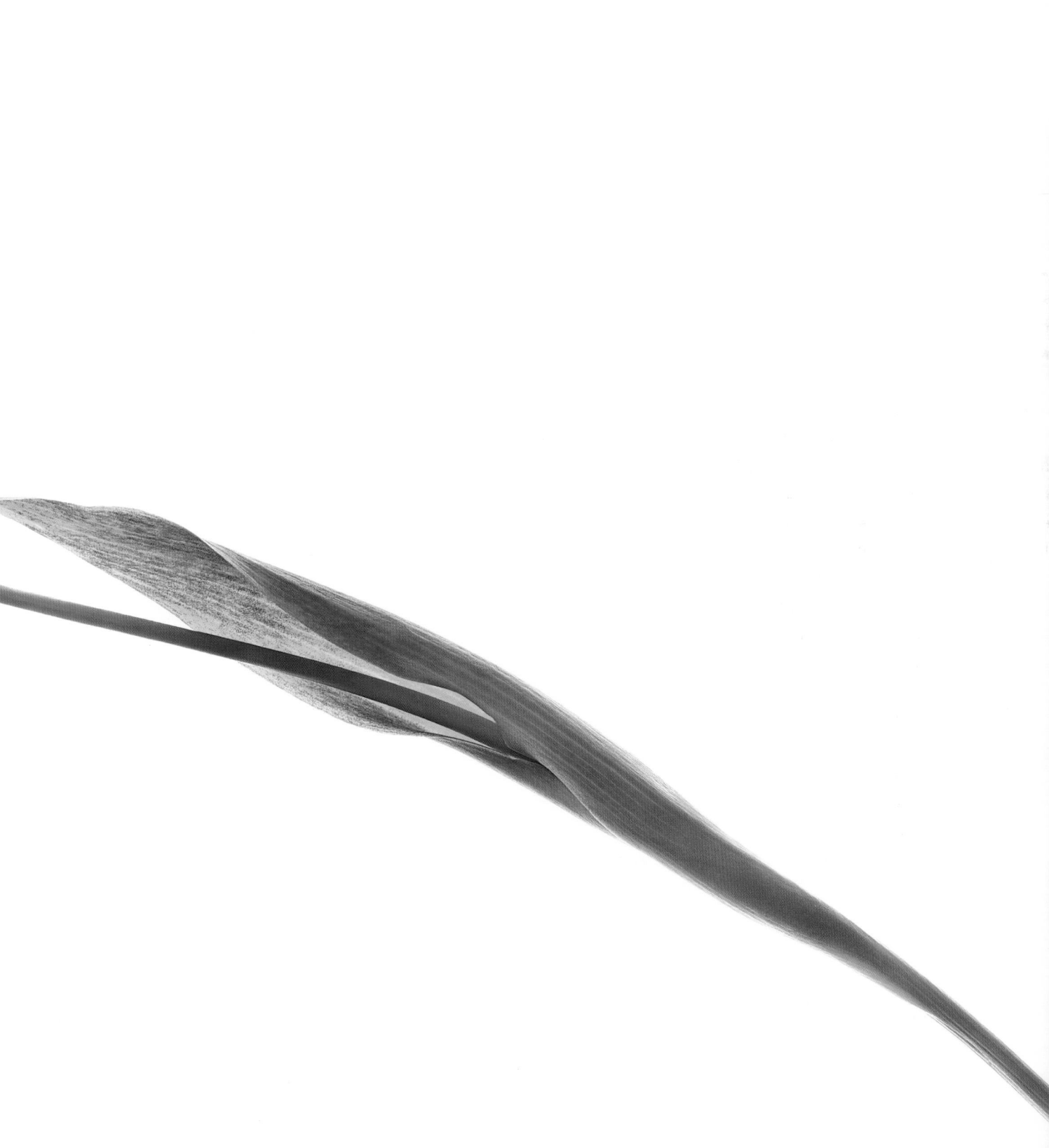

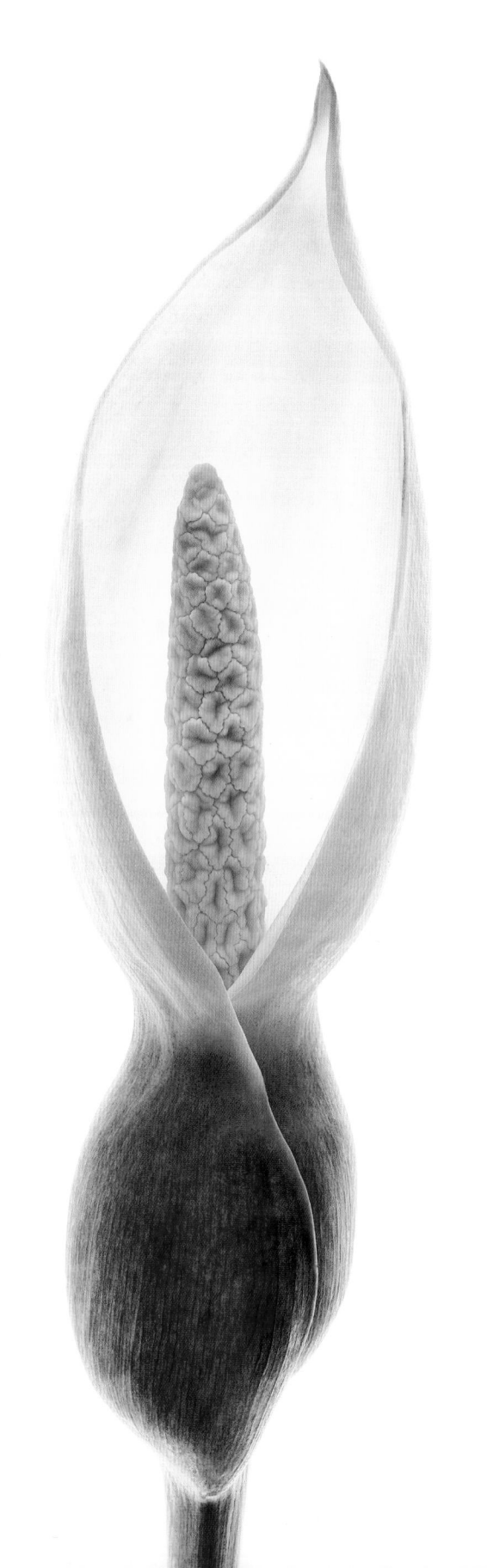

extent and fleshy texture attest to their confidence that water is not in short supply when they are growing in situ or are well cared for in a damp-soil garden setting. In other words, these leaves are not conservative. Skunk cabbage leaves are big, dark green, and, well, cabbage-like. Taro leaves are shaped like hearts and may be as broad as 3 feet across, covered in tiny silken hairs that shed water like lotus leaves. Many are burly, broad versions of an arrowhead or spear shape, such as the leaves of calla lilies, lords and ladies (*Arum*), and *Dracunculus*.

But it is the flowers, when they occur, that gain the most admiration—or, at least, inspection. They are a unique, two-part, spathe-and-spadix arrangement that invites a closer look, particularly when they are hidden from immediate view under the leaves. Jack-in-the-pulpit, for example, has compound leaflets of three leaves each that hover over the strange-looking inflorescence and shelter it like an umbrella. You have to crouch down and move them aside first. The spathe, or pulpit, is a leafy outer covering; the jack within is a thin column called a spadix, which is a fleshy stem covered with numerous tiny flowers or florets. In calla lilies, the yellow spadix is easier to view, wrapped loosely in a single elegant petal that is usually cream, pink, or yellow. That petal, of course, is technically a spathe. The flowers of taro look like huge calla lilies. In *Dracunculus*, both spathe and spadix are a lurid shade of purple.

In nature, bees, flies, and beetles accomplish pollination. The flowers can smell softly sweet, as with callas, while others—like the aptly named skunk cabbage and foul-smelling *Dracunculus*—reek. Sometimes the flowers are hermaphroditic, deciding at some point which sex to be. Female plants tend to be larger because they need enough energy to produce fruit and seeds. Sometimes the flowers exist in separate female and male zones on a monoecious inflorescence. Sometimes there is a sterile region. In any event, female parts have the odd tendency to be receptive *before* the male pollen is released. Meanwhile, willing insects may arrive only to drown in water collected at the base of the spathe. If pollination succeeds, fruits develop along the spadix as tight clusters of tiny berries. In some genera, these turn a bright color when ripe, attracting notice (from humans, at least) that the flowers may have evaded. Botanists continue to study the intricacies of how and why all this works toward the ongoing survival of the plants. Suffice to say, it's complicated.

Lords and ladies, *Arum maculatum*, owes its evocative common name to its classic spathe-and-spadix bloom arrangement. The spathe is a leafy covering for the true flowers on the erect spadix within. In Elizabethan times, starch from the root was used to stiffen the ruffed collars that wealthy men wore; this practice may also figure in the name.

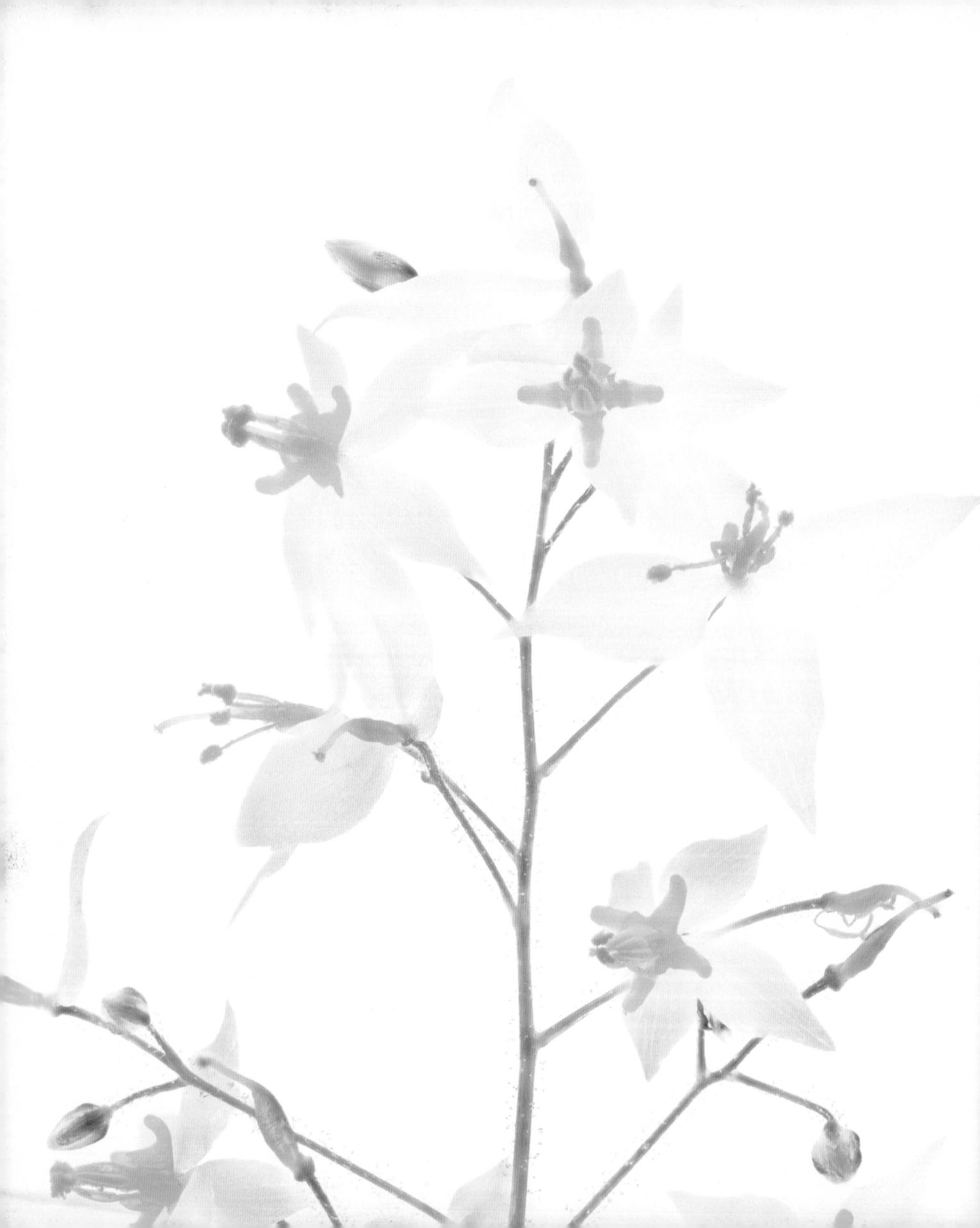

THE BARBERRY FAMILY

Berberidaceae

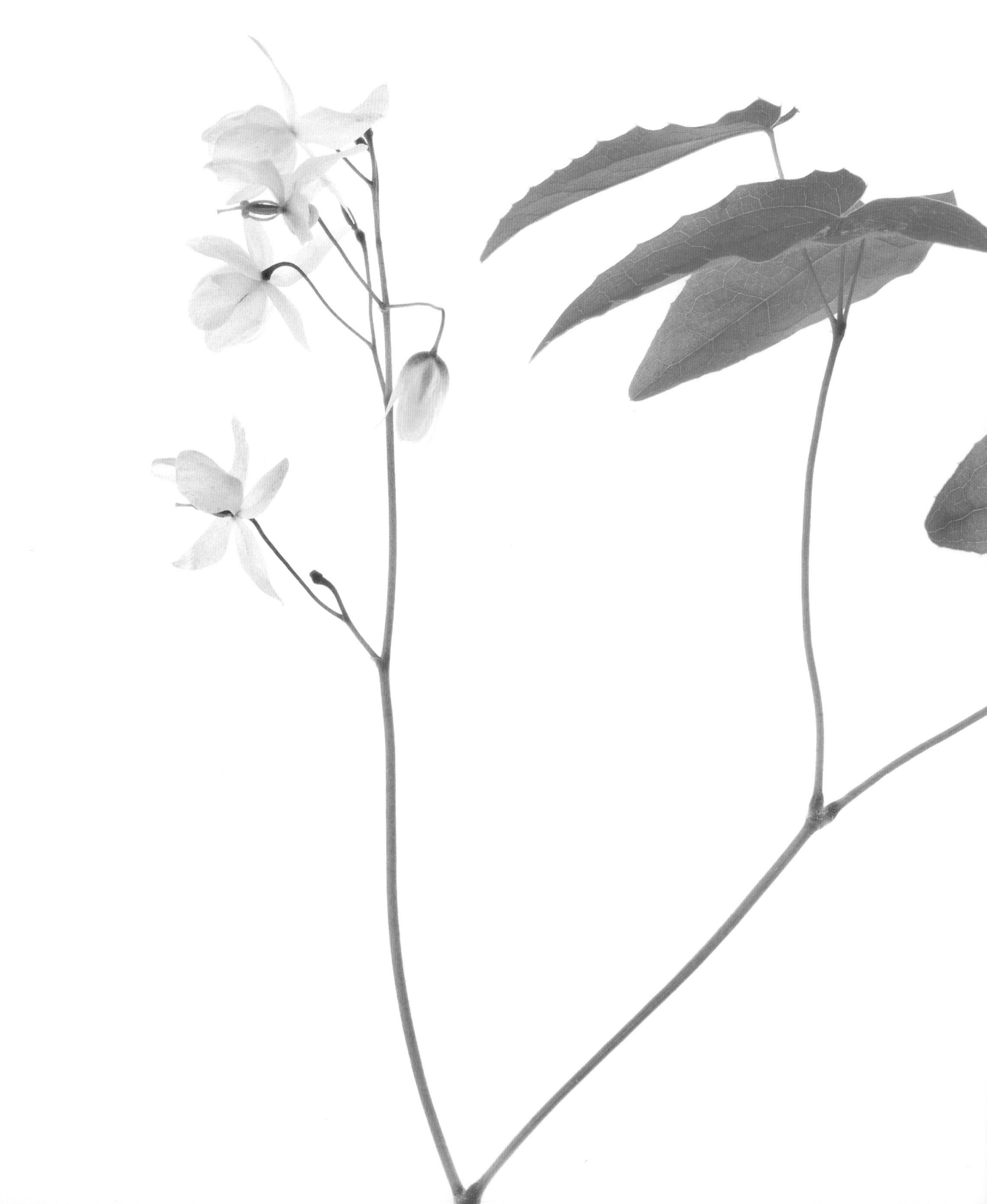

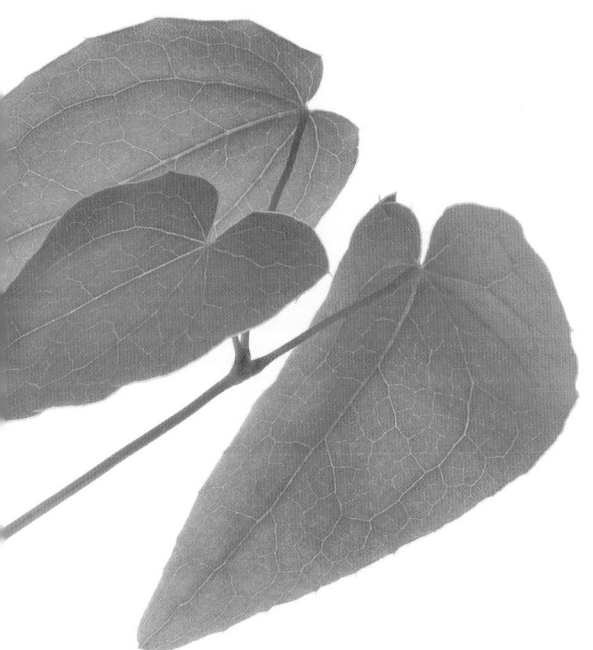

"Some persons have wondered why we have many more wild spring flowers than summer blooms. These dainty gems must hasten to produce their blossoms before the trees overhead shade them too much. . . . Every garden of any size should make room for a bed of wild flowers. . . . Flowers are never more enjoyable than early in spring. After the winter we look forward to these denizens of the woods."

ALFRED C. HOTTES, *A Little Book of Perennials*, 1923

LEFT AND PAGE 34 Bewitchingly pretty *Epimedium xyoungianum* has soft white flowers that are spurless, or nearly so, giving them a more conventional look than some of their near relatives.

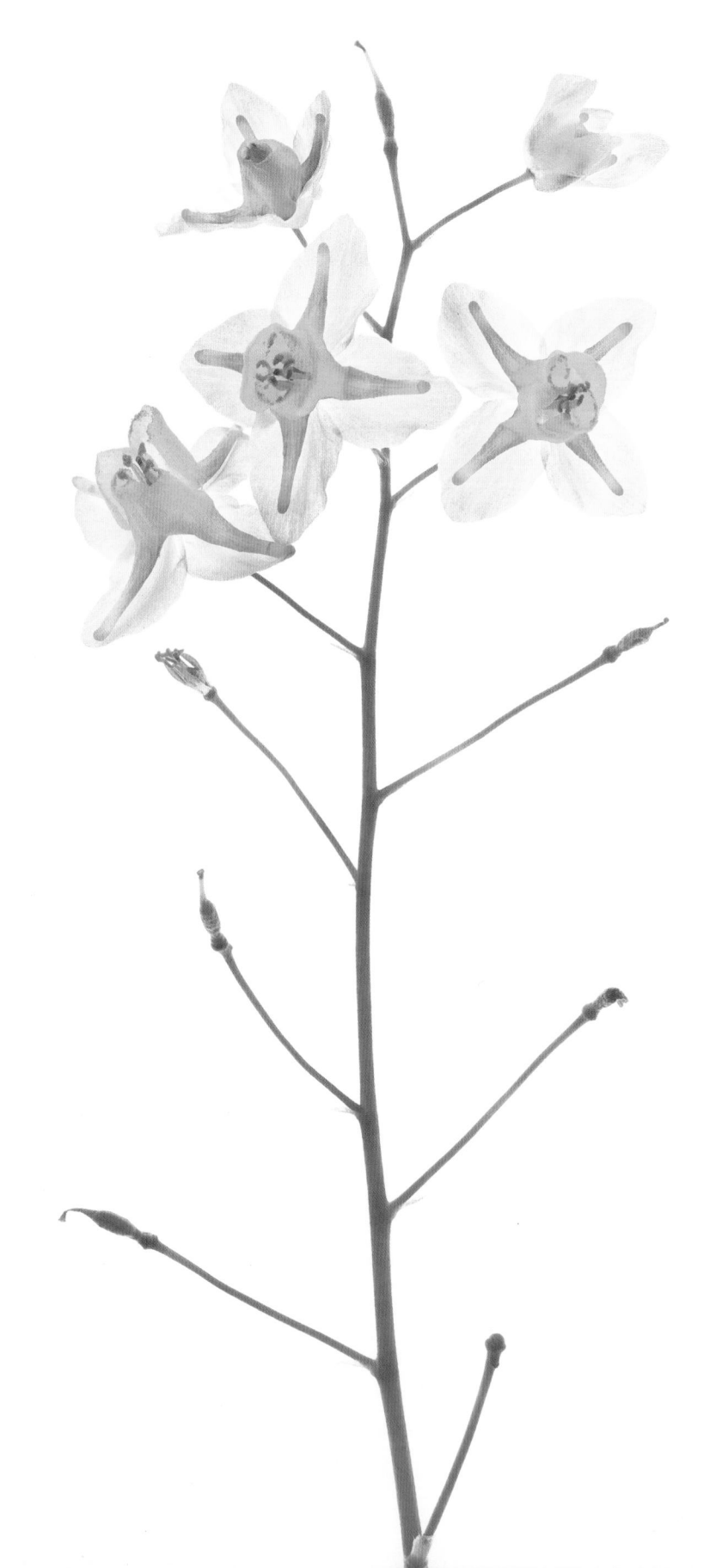

When Captain George Vancouver, botanist Archibald Menzies, and crew sailed up the coast of present-day Washington State and British Columbia in the 1790s in a ship called *Discovery*, one of their goals was to catalog the plants of the damp, wooded wilderness on the mainland and many islands. The sweet little wildflowers *Vancouveria hexandra*, or Vancouver fern (not a fern at all, although to some eyes it resembles a robust maidenhair; the *hexandra* actually refers to the six stamens in the tiny blooms), and similar *V. planipetala* and *V. parviflora*, inside-out flowers, bear the captain's name.

Other related wildflowers are mayapple, *Podophyllum peltatum*, and twinleaf, *Jeffersonia diphylla*. These also bloom in spring before the overhead trees are fully leafed out, bear little white solitary flowers, and carry their leaves in sets of two. You may encounter them in the woods; they are rarely grown in gardens.

But you are probably familiar with a close relative of all of these, *Epimedium*, or barrenwort or Bishop's hat. A genus of dozens of species, these have dainty leaflets borne on wiry stems. Mostly basal, the foliage is usually pinnately divided into triangular or angel-wing leaflets. Sometimes the leaves are veined, rimmed, or mottled in red, and they turn a handsome shade of bronze in fall's cooler weather. The springtime flowers, like vancouveria, are carried in airy sprays. In *E. grandiflorum*, flower color varies from white to shades of purple, and cultivars have been bred to display brighter and larger flowers, as with the enduringly popular 'Rose Queen'. The hybrid *E.* ×*versicolor* (a result of crossing *E. grandiflorum* with *E. pinnatum* subsp. *colchicum*) is widely grown as well, and has yellow-hued flowers. Those of *E.* ×*youngianum* (a cross between *E. diphyllum* and *E. grandiflorum*) are white and spurless or nearly so, and are delicate beauties.

It is fairly easy to see that these plants are related. In addition to favoring humusy soil under tall trees, they share physical characteristics. Their foliage is low, sprawling, and evergreen. Their tiny, generally cup-shaped flowers are certainly distinctive, sporting two rows, or layers, of sepals and petals. Often the petals are modified into showy spurs or pouches, hence epimedium's other, whimsical common name: Bishop's hat.

What unites these plants, besides springtime bloom and a preference for shady growing conditions, is their paired leaves. Large or small, these are carried jauntily aloft, like small umbrellas that could shelter little woodland faeries.

It is genuinely surprising, then, that they all belong to the family Berberidaceae. Yes, the family of barberry shrubs, *Berberis thunbergii*, and its many cultivars and related species, which also includes heavenly bamboo, *Nandina domestica*, and Oregon grape, *Mahonia aquifolium*. All are substantial bushy plants you may know best for their attention-grabbing, brilliant red to purple foliage every autumn. Their leaves are a far cry from umbrellalike, but rather small and compact, with heavy, lustrous, even leathery texture. Nor does *graceful* describe these woody relatives—barberry branches are spiny, heavenly bamboo gets quite tall and leggy, and Oregon grape can be either dense or stoloniferous.

How did this happen? The taxonomists were not thinking of the leaves or plant habit; they classified by the flowers. Typically members of this family sport four to six true sepals and four to six petals, plus four to six stamens. The fleshy or dry fruit that follows—whether mayapple or barberry, epimedium or Oregon grape—is a single-compartment berry or drupe, sometimes touted as edible (albeit sour). Genetic research has confirmed the kinship of these seemingly diverse plants, leading us once again to marvel at how relationships exist even when not immediately obvious.

The short-spurred yellow blooms of *Epimedium* ×*versicolor* 'Sulphureum' are carried in dainty racemes and composed of bright yellow petals and spurs, plus pale yellow sepals. Gardeners appreciate the way it prospers even in dry-shade settings.

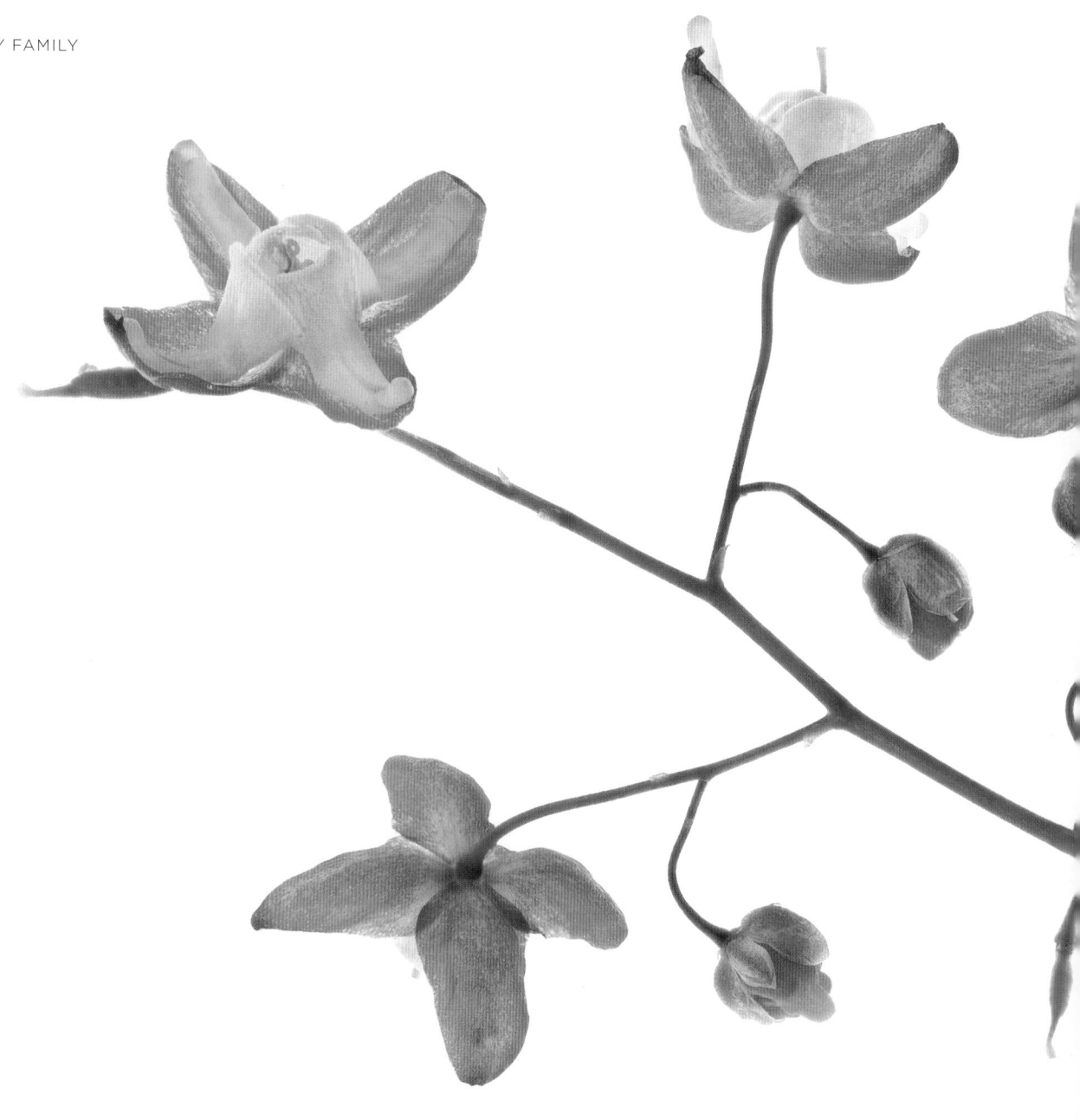

Red barrenwort, or *Epimedium ×rubrum*, gets its name from the ruddy color of its splashy little blooms. Notice that the flower form is uncomplicated, however, with four soft yellow petals and four red sepals each.

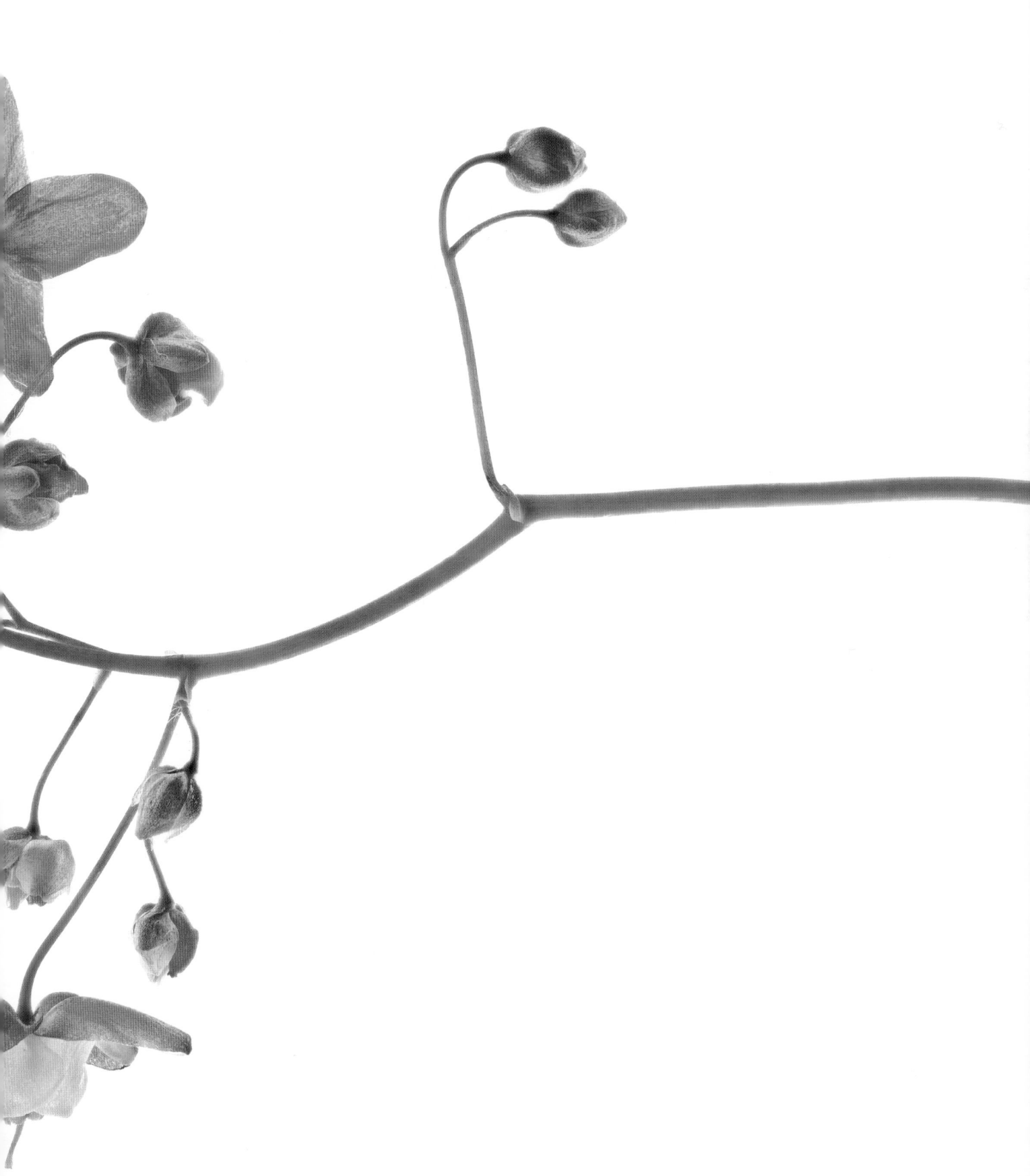

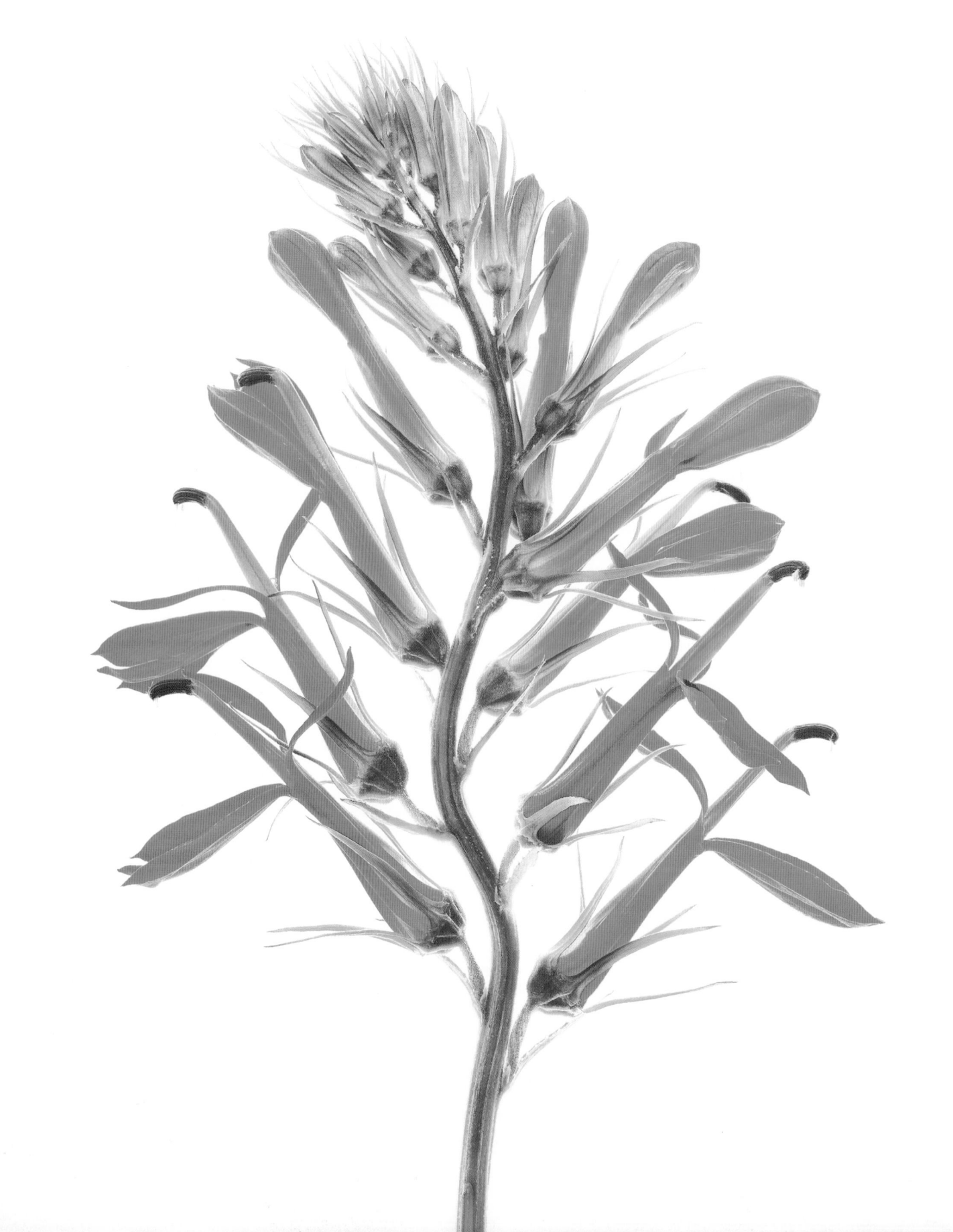

THE BELLFLOWER FAMILY

Campanulaceae

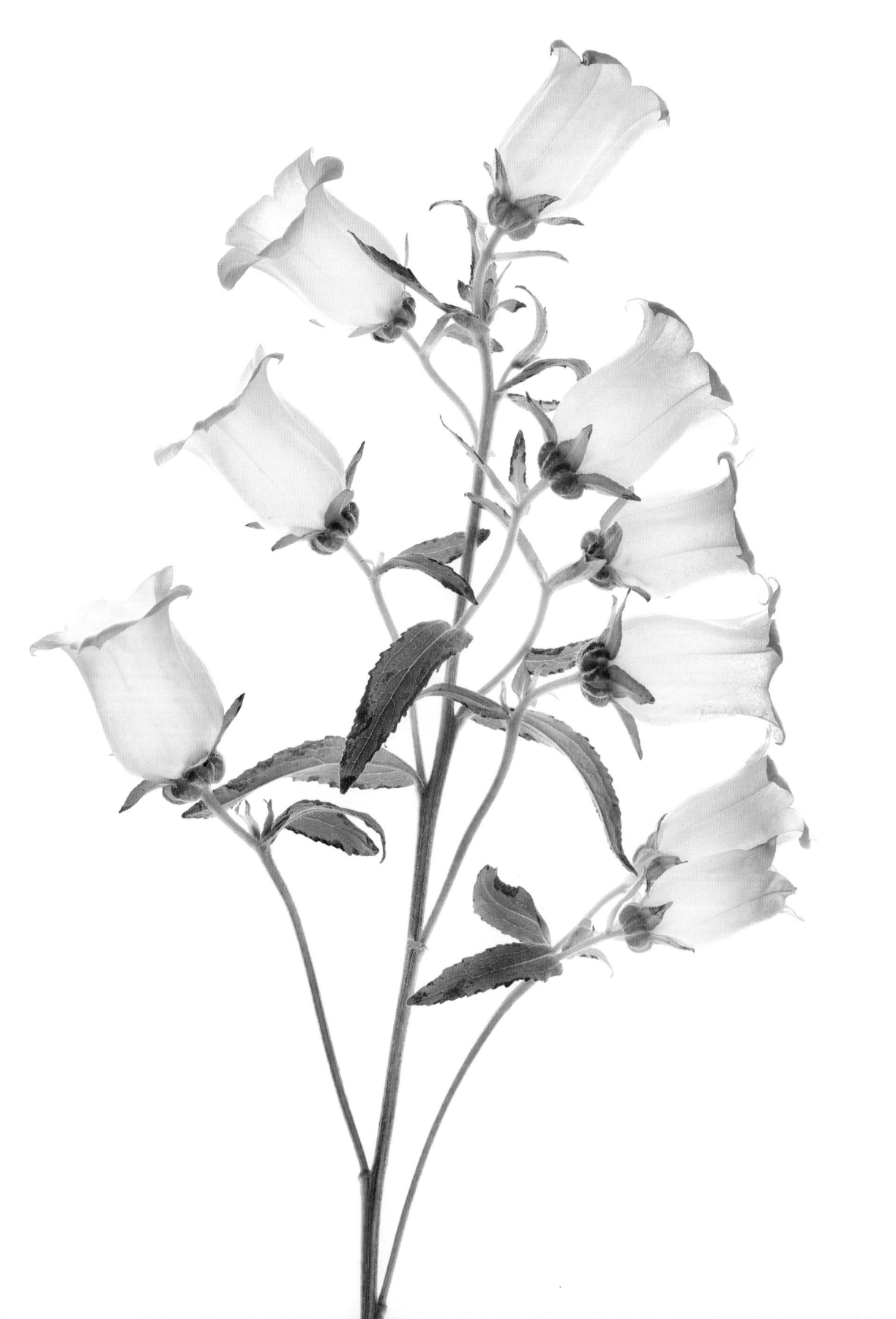

"A fair bell-flower

Sprang tip from the ground;
And early its fragrance

It shed all around;
A bee came thither

And sipp'd from its bell;
That they for each other

Were made, we see well."

JOHANN WOLFGANG VON GOETHE, "Like and Like," 1814

Canterbury bells, *Campanula medium,* usually behaves as a biennial, but the blossoms are worth waiting for, as they are vividly colored and long lasting. Each "bell" is actually made up of five fused petals.

PAGE 42 At first glance, the slender, more elongated blooms of cardinal flower, *Lobelia cardinalis,* do not seem related to the bellflowers. But these also have the five petals characteristic of the family, two pointing up and three directed down.

The aptly named bellflower family, or Campanulaceae, is viewed with great affection wherever its delightful members are seen and grown. Although there are about seventy genera included in the tribe, the most familiar ones are the campanulas. These go by such evocative common names as harebell, bluebells, Canterbury bells, Coventry bells, wild hyacinth, and Our Lady's nightcap. Some bells are short and round in overall shape, while others are long and tubular. Usually blue flowered, these plants range from low-growing, mound-forming favorites like *Campanula carpatica* to lanky *C. lactiflora* and the towering chimney bellflower, *C. pyramidalis*.

Also in this family and similar in looks and habits is the balloon flower, *Platycodon grandiflorus*. The flowers in bud are small, chubby, inflated balloons that eventually spring open. The cup-shaped blooms are much like bellflowers, although a touch shallower. The flowers of the species tend to be darker hued than most campanulas, closer to purple-blue. This species has rather weak stems, so two shorter, stouter selections, 'Mariesii' and 'Sentimental Blue', have been bred. They are 1 to 2 feet tall and a mere 6 to 9 inches tall, respectively.

A bell of these lovelies is created by five fused petals that curve ever so slightly outward, like an elfin cap, with five long, pointed sepals right behind them. Peering into the interior, you can observe five jaunty white stamens, closely packed or fused together so they form a minute tube. This tube becomes laden with (male) pollen grains. Meanwhile, the pistil in the flower's center, consisting of three fused carpels where seed eventually develops, curls back from the tips. As pollen accumulates, the (female) style expands and nudges the grains ever upward in expectation of a helpful bee so pollination can commence. It's an ingenious system. The grains land on a bee, which conveys them away to a nearby flower. Yes, this means members of this family cannot self-pollinate. Not until all the pollen has been ejected does the (female) style mature at last, splitting open to expose a sticky surface that can receive pollen from a different flower.

In the campanulas, you might also detect very tiny, downward-pointing hairs along the outside of the petals. Evidently these discourage ants from climbing up the blooms. It seems their services are not wanted, as they would not be very efficient pollinators.

Unique to this family is another protective feature: The ovary is tucked below where the petals attach, so insects visiting the flower cannot dine on the immature seeds. It appears that members of Campanulaceae prefer only bees as pollinators, and all other visitors are less welcome.

However, this turns out not to be completely true. Also included in Campanulaceae—surprisingly, at first glance—are the lobelias. Look closely. Yes, the flowers have five petals. But unlike their cousins, two point up and three point down. Those of cardinal flower, *Lobelia cardinalis*, feature an especially long tube with the pollen tube projecting beyond it. This configuration, and its usually dark red color, entices completely different, larger pollinators: hummingbirds and daytime moths with long mouthparts. These visitors are undaunted by the poor (that is, short) landing area and instead hover before the flower sipping nectar from deep in the tube. In the process, pollen rubs off on their foreheads.

Many flower lovers admire balloon flower, *Platycodon*, for the fun way its inflated blooms finally burst open (to look quite a bit like their cousins, the campanulas). Curiously, those of the cultivar *P. grandiflorus* 'Komachi' never do.

NEXT SPREAD So common is the (usually) blue-flowered lobelia, *Lobelia erinus,* and so small are its profuse flowers, that we may never pause to look more closely. And yet the individual blooms are worth marveling at, with their fan-shaped lower lips and frequently contrasting-color centers. Such a form is irresistible to pollinators, including butterflies.

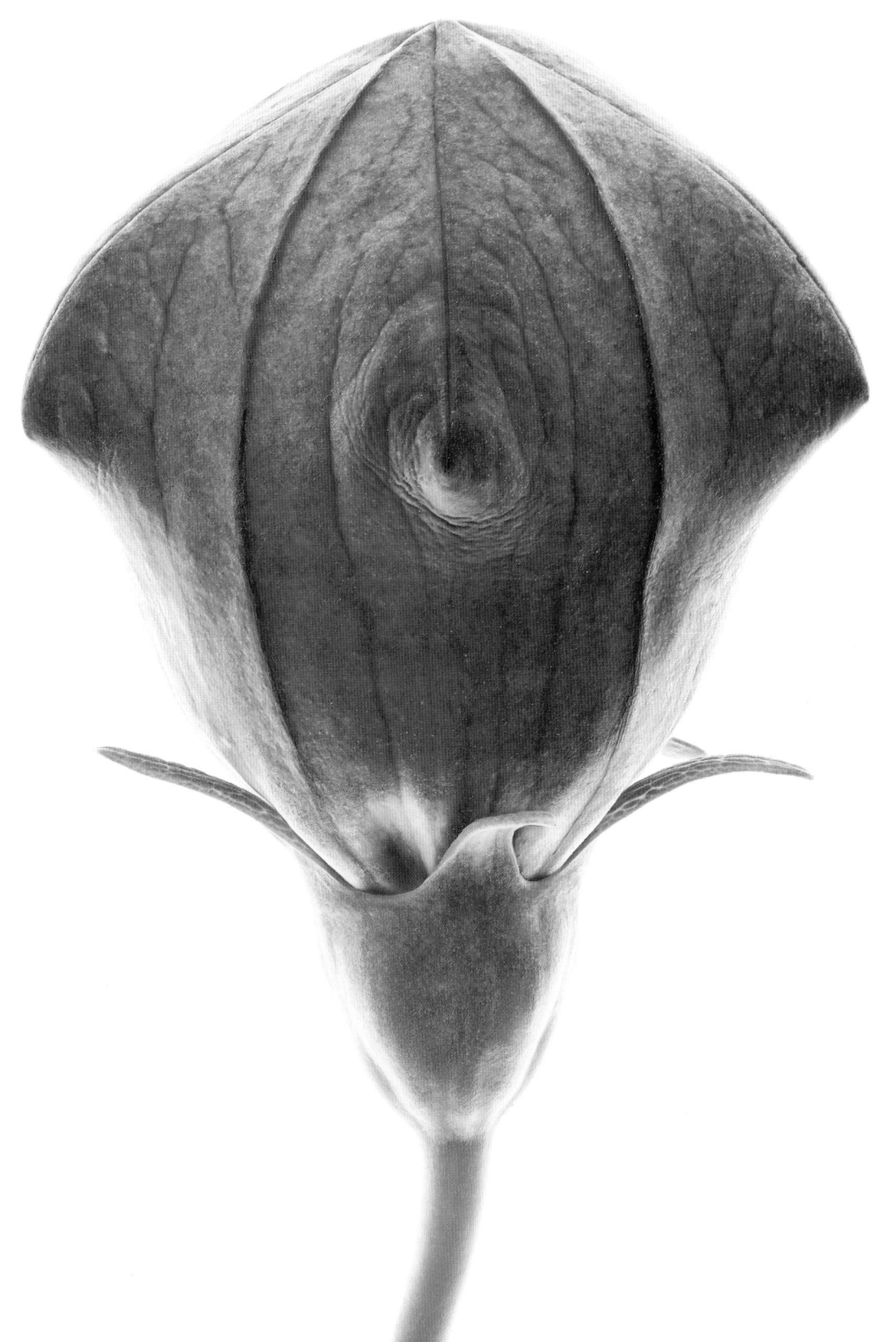

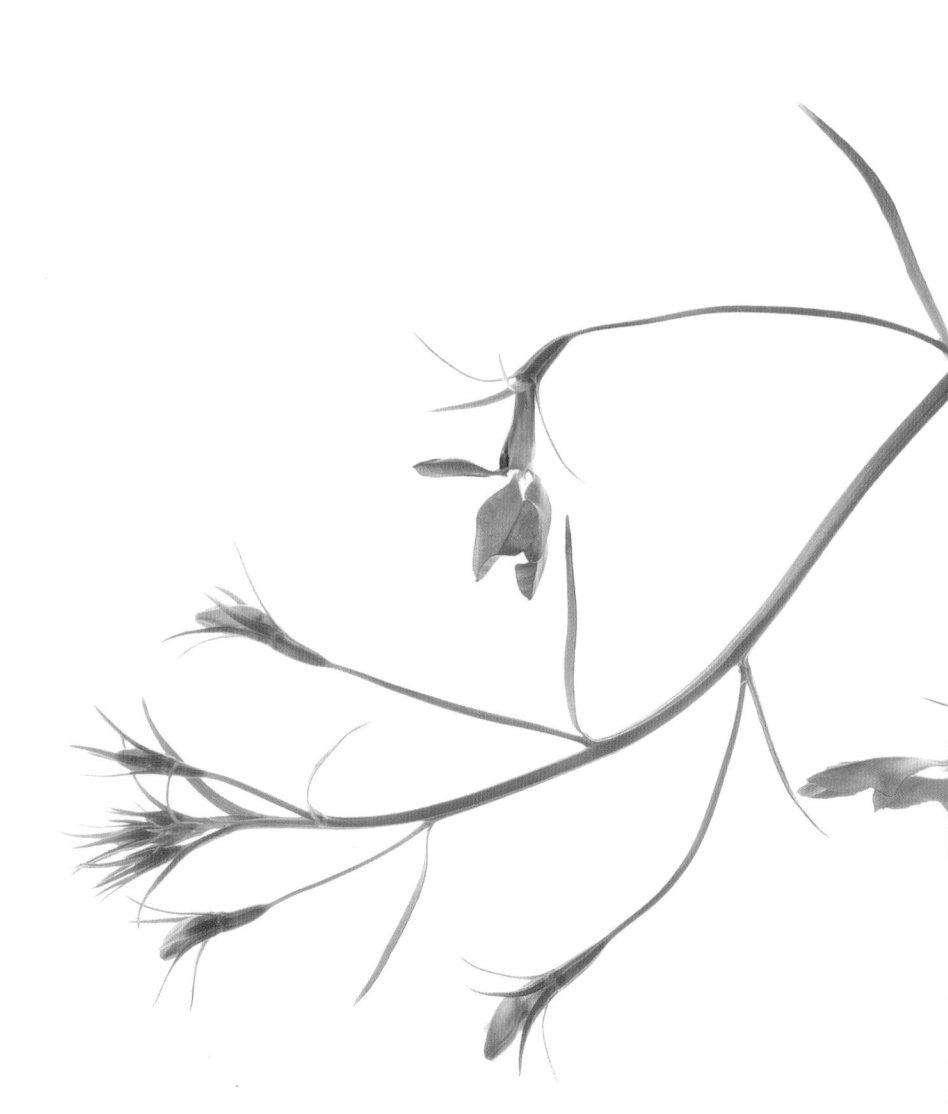

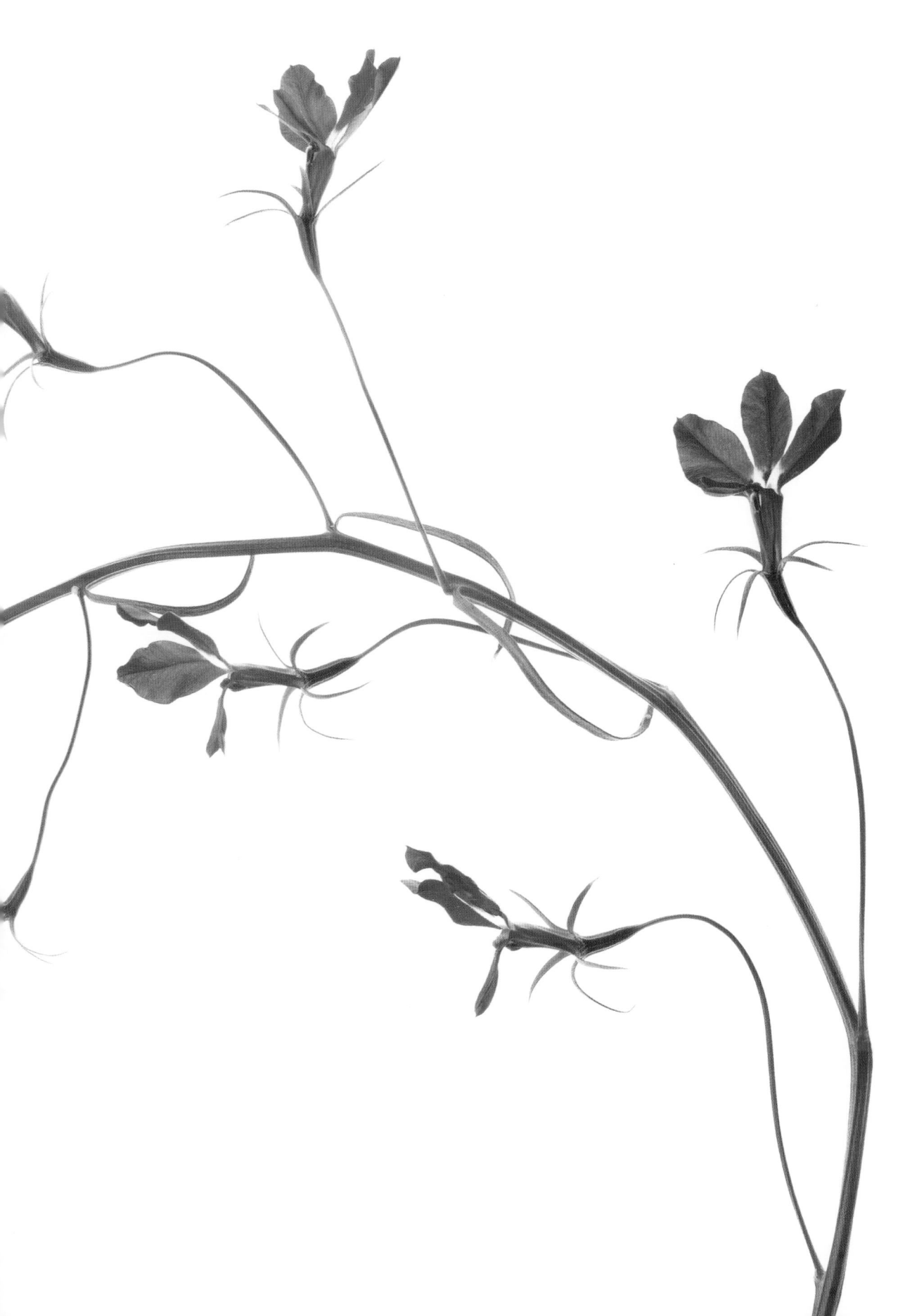

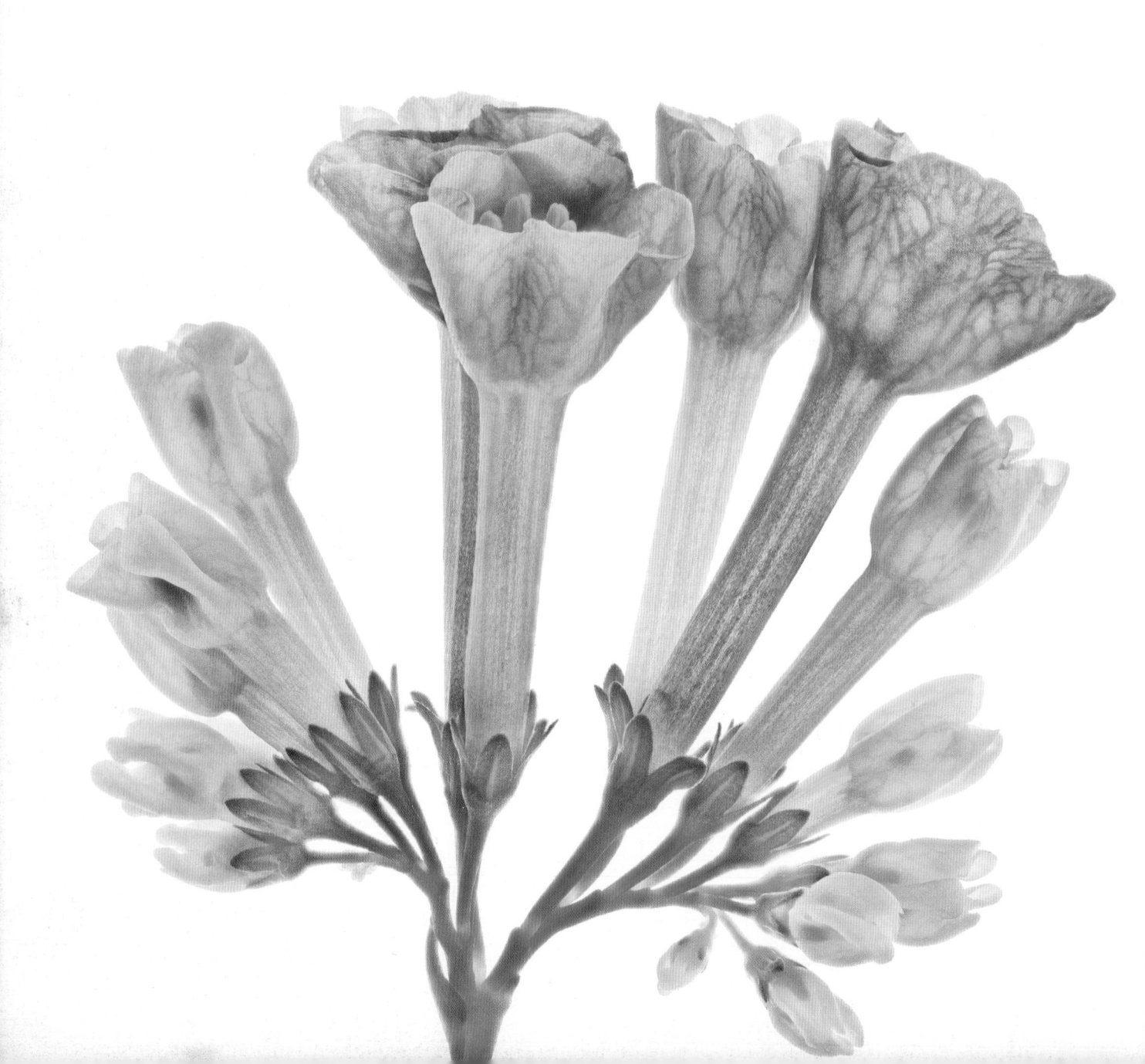

THE BORAGE FAMILY

Boraginaceae

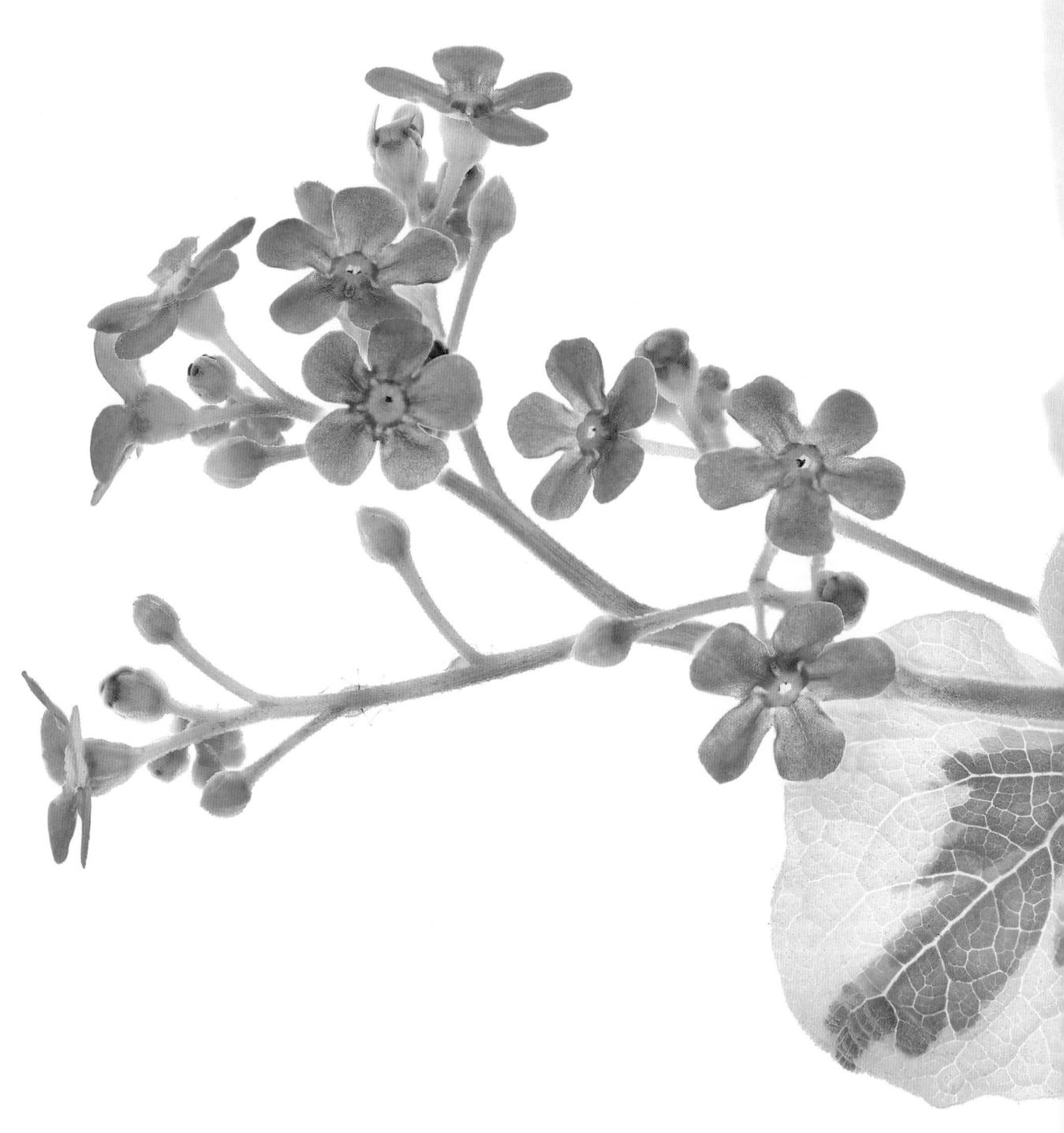

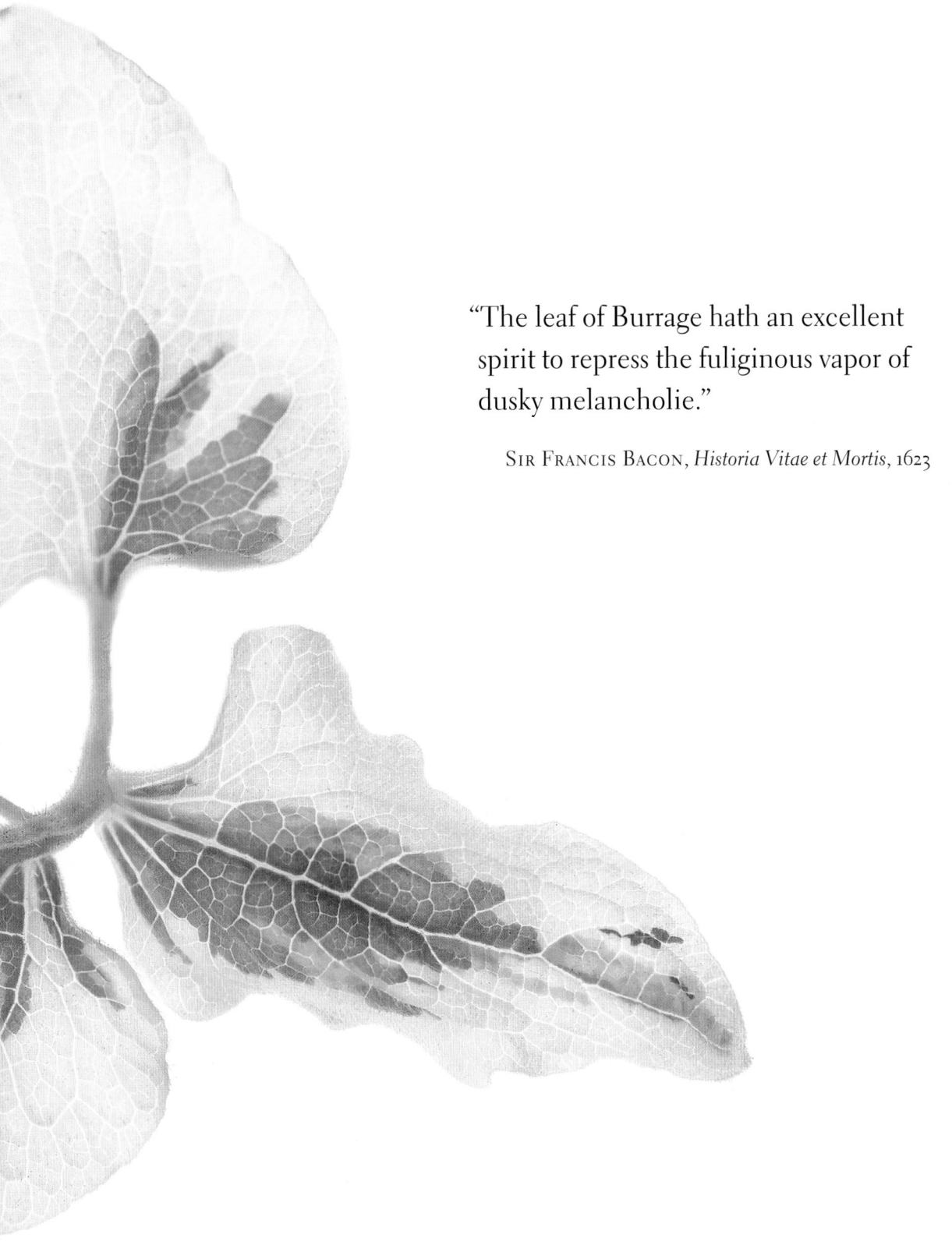

"The leaf of Burrage hath an excellent spirit to repress the fuliginous vapor of dusky melancholie."

SIR FRANCIS BACON, *Historia Vitae et Mortis*, 1623

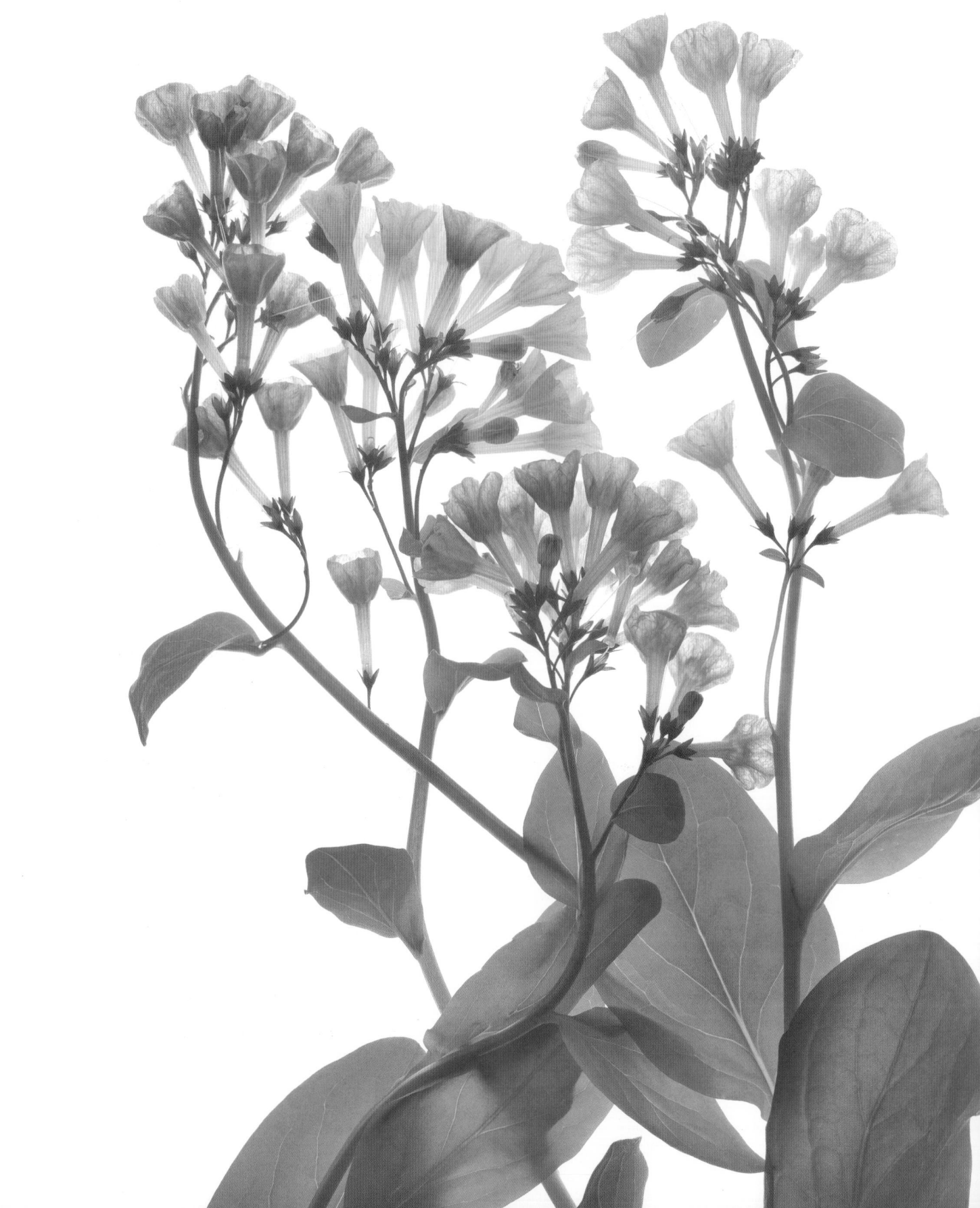

W hile it is not a showy or attention-grabbing group, members of the family Boraginaceae have plentiful virtues that have endeared them to ornamental and herb gardeners alike. The namesake itself, borage (*Borago officinalis*), is a long-popular herb credited with various beneficial and healing properties. Other relations include comfrey (*Symphytum*), Virginia bluebells (*Mertensia virginica*), forget-me-nots (*Myosotis* species), lungwort (*Pulmonaria*), Siberian bugloss (*Brunnera macrophylla*), and viper's bugloss (*Echium vulgare*).

Detect a theme? All these plants have flowers in shades of true blue, which is not always easy to find in the plant world. In some of the species, the presence of anthocyanins makes the distinctly red or pink buds open light purple and gently age to blue. The little blossoms are borne in clusters, so the plants become spangled with a charming contrasting, two-tone look. Botanists believe the color change is a signal to pollinators (mainly bees) that the nectar and pollen has become depleted, a chemical "no vacancy" sign.

Members generally share other distinguishing characteristics. The plants emerge from a basal rosette. Hollow stems arise from the center. Often both stems and leaves are fuzzy. These little hairs are sticky and at times even irritating to the touch, which may deter chewing insects.

Borage, an annual herb, has been in cultivation all the way back to medieval Europe. This flagship species for the family has earned notoriety, or at least familiarity, because it is widely distributed and naturalized on both sides of the Atlantic. Early herbalists credited it with invoking courage. Soldiers evidently believed that drinking borage-flavored wine would give them courage in battle, and maidens were urged to serve their suitors this wine if they wanted a marriage proposal (if helpful, one wonders whether it was the wine acting).

Healers prescribed borage not only as a mood enhancer but also as a diuretic, an emollient, and a treatment for diarrhea. Of all these uses, the only one modern science supports is the last. However, fuzzy leaves or not, borage is perfectly safe to consume and has a light but pleasant cucumber flavor. Indeed, it has made an appearance in a trendy modern-day dish; *Saveur Magazine* popularized its use in herb-and-cheese-filled ravioli. (However, the recipe allows that you could substitute dandelion leaves, Swiss chard, or spinach.)

Despite its abundance of ¾-inch blue star-shaped flowers, borage itself has not attracted a lot of attention as an ornamental plant. At 2 to 3 feet high and wide, with a lax growth habit, it is a little too sprawling for most flower gardens, although it certainly brings welcome color to an herb garden.

Its cousin comfrey, on the other hand, gets more credit as a useful or medicinal herb (for external use only). Comfrey contains allantoin, a fascinating compound that affects tissue growth. Poultices are used to heal wounds, bruises, and various skin problems, and it is employed in cosmetic lotions and creams. It once enjoyed popularity as a salad green (it is a rare plant source of vitamin B_{12}, and it was once billed as "the world's fastest protein-builder"), until research in the early eighties showed it to be possibly carcinogenic. Alas, like borage, it is not much valued as an ornamental

LEFT AND PAGE 50 The enchanting native American wildflower Virginia bluebells, *Mertensia virginica*, blooms generously every spring. Pink buds in coiled racemes elongate and open to little blue trumpet-shaped flowers. Look closely: Each one houses five stamens and a lone pistil.

PREVIOUS SPREAD Despite its clunky common name of Siberian bugloss, *Brunnera macrophylla* brings a gracious demeanor to shade gardens, thanks to its tidy clumps of large heart-shaped leaves. Sprays of diminutive, five-petaled blue flowers add to the show in spring. The cultivar 'Jack Frost' has silvery leaves.

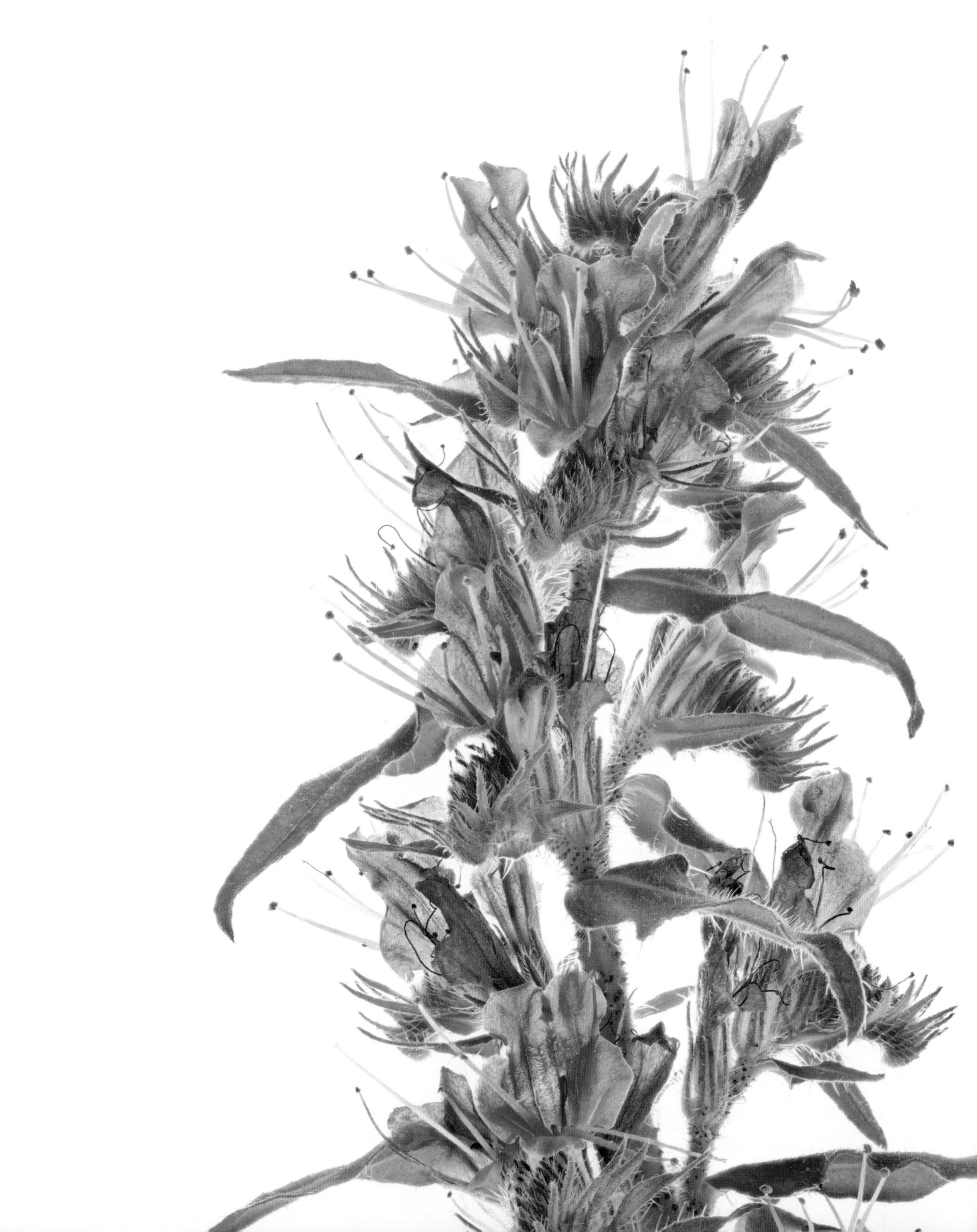

plant because of its clumsy, rambling profile—up to 5 feet tall and wide.

Another bulky borage family member, however, has become a standby in many California and other mild-climate gardens. Pride of Madeira, *Echium fastuosum*, a native of the Canary Islands, grows thickly to as much as 6 feet high, laden with spiky foliage and impressive bluish flower spires. A mature plant is truly spectacular, and often seen anchoring mixed borders or flanking garden gates.

Meanwhile, in cooler climates, modest-size Virginia bluebells and low-growing forget-me-nots are enduringly popular for springtime color. They make good companions for spring bulbs, especially tulips, and hang around to distract from the dying-down foliage. The latter can even become a self-sowing nuisance.

Pretty *Pulmonaria* is probably the borage family member that has enjoyed the most horticultural refinement. This one can be found at many garden centers in various hues of blue, as well as in pure pink- and white-flowered selections ('Sissinghurst White' is a classic). Its signature spotted leaves provide a handsome backdrop for its own flowers, as well as others nearby.

The truth is, blue in any hue is always welcome in gardens and bouquets. It mingles well with other pastels, especially pink flowers, and is equally terrific alongside orange or yellow. The gift of this family is a range of easygoing plants that will supply plenty of cheery blue, no matter what sort of climate your garden is in.

Look past the blue petals of viper's bugloss, *Echium vulgare*, and you'll see that the flowers also feature red stamen filaments and blue pollen. These help them to stand out in form as well as color to pollinators—and to us.

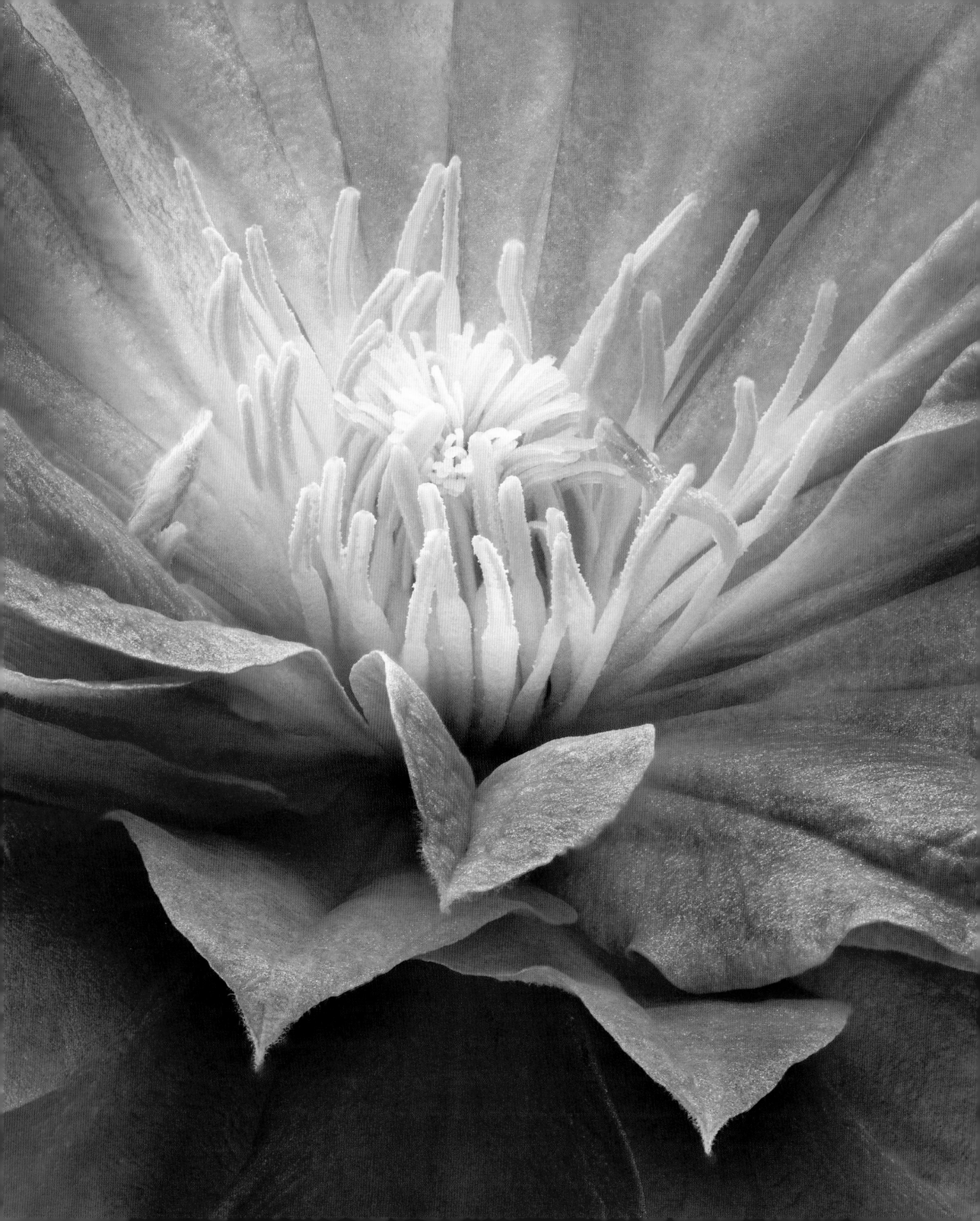

THE BUTTERCUP FAMILY

Ranunculaceae

"There is a Flower, the Lesser Celandine,
 That shrinks, like many more, from cold and rain;
 And, at the first moment that the sun may shine,
 Bright as the sun itself, 'tis out again!"

WILLIAM WORDSWORTH, "The Small Celandine," ca. 1803

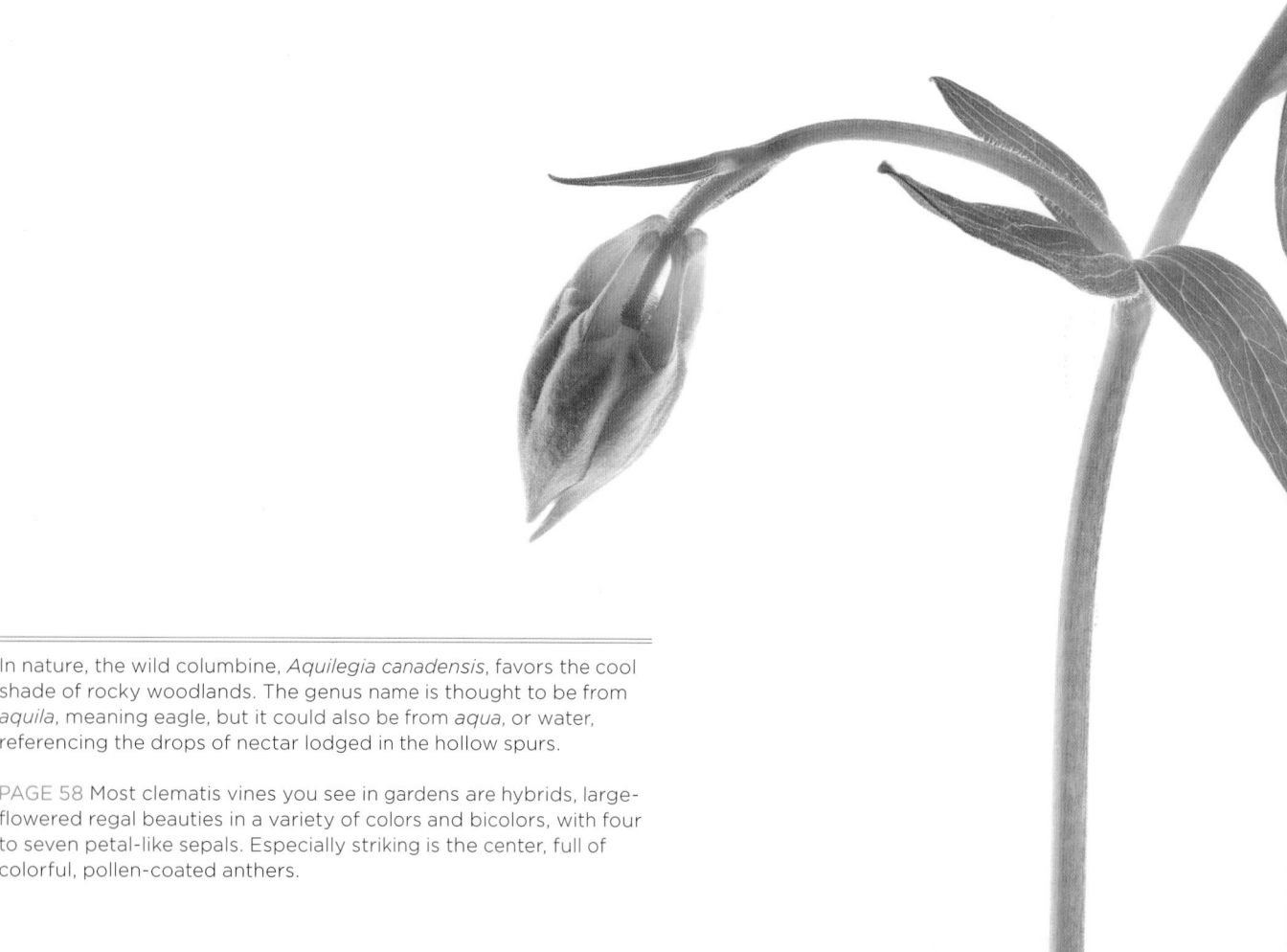

In nature, the wild columbine, *Aquilegia canadensis*, favors the cool shade of rocky woodlands. The genus name is thought to be from *aquila*, meaning eagle, but it could also be from *aqua*, or water, referencing the drops of nectar lodged in the hollow spurs.

PAGE 58 Most clematis vines you see in gardens are hybrids, large-flowered regal beauties in a variety of colors and bicolors, with four to seven petal-like sepals. Especially striking is the center, full of colorful, pollen-coated anthers.

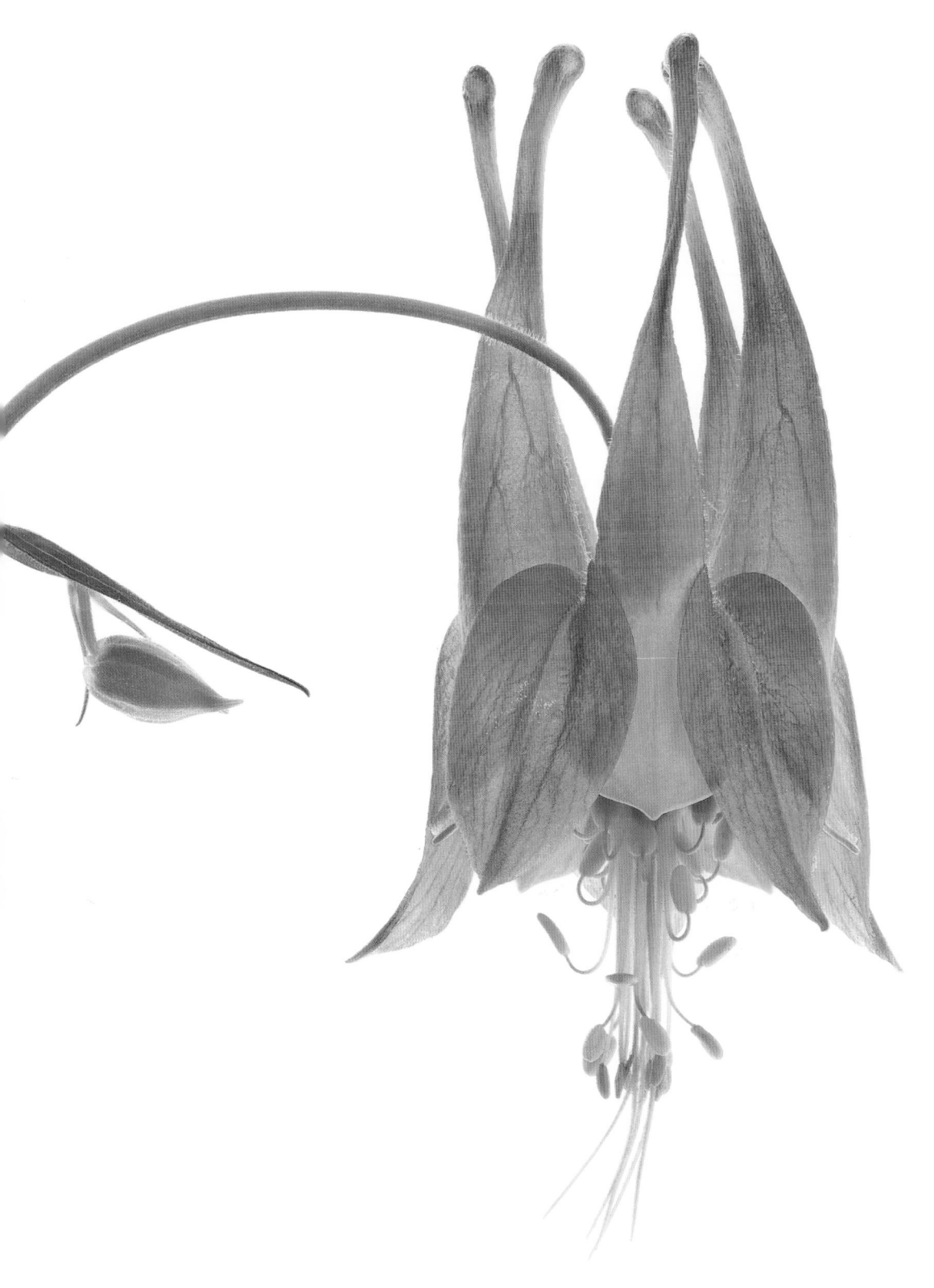

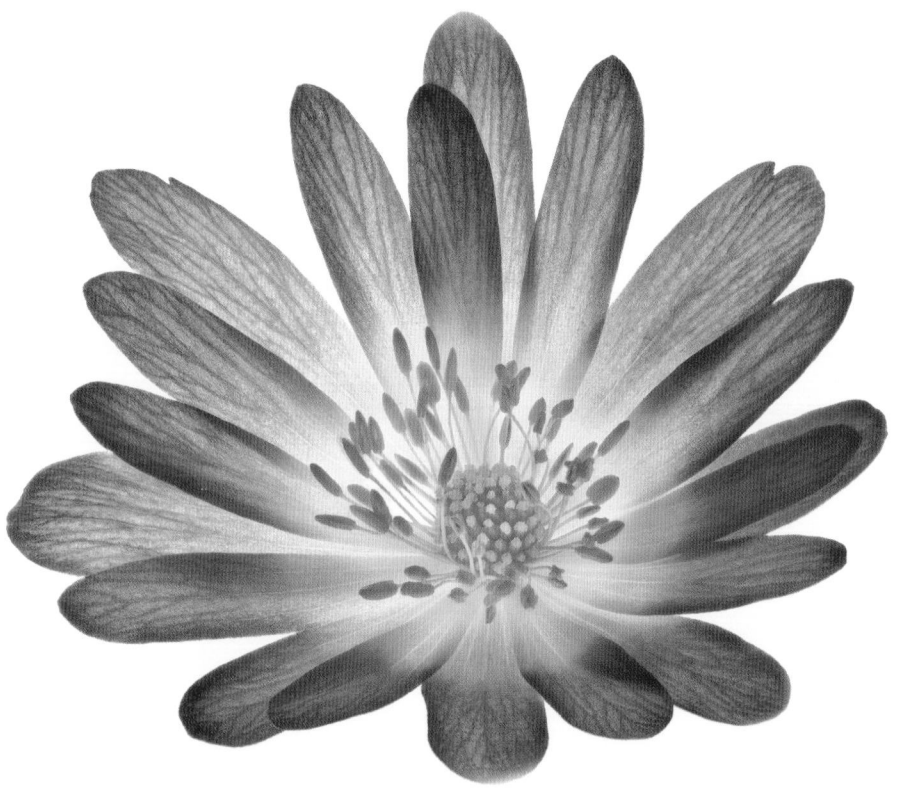

From columbines to clematis, delphiniums to hellebores, anemones to winter aconites, the family Ranunculaceae (also known as Crowfoot or Buttercup) encompasses a broad range of handsome, mainly herbaceous plants. It is a substantial and diverse group that includes favorite garden perennials, flowering bulbs, and even some weedy wildflowers: Botanists place over 1700 species here. In an era when molecular biology determines plant taxonomy, resulting in many revisions and splitting, it is remarkable that this family continues to contain so many members.

The family name has a folksy origin. In Latin, *ranunculus* means "a little frog." Celebrated Roman botanist Pliny evidently bestowed this name long ago because all the species he knew seemed to grow near ponds and wetlands, where frogs also reside. Many species prefer a moist habitat, especially the celandines, although the group is so large that it's hard to generalize.

In any event, with very few exceptions, the flowers of Ranunculaceae plants have five sepals (which may be called, or appear to be, petals). Stamens are many, and fused carpels are common—peer inside columbines or delphiniums, for example. Leaves are generally lobed

ABOVE Native to southeast Europe, including Turkey and the island of Cyprus, the plucky Grecian windflower, *Anemone blanda*, grows from a small tuber or corm and has daisylike flowers. Like other members of the buttercup family, however, the petals are actually colorful sepals, and abundant stamens flank fused carpels.

RIGHT Among the finest perennial garden flowers, delphiniums tower over many others, and *Delphinium elatum* hybrids like this one often need staking. Individual single flowers have five petal-like sepals around four true petals; doubles, of course, are fuller. Butterflies and bumblebees are the pollinators of these lovelies; resulting seeds are tiny, shiny, and black.

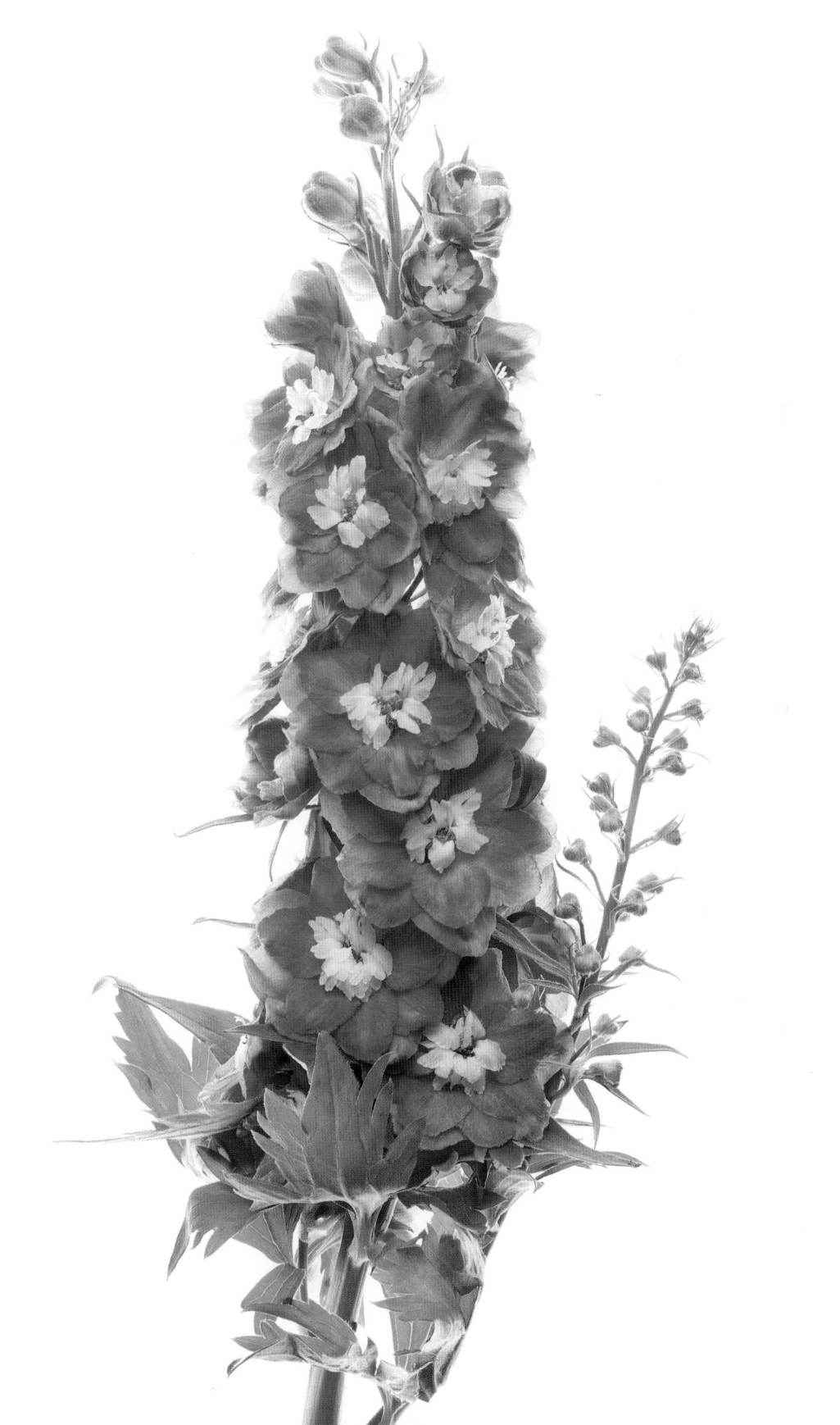

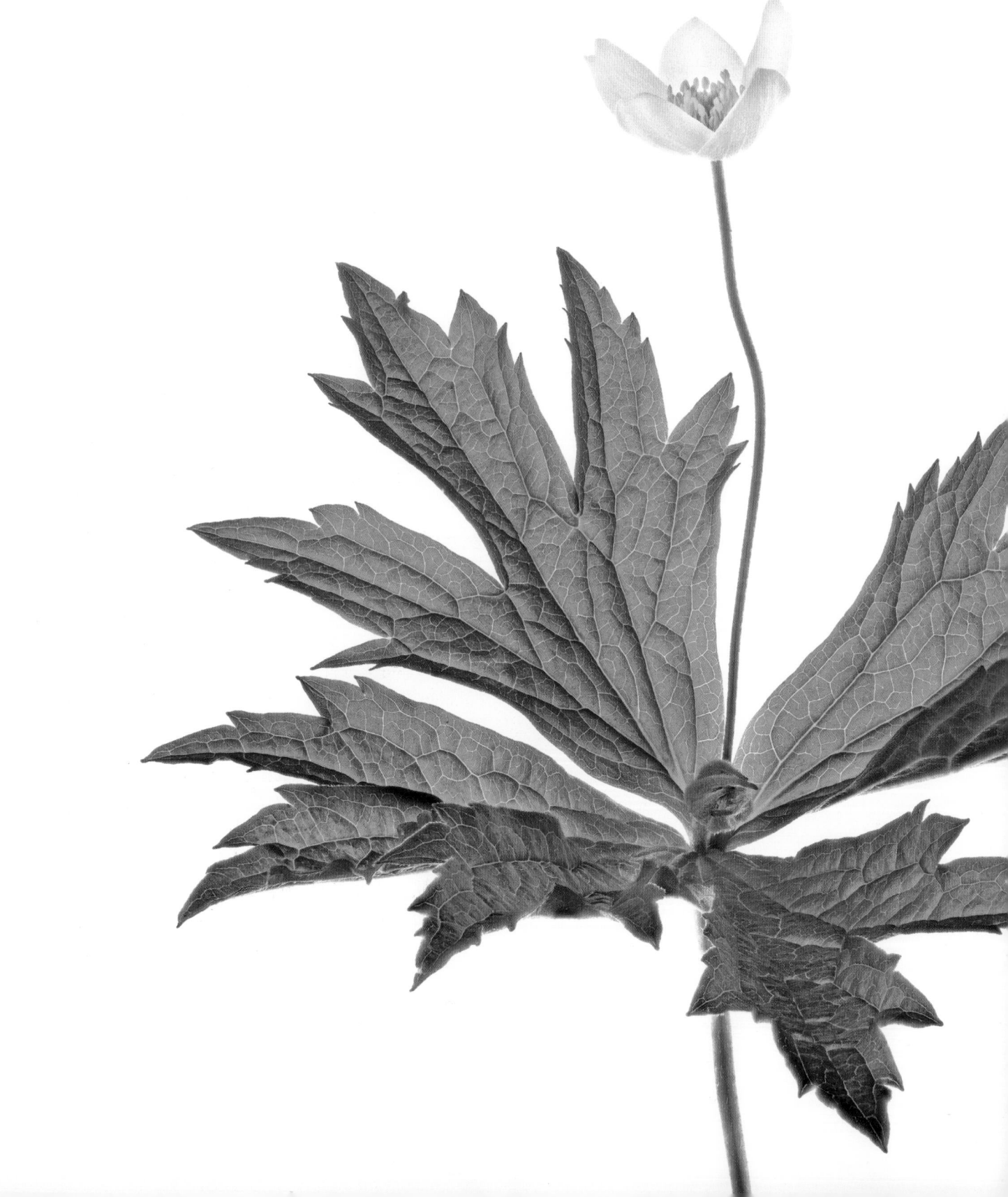

and/or deeply cut or divided, as with anemones, buttercups, and celandines. The "crowfoot" common name refers to such leaves.

Leaves and stems, and often all other plant parts, house a bitter, acrid juice that contains toxic compounds. Among them are protoanemonin, alkaloids, and glycosides, all of which can be harmful to humans and animals if consumed and are capable of causing blisters on exposed skin. Livestock avoid grazing on members of this family, and wild nibblers such as deer and rabbits also tend to eschew Ranunculaceae.

On the other hand, the chemicals may be useful if deployed with great care medicinally, as with the controversial herb goldenseal (*Hydrastis canadensis*). This native American plant was a favorite of Cherokee healers, who relied on its root for treating a wide range of ailments. A turn-of-the-century self-taught country doctor, Jethro Kloss, declared it "one of the most wonderful remedies in the entire herb kingdom" and promoted its use far and wide. Over the years, goldenseal has been credited with easing everything from insect bites to an upset stomach, preventing morning sickness, and lowering blood sugar. During its heyday, as native stands were plundered for profit, it eventually became expensive and hard to find. It remains controversial, as some medical professionals and herbalists have branded it harmful and poisonous. It should never be taken internally or used inadvisably.

Casual handling or contact with plants in this family, including in your garden, should not be risky if you are prudent. Indeed, people love to pluck a blooming stem of petite wild buttercup, hold it under their chins, and look for the yellow reflection that reveals that they like butter. Cautious perennial gardeners are careful to steer curious toddlers and browsing pets away from their

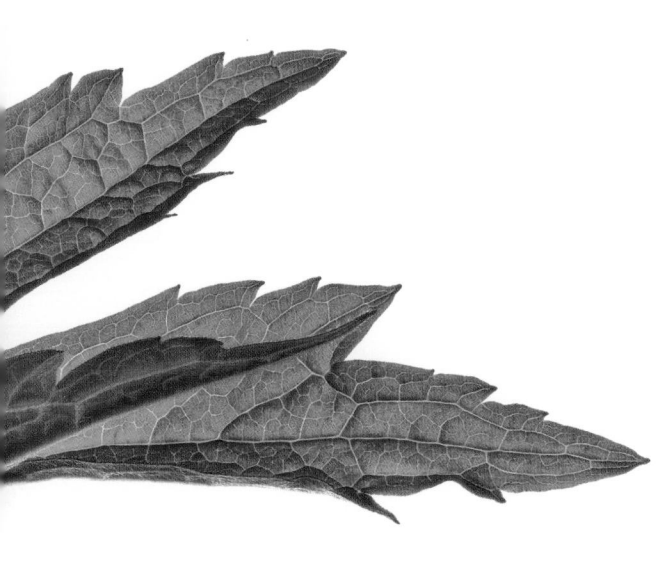

The starry white flowers of the meadow anemone, *Anemone canadensis*, are composed of five sepals, as is typical of most plants in this family.

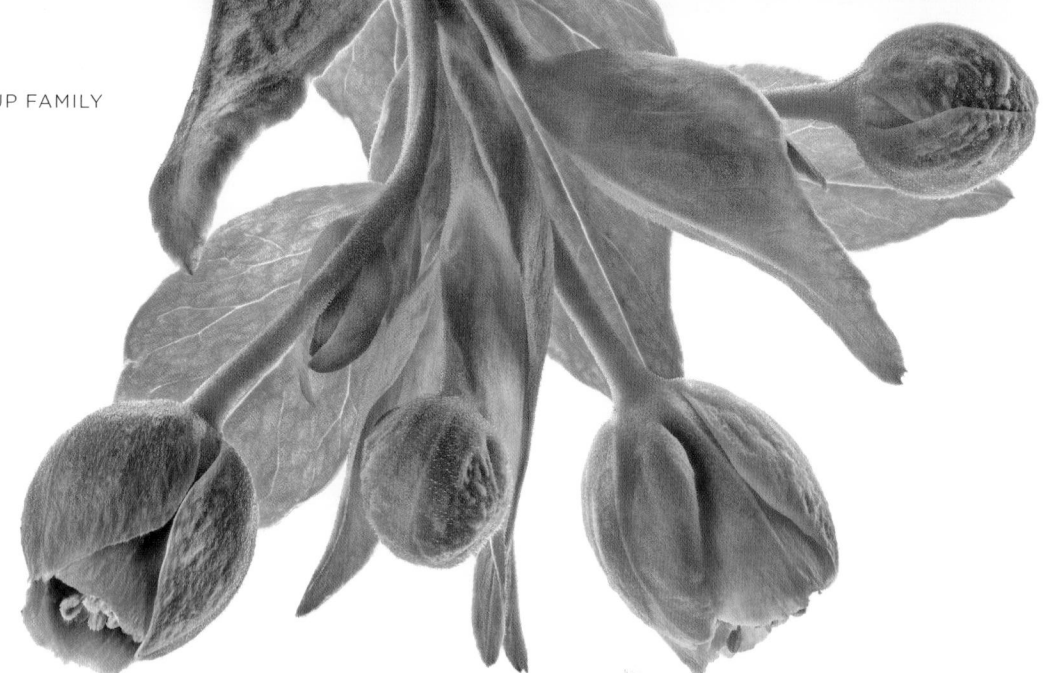

monkshood (*Aconitum*) and delphinium plants. Since both of these are tall, spiky growers, your best bet is to prevent trouble: Keep them shielded or beyond reach with a buffer of shorter, harmless plants (which also looks nice).

The greasy sheen on buttercup petals is neither imagined nor magical. Small and slim though they may be, these petals have three parts: a waxy layer on top, a layer of yellow cells, and a layer of white cells that gives the yellow a boost in brilliance. Close inspection reveals, however, that the white layer does not extend all the way into the very center of the flower. Also deep in the center of a buttercup flower—in addition, of course, to the signature spray of tiny pistils and stamens—is a little flap at the very base of each petal, called a nectar scale. Here, nectar is produced and stored for the pollinating insects, including but not limited to bees.

The wide array and attractiveness of the flowers in this family is hard to resist. Horticulturists have selected or developed varieties that gardeners seek out and cherish. From the original, dainty red-and-yellow wild columbine, *Aquilegia canadensis*, and the gorgeous state flower of Colorado, blue-and-white *A. caerulea*, have sprung numerous colors and bicolors, as well as double flowers. Breeders have also brought us bigger columbine flowers, which is welcome indeed—the multicolored McKana hybrids, for example, are enduringly popular. Meanwhile, more sumptuous hues of hellebore (*Helleborus* cultivars and strains) come to market practically every year. Delphiniums, ranunculus, larkspur, clematis—all of these Ranunculaceae, and more, continue to be offered in a broad range of colors, forms, and sizes. Dicey chemical content notwithstanding, no garden should be without a few, for their beauty is so compelling.

ABOVE In late winter or earliest spring, green hellebore, or *Helleborus foetidus*, arises to display clusters of maroon-edged, pale green flowers atop sturdy stems. It has the five petal-like sepals typical of this plant family.

RIGHT The blossoms of columbines, *Aquilegia*, may look complicated, but they still conform to the five-part form of many of their kin. Five long, hollow spurs on the petals extend upward, ending in tiny bulbous tips that shelter the nectaries. Sepals, colored like the petals, alternate with them. The long style and numerous stamens protrude from the corolla. When a flower begins to fade, the downward-curving pedicels straighten to hold the distinctive ripe fruit erect. At this point, it is easy to discern five parallel, tube-shaped follicles full of seeds.

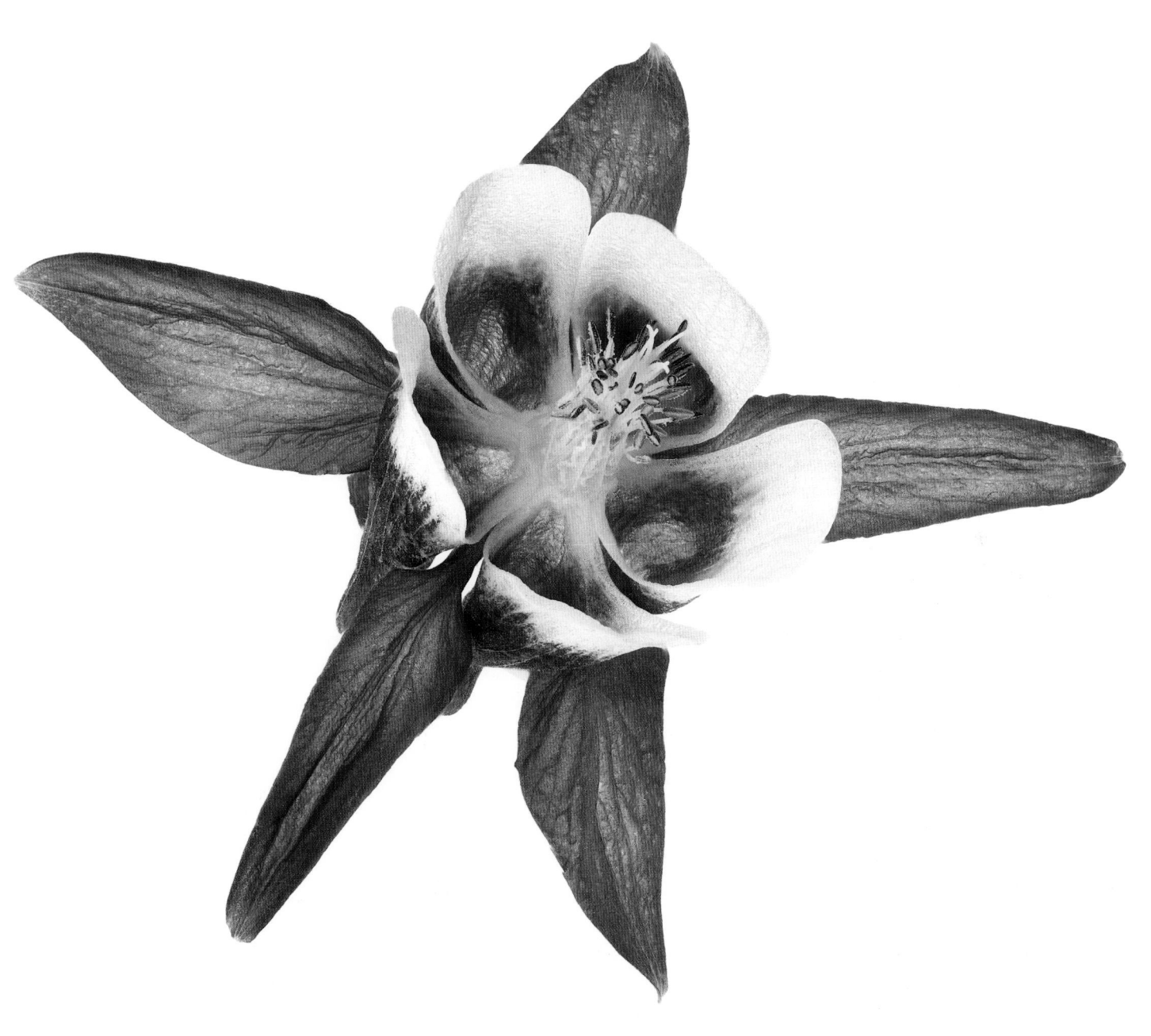

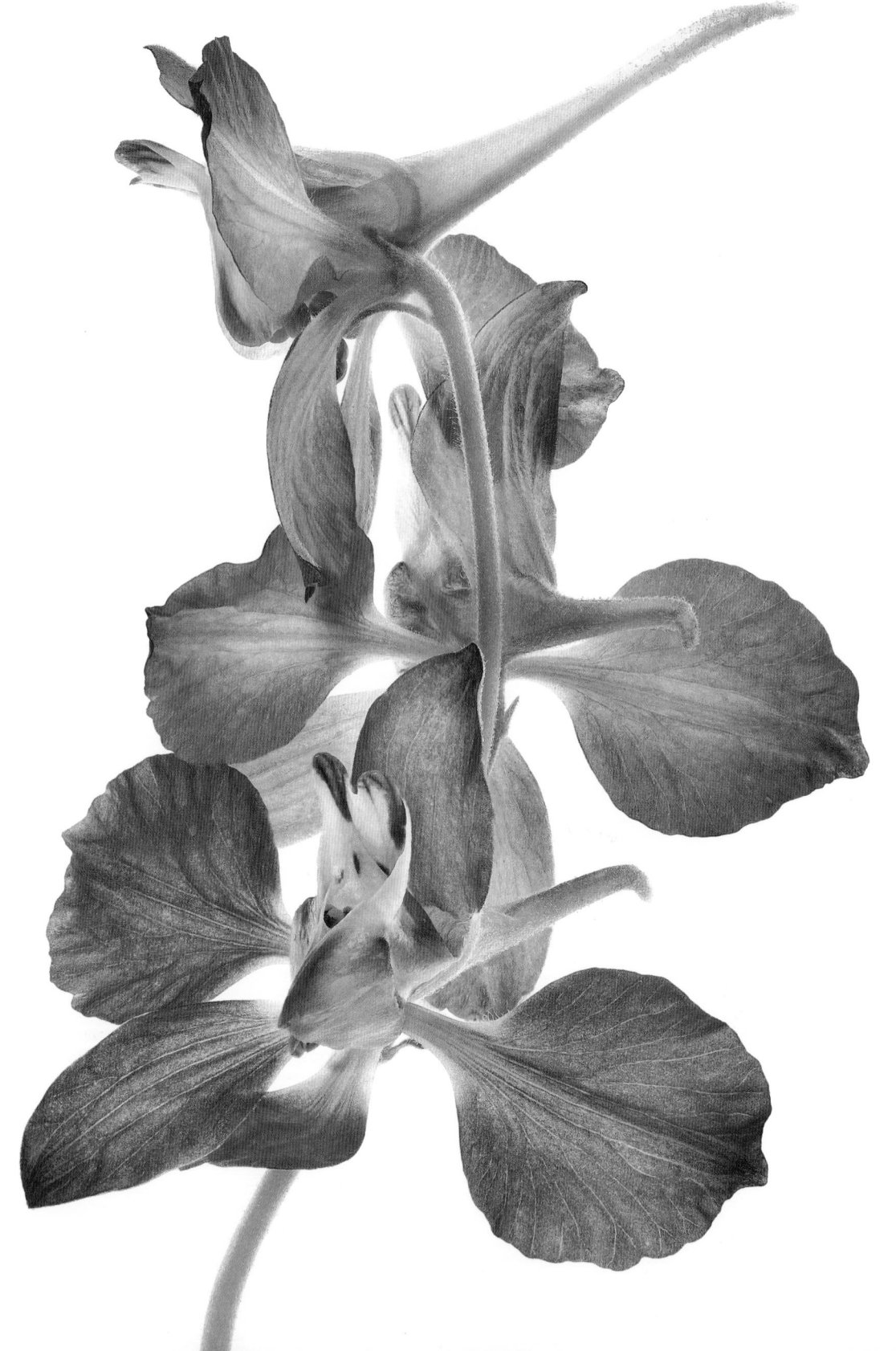

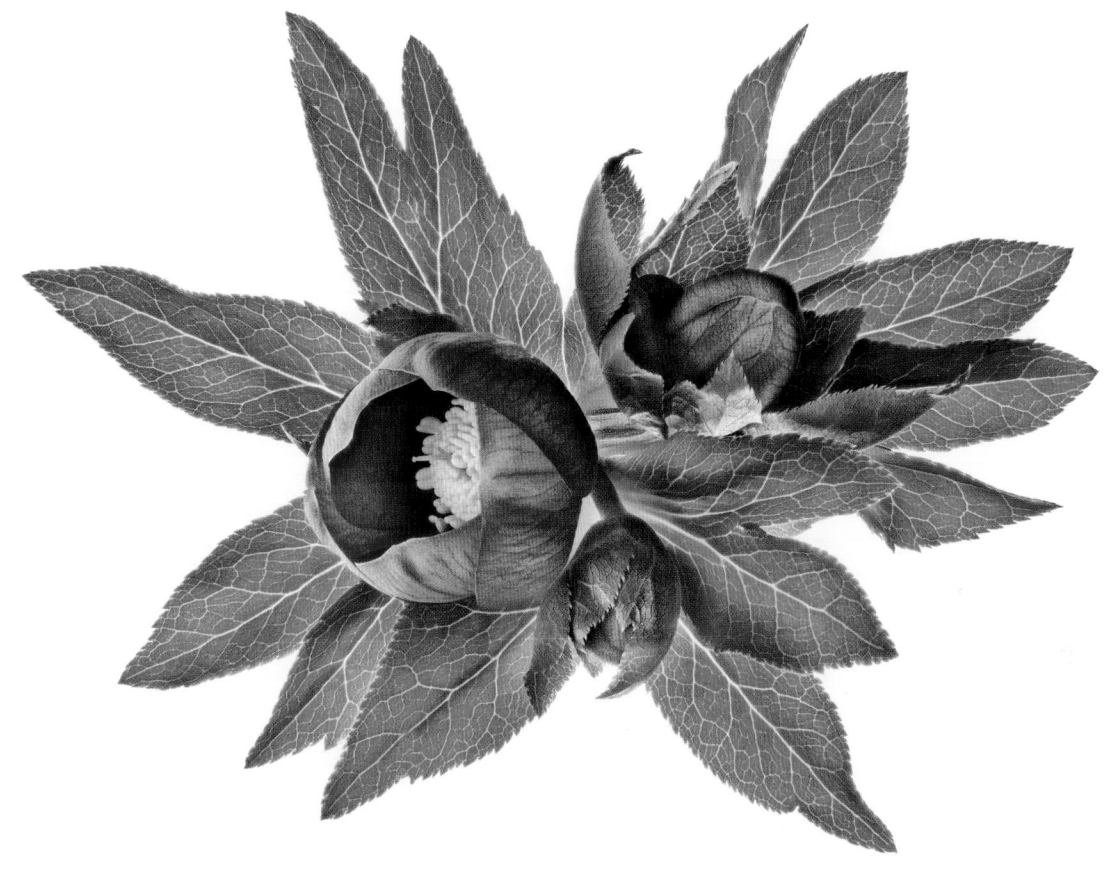

LEFT The blooms of larkspur, *Consolida ambigua (C. ajacis)*, are carried in loose racemes. They differ from delphiniums in that individual larkspur flowers are shaped more like stars, and the little spur on the back of each tends to be smaller and slimmer than those on delphiniums, or may be lacking entirely. Also, when a larkspur goes to seed, you can detect that the ovary is not divided, whereas delphinium ovaries have three or four follicles.

ABOVE Oh, the delectable Lenten rose! Known as *Helleborus ×hybridus* or *H. orientalis*, and encompassing singles, doubles, semidoubles, picotees, bicolors, and more, these luscious early bloomers have mixed up or debated genetics. Seed-grown ones are notoriously variable. Heavy texture allows them to stand up to wind and weather.

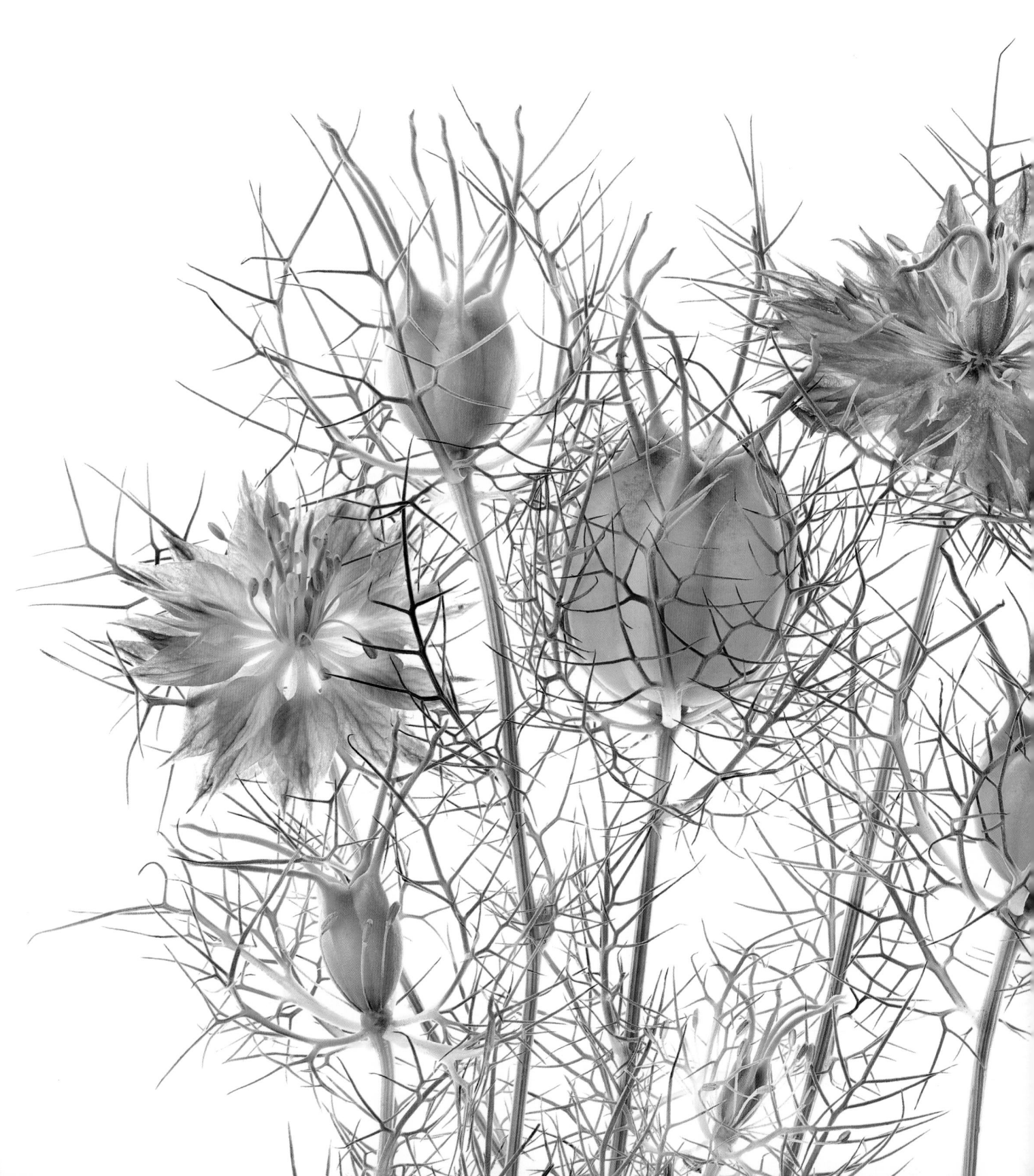

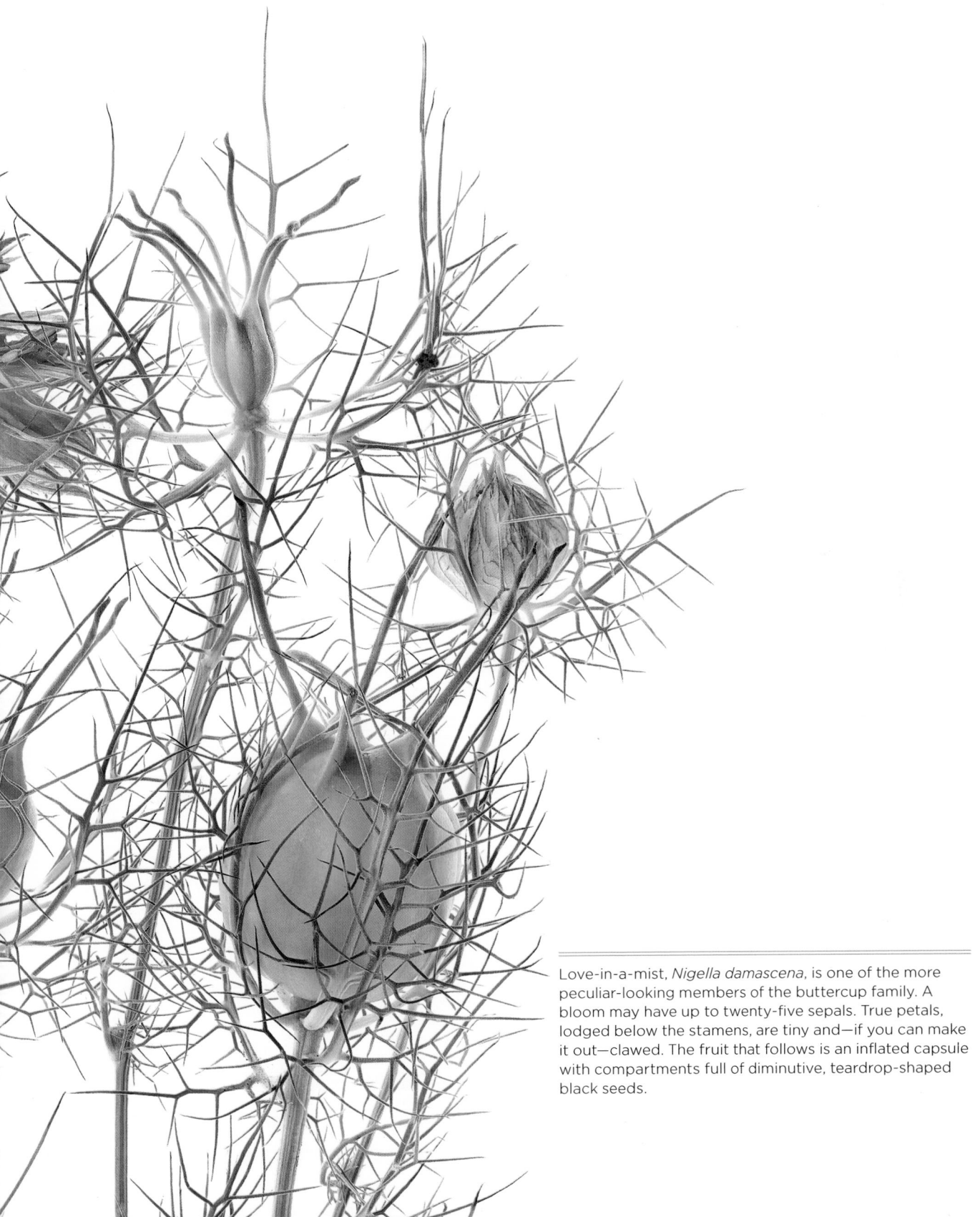

Love-in-a-mist, *Nigella damascena*, is one of the more peculiar-looking members of the buttercup family. A bloom may have up to twenty-five sepals. True petals, lodged below the stamens, are tiny and—if you can make it out—clawed. The fruit that follows is an inflated capsule with compartments full of diminutive, teardrop-shaped black seeds.

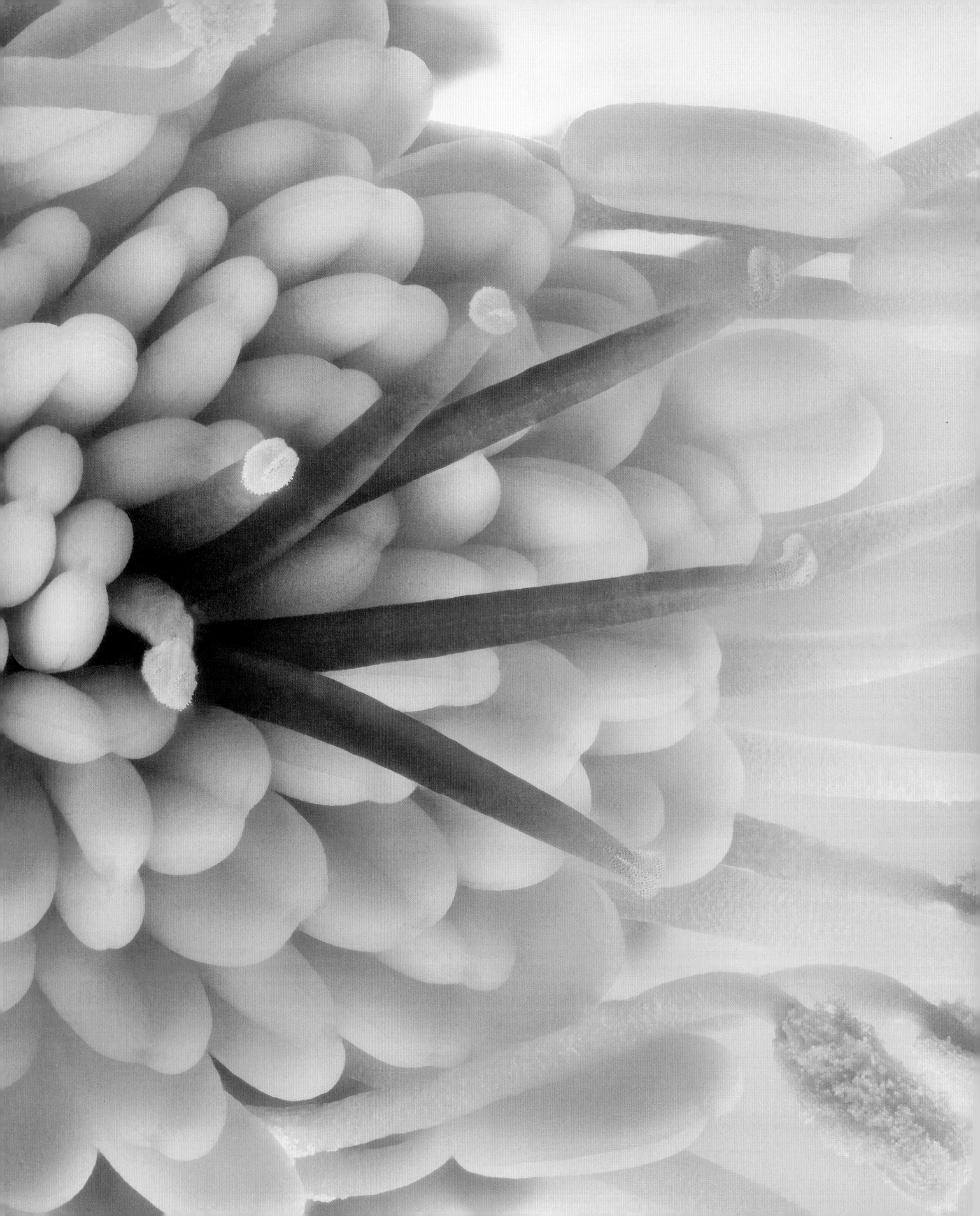

LEFT A hellebore flower consists of five durable outer sepals. Unlike true petals, they stay on the flower long after fertilization and never really fall off, or dehisce. In the center, a crowd of pollen-topped anthers surrounds the stigma.

BELOW Ever-gorgeous peonies, *Paeonia* hybrids, traditionally in Ranunculaceae, have been shifted to the family Saxifrage, and even been granted a family of their own, depending on which botanical source you consult. The plush, petal-laden ones are favorites, but the single ones allow you to study flower form more easily. Here we find a row or more of at least five wide "guard petals" sheltering a boss of (usually) golden, pollen-laden stamens and a cluster of carpels in the very center. Note that double flowers tend to be sterile, as they have transformed their pollen- and seed-bearing structures into petals.

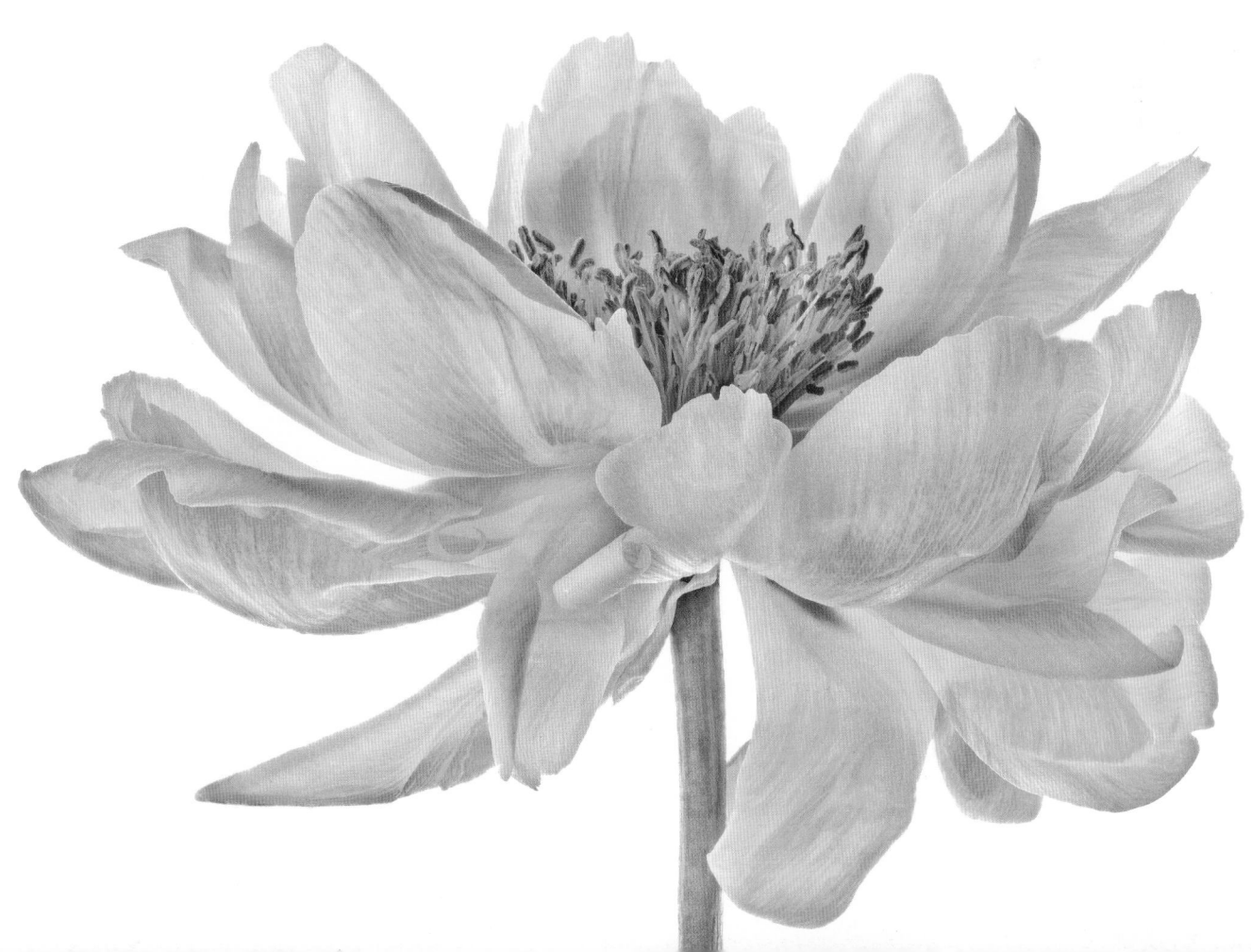

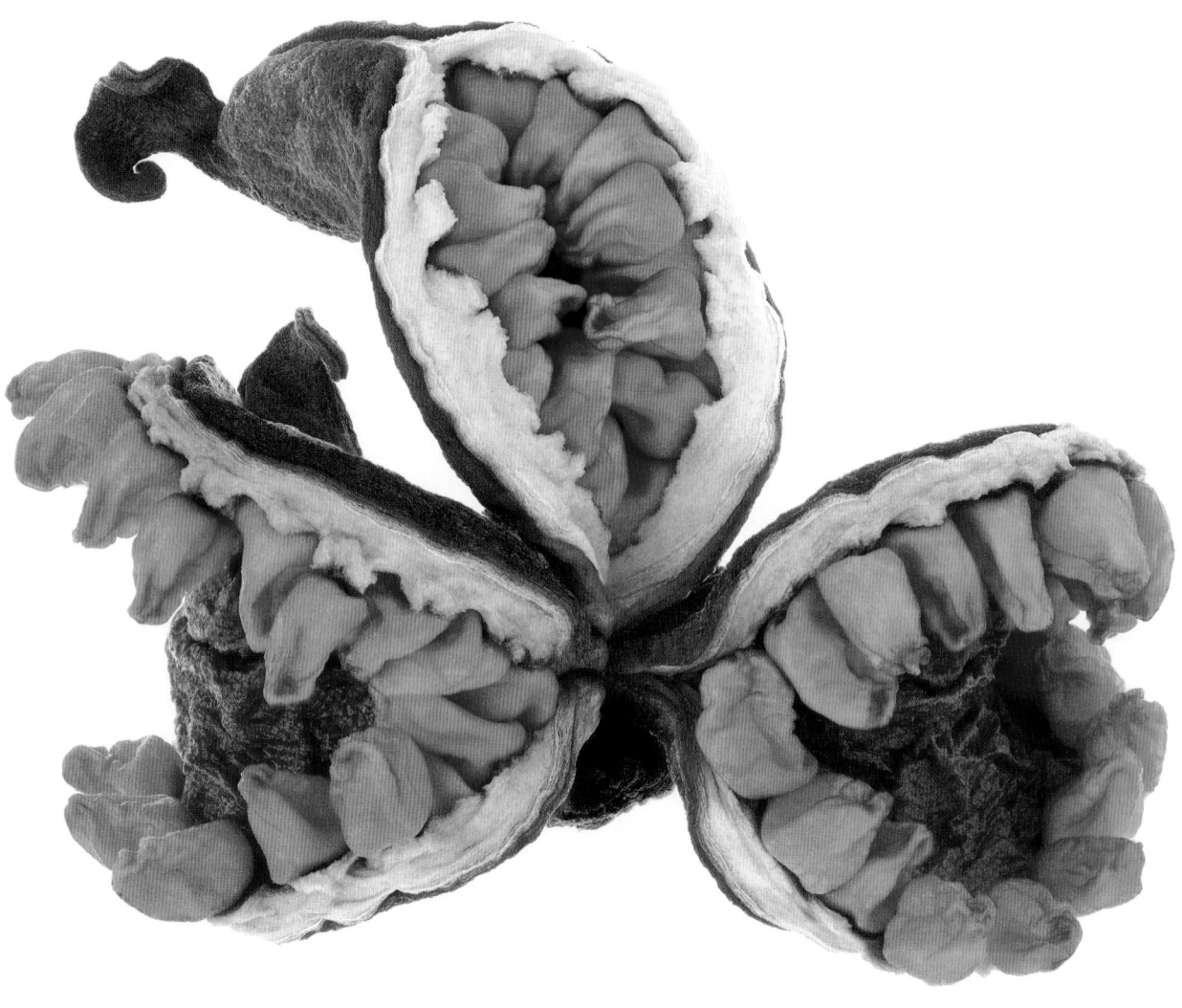

ABOVE A successfully pollinated peony flower will slowly ripen seed over the summer months. When ready, a segmented pod splits open to reveal hard, dark-colored seeds the size of peas.

RIGHT Frothy-flowered yellow meadow rue, *Thalictrum flavum*, actually lacks true petals. Instead, you see clusters composed of pale yellow sepals and stamens. Bees relish it.

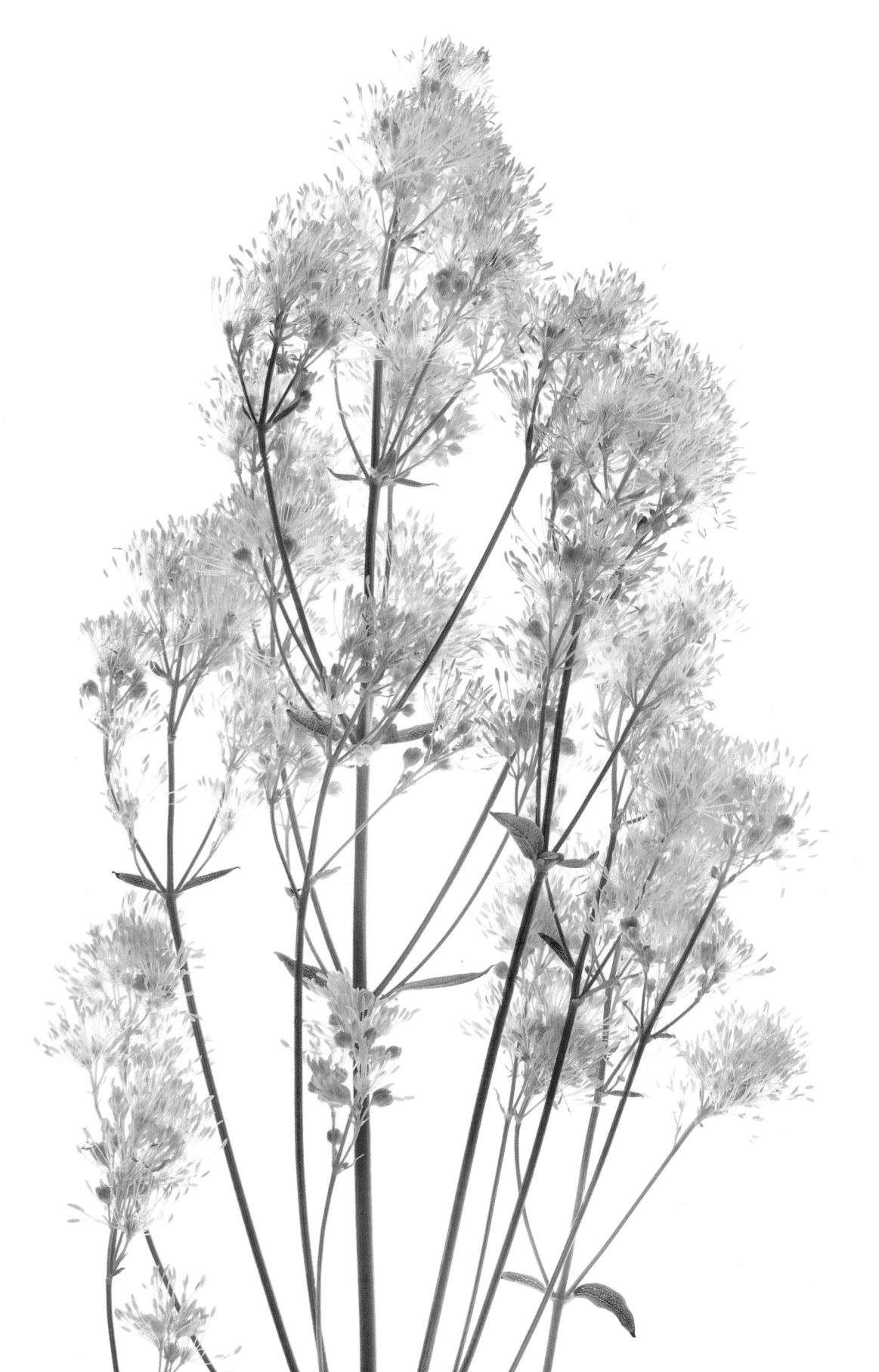

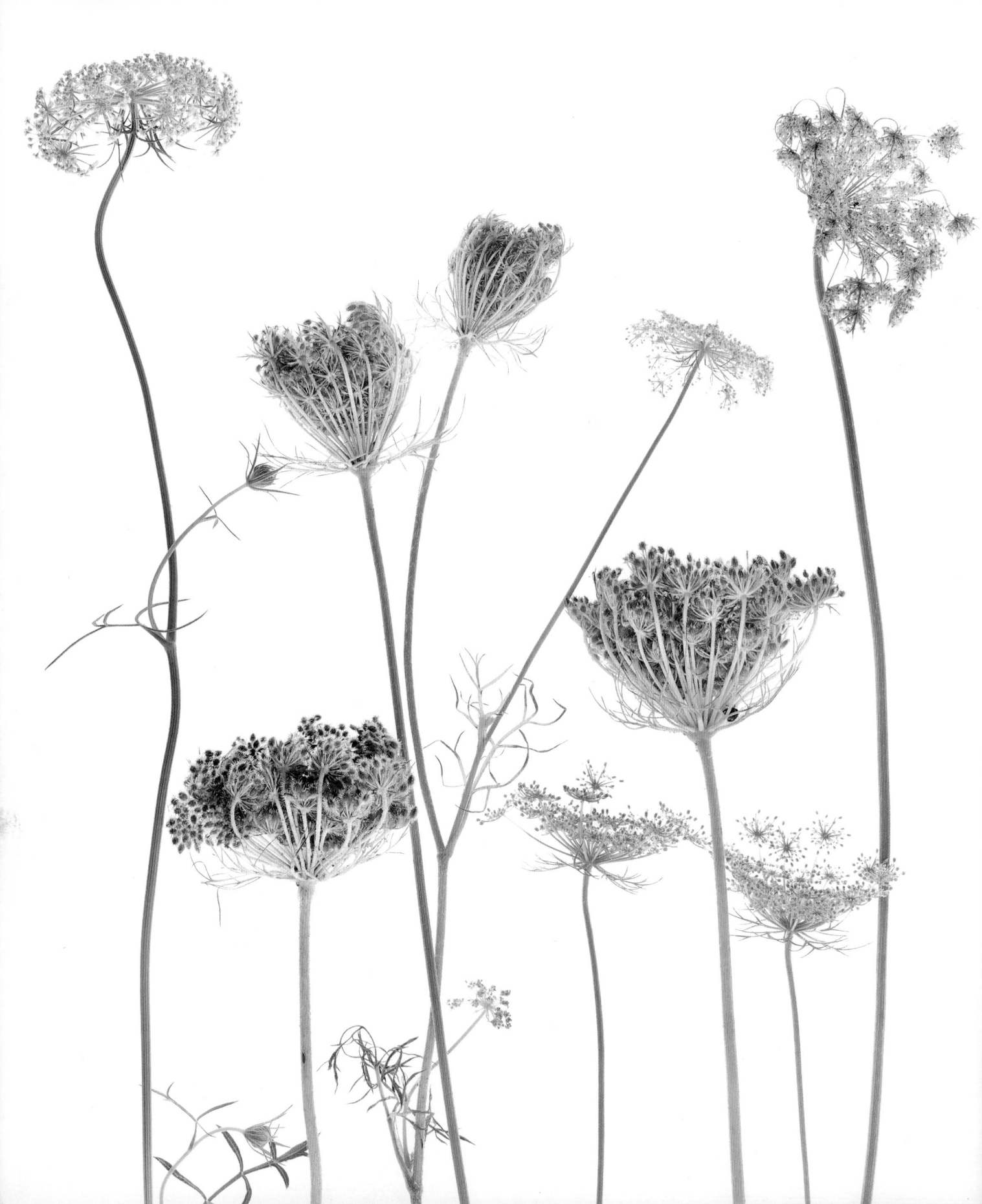

THE CARROT FAMILY

Apiaceae (or Umbelliferae)

"Her body is not so white as
    anemone petals nor so smooth — nor
    so remote a thing. It is a field
    of the wild carrot taking
    the field by force; the grass
    does not raise above it.
    Here is no question of whiteness,
    white as can be, with a purple mole
    at the center of each flower."

WILLIAM CARLOS WILLIAMS, "Queen Anne's Lace," 1921

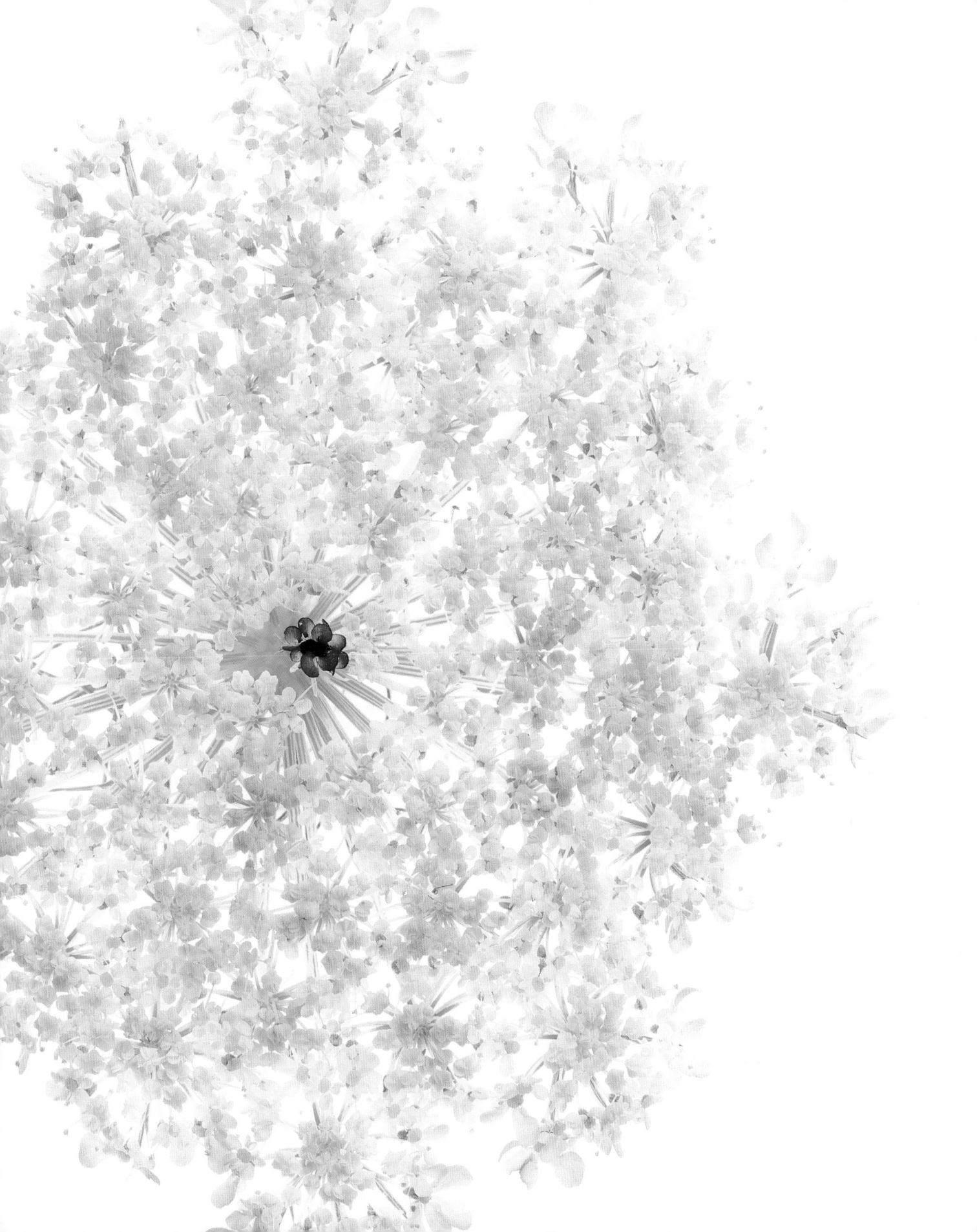

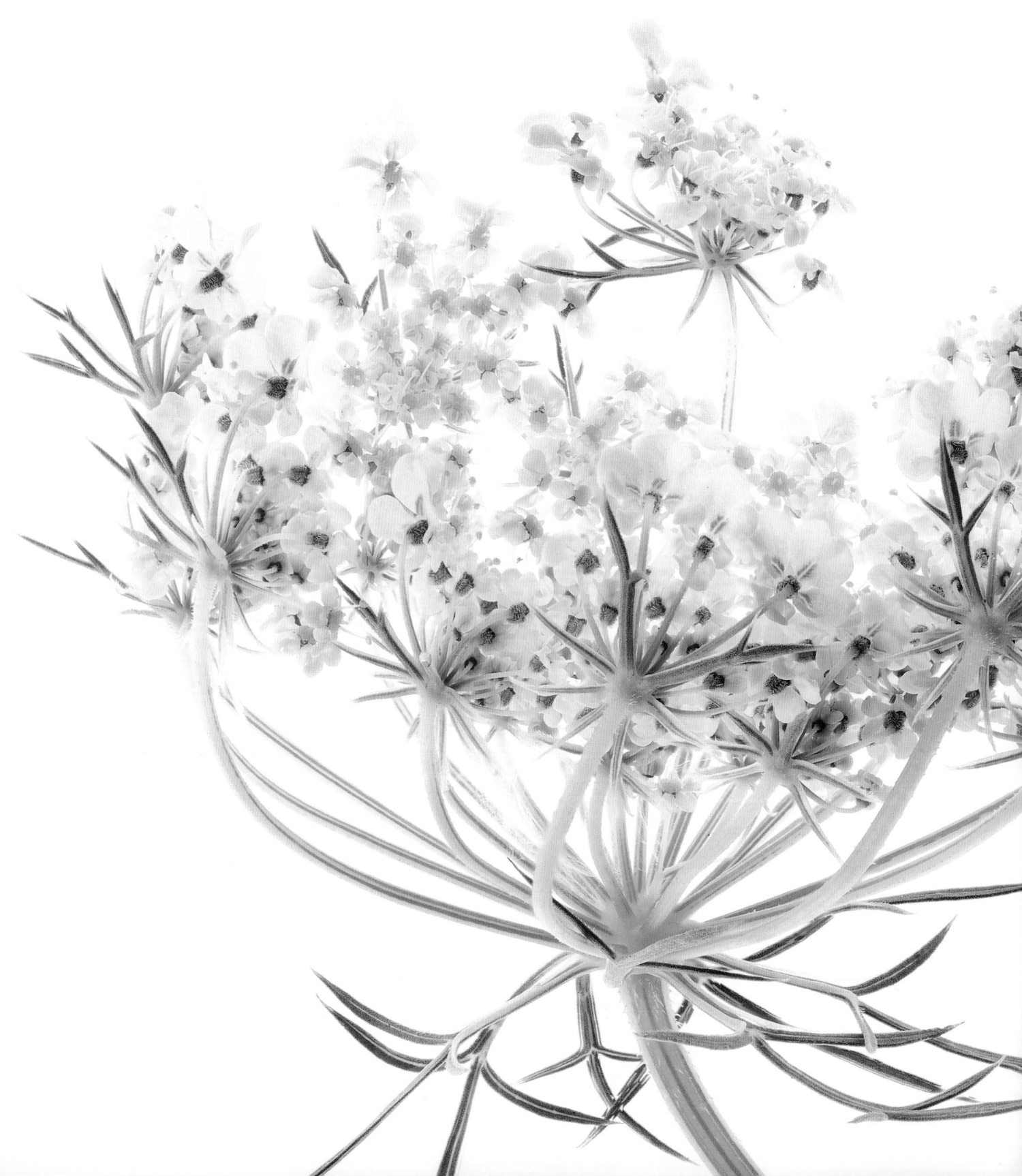

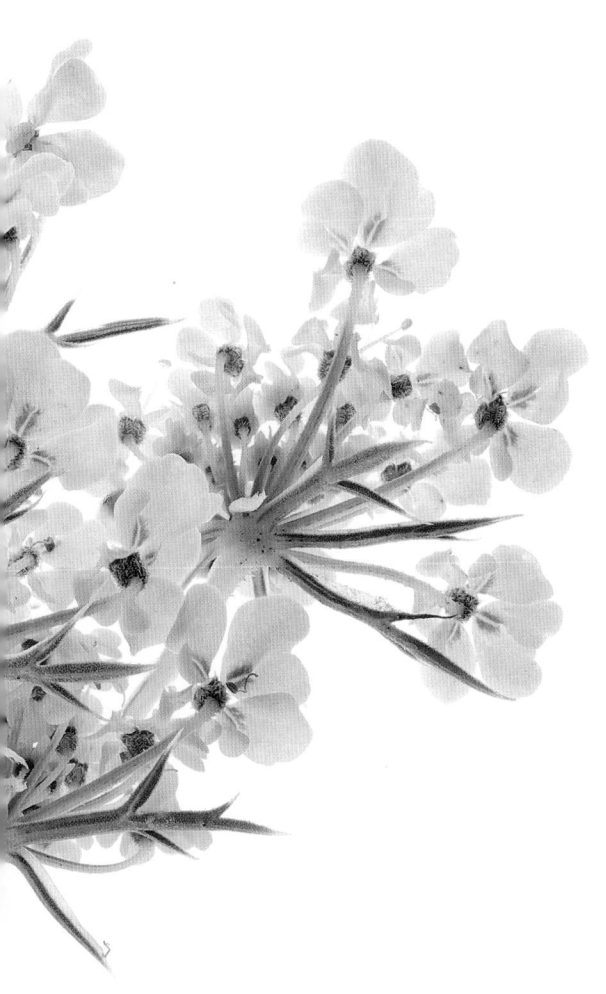

Although it is well known from meadow and roadside, Queen Anne's lace, *Daucus carota*, is not a simple or dull weed. Its history is long and colorful, and its characteristics are quite intriguing. As you can deduce from the botanical name, it is indeed related to the carrot of our vegetable gardens, *Daucus carota* var. *sativus*. Both are members of the carrot or parsley family, Apiaceae or Umbelliferae, which also includes many favorite herbs and vegetables: parsley, cilantro (also known as coriander), anise, caraway, fennel, parsnip, celery, dill, chervil, and lovage. Some are annuals, while others are biennials (that is, they form a rosette their first season and bloom their second).

The majority are edible in root, leaves, fruit, and/or seed. (You could eat the skinny white roots of Queen Anne's lace.) Other wild strains of this plant, with yellow, purple, and occasionally orange roots, apparently took a 5000-year journey from Afghanistan to southern Europe to Holland to North America, eventually becoming the thick, sweet, orange roots, so rich in vitamin A, that we prize today.

Yes, the common contents of the vegetable crisper in your refrigerator, carrots and celery, are in the same family. You would recognize their kinship if you saw them growing in a garden or at a farm, as their flowers are generally carried at the top of a slender stem that sways in summer breezes. Most often white, flowers in this family also come in green, yellow, or even pink, depending on the species. As the name Umbelliferae suggests, these are technically umbels—broad, round clusters that are flat topped like a parasol during part of their blooming cycle. Typically the inflorescence is

LEFT, PREVIOUS SPREAD, AND PAGE 76 A familiar wildflower of field, roadside, and garden edges. Queen Anne's lace, *Daucus carota*, is easily distinguished from its similar-looking relatives. It is the only umbellifer that has a lone (sterile) purple floret in the center of nearly every bloom.

slightly rounded when it first opens and flat during fertilization, ultimately becoming concave or curled inward while ripening seed.

Other telltale identifiers for this family include finely divided, feathery leaves, and fruits rich in fragrant essential oils, or terpenoids, such as carvone, limonene, linalool, and pinene. Some are grown for both, such as cilantro, with its delicious lemony leaves; when this plant's spicy seeds are harvested for kitchen use, they're known as coriander.

Not all members of this family are edible and benign, however, so be exacting about identification, especially for plants you may encounter growing wild. Poison hemlock, *Conium maculatum*, is indeed deadly—legend has it that this plant killed Socrates (rather than the common hemlock tree, which is another family and species altogether). Fool's parsley, *Aethusa cynapium*, must also be avoided. Both have smooth stems and foul-smelling leaves, a clue to their nefarious personalities, whereas Queen Anne's lace has fuzzy stems and leaves and smells pleasantly of carrot greens.

The distinctively lacy look to this family's inflorescences is created by scores of tiny flowers. Peer closely at an individual one and you'll always find five petals, five sepals, and five stamens. Observe how the petals are actually irregularly shaped, like wee mittens. All have male and female parts, although only the male parts on the inner ones may be functional. Flowers on any given cluster may be self-fertile, but also able to pollinate their nearby neighbors.

Much has been made of the small purple, or deep red flower often lodged in the center of a Queen Anne's lace inflorescence (other Apiaceae lack this tiny flourish). The story is that the queen pricked her finger during the delicate, arduous work of making lace, and the flower represents a drop of royal blood. Extracting and chewing on this flower was once believed to prevent epileptic seizures. Actually, it appears to have a practical purpose, although its draw is thought to be purely visual, as it is sterile. A head shot of an umbel with black-and-white or infrared photography shows a target format, with bigger flowers to the outside, smaller ones around the middle, and that lone colored flower in the bull's-eye position. Botanists believe this creates an irresistible, bold beacon for passing insects.

What pollinates these flowers? If you observe a member of this family at the height of its growing season, you have a ready answer: practically any insect that happens by. This includes a variety of flies, gnats, and wasps, and also moths, butterflies, and bees—and sometimes even beetles. Plentiful nectar and a nice flat landing area are an irresistible combination.

Once an umbel has been successfully pollinated, it closes in on itself and starts to dry out. At this stage, the concave, clawed form resembles a small, unruly bird's nest, which is an alternate common name for Queen Anne's lace but certainly applicable to related plants. Because the bloom is sensitive to moisture, you can observe the outer parts bending in on moist or humid days and splaying outward on dry ones. Seeds are thus dispersed on dry days, hopefully with a nice breeze to send them on their way. Individual seeds have tiny rows of spines to help with additional distribution via animal fur.

The seeds of Queen Anne's lace are safe to eat, but nothing special. The seeds of some of its relatives, however, contribute unique and delectable flavor to recipes in many cultures. Aromatic dill seeds are added to apple pie, herb butter, relish, breads, cookies, and cakes. Caraway seeds contribute a nutty-licorice flavor to rye and Irish soda bread, as well as cookies, crackers, and sauerkraut and other cabbage dishes. Fennel seeds are

Once a Queen Anne's lace plant has been pollinated, the flowerhead curls inward and seeds form and ripen.

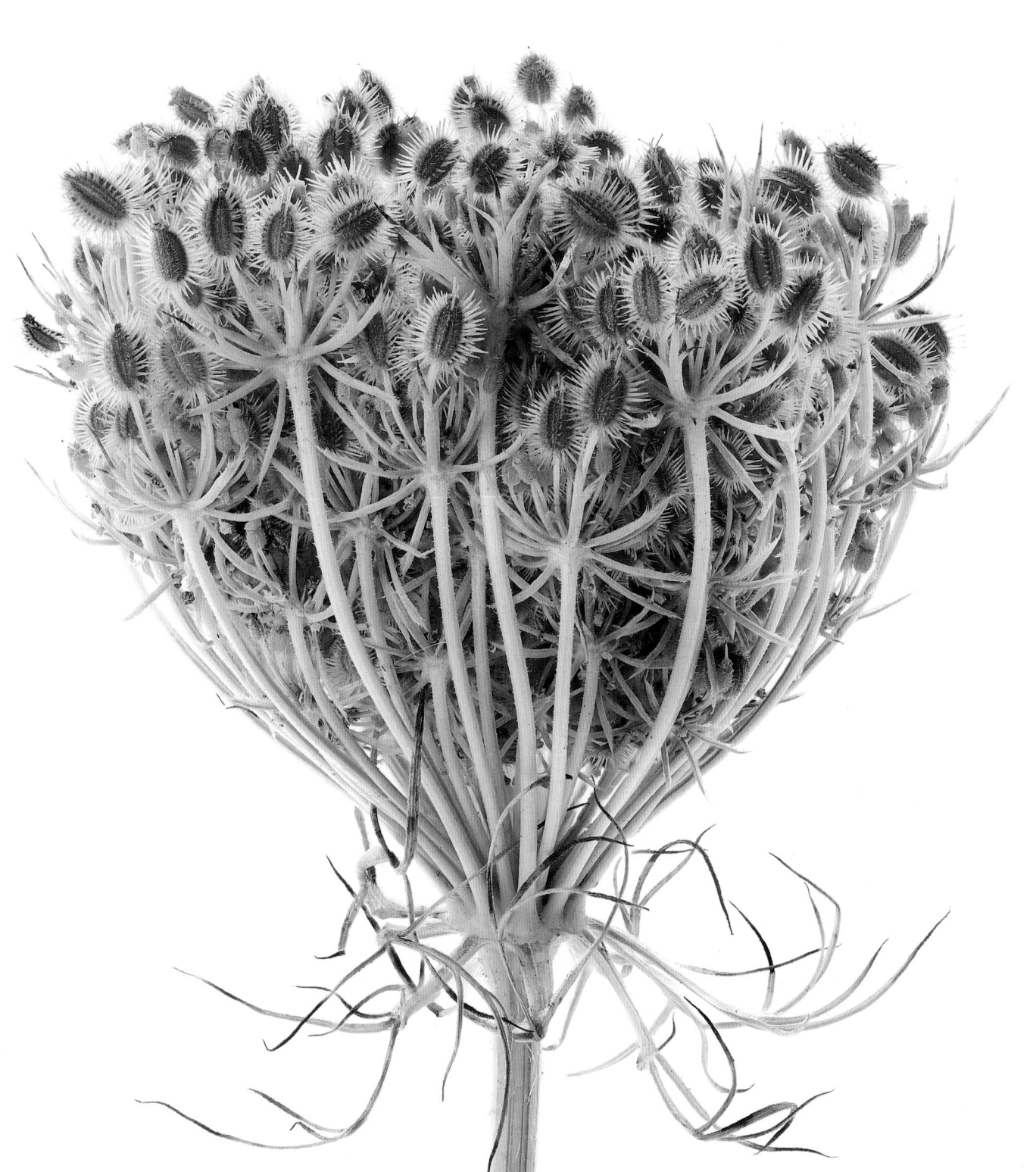

cherished for a bracing, almost minty flavor that enhances everything from fish dishes to homemade sausages.

Although easygoing, hearty, and long blooming, these plants are not generally grown for their ornamental beauty. Their rangy, lanky profiles and untidy or dainty flowers are fine in a vegetable or herb garden, perhaps okay in your flowerbeds if they are informal affairs. With the main exception of carrots, no one has lavished effort on creating and nurturing a broad range of improved umbellifers.

However, over the years the nursery trade has brought forth a handful of improvements that focus on culinary use. If you prefer to grow fennel not for its seeds but its crunchy, sweet stems, seek out the selection or variant known as Florence fennel, *Foeniculum vulgare* var. *dulce*, which is broad or swollen near the base. And 'Dukat' and 'Superdukat' dill are intensely flavorful dwarf, bushy varieties suitable for a smaller garden or container culture; they are slow to go to seed, prolonging the tangy, leafy harvest. 'Giant of Italy' flat-leaf parsley is marketed as the best-tasting choice, with lush plants laden with rich green leaves. Overall, from root to leaf to flower, the species in this family give the best they've got, and gardeners, flower arrangers, and cooks can choose to decline or accept them as they are.

The popular herb dill, *Anethum graveolens*, has classic flowers for its family: flat umbels made up of many small yellow flowers. If you get close, you'll observe that the wee petals are rolled inward from either side.

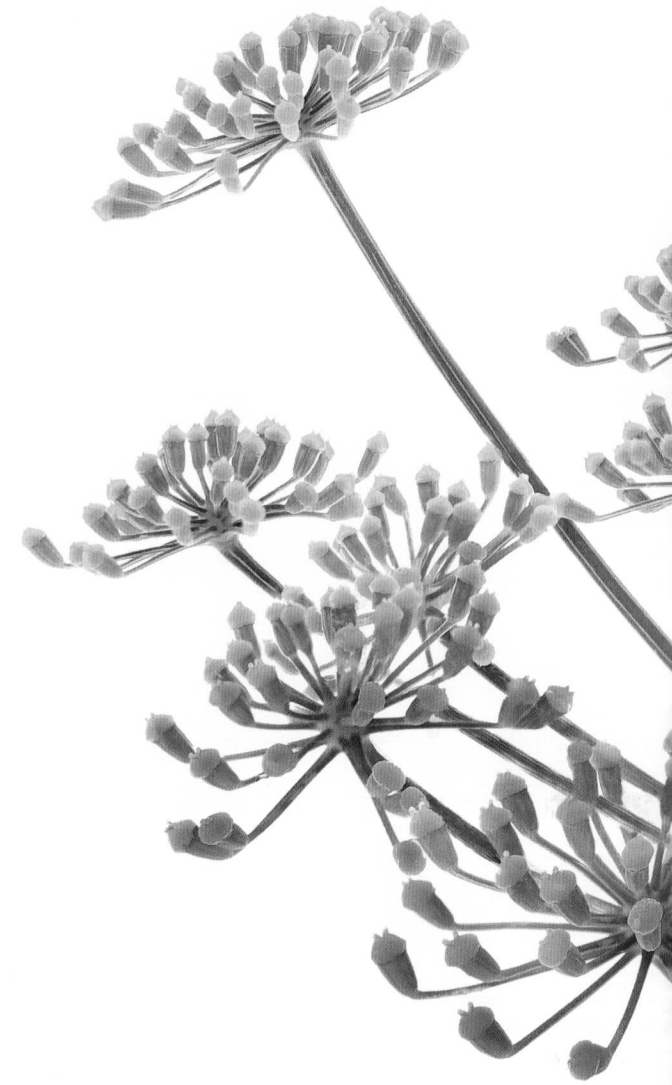

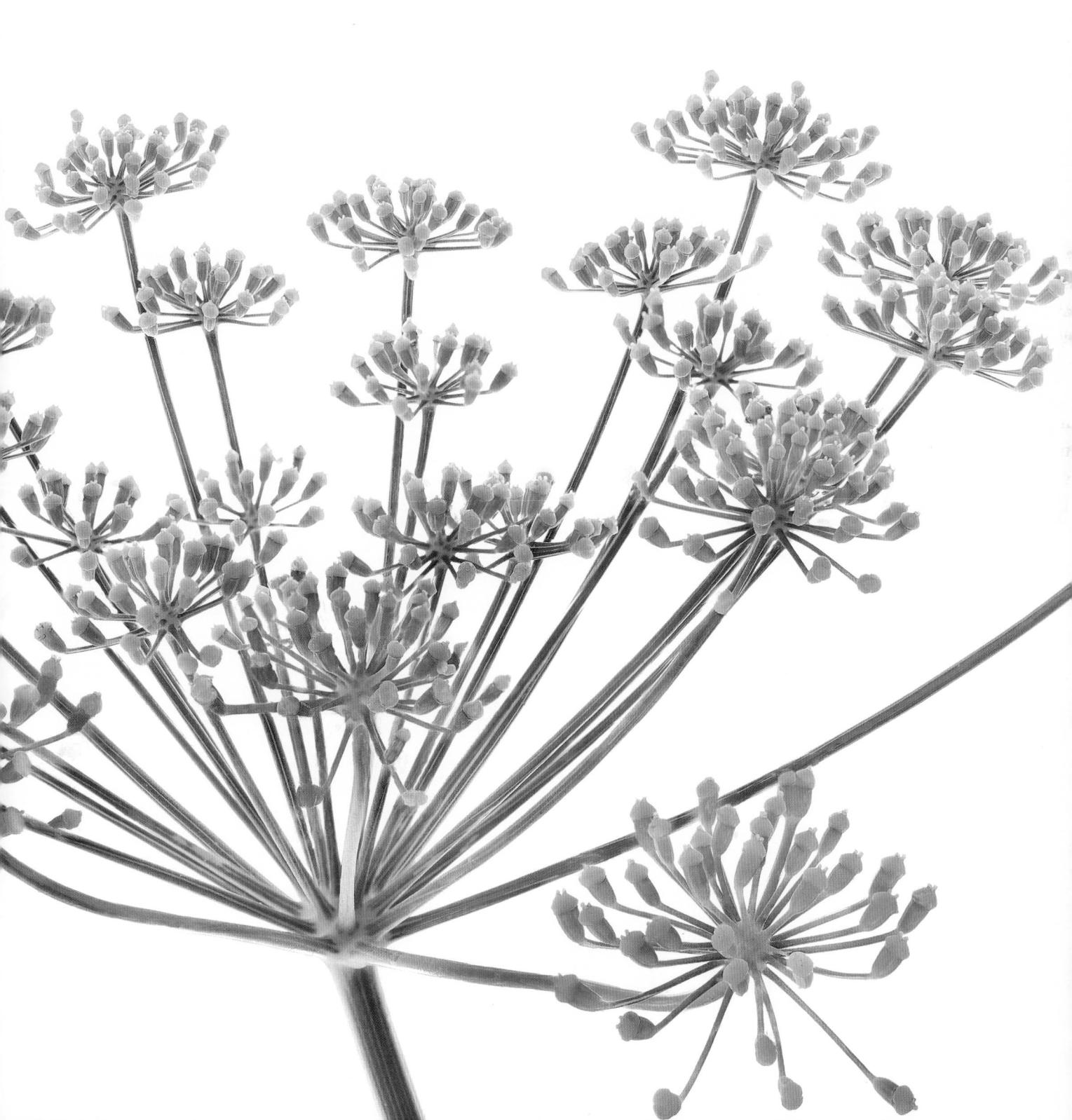

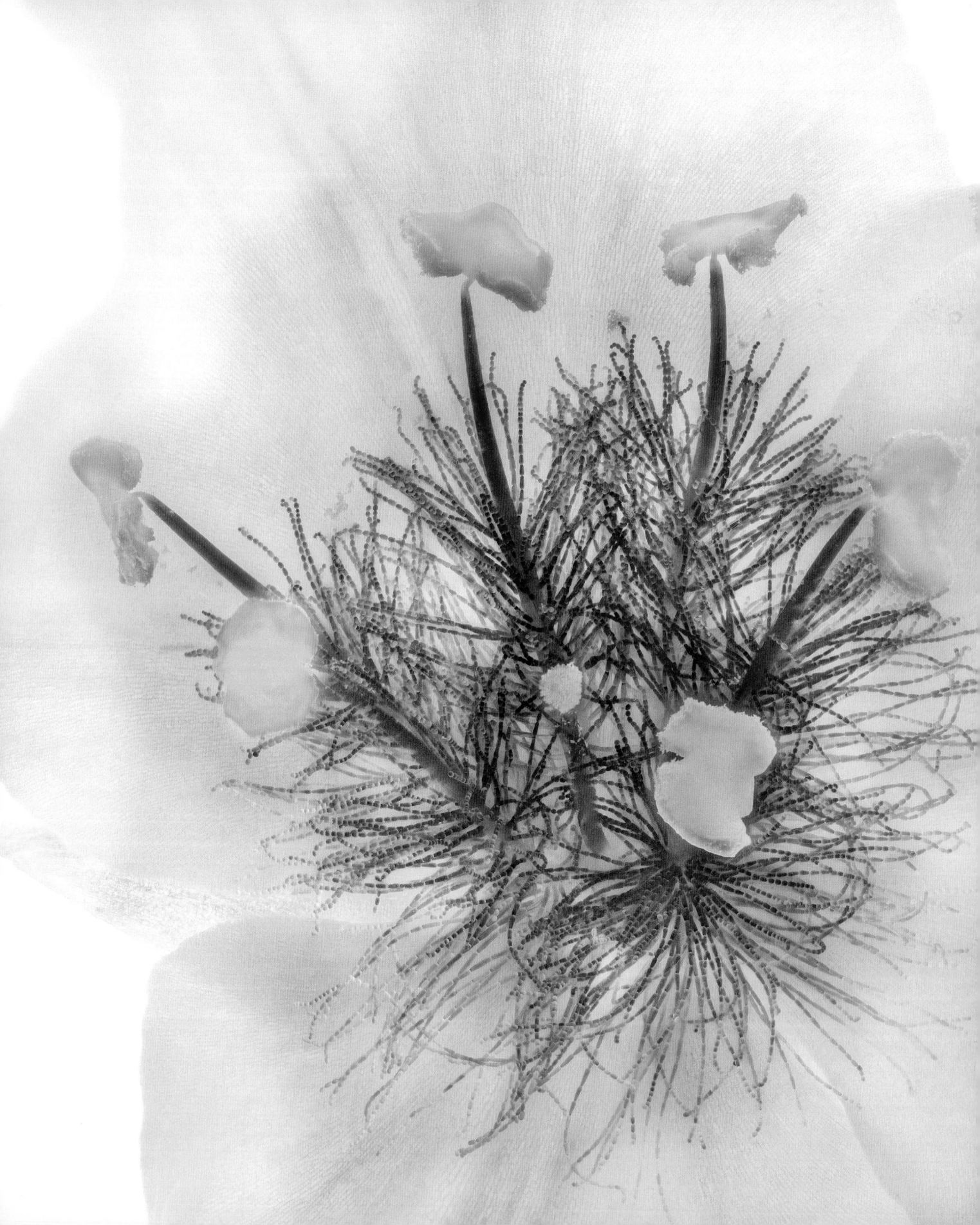

THE SPIDERWORT FAMILY

Commelinaceae

"Night is here.
The wind and the rain announce the news
that spring is coming.
Still I sleep alone, my dream not yet realized.

Flower petals falling…
seem to understand my dreams and aspirations.
They touch the ground
in perfect silence."

NGUYEN BINH, as quoted by Thich Nhat Hanh
in *Cultivating the Mind of Love*, 1996

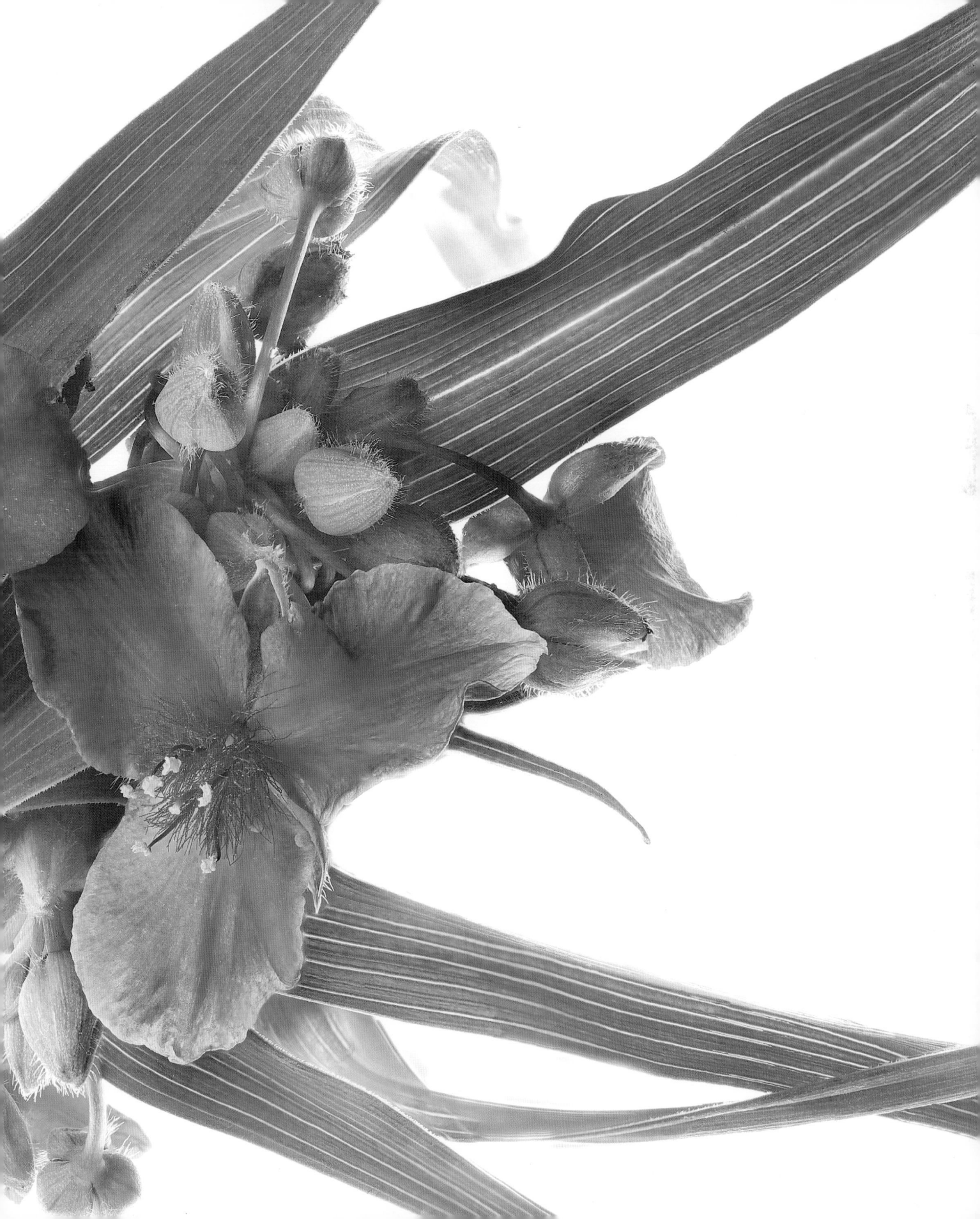

The lush, gangly tresses trailing over the edge of a hanging basket contain a plant so familiar, we scarcely give it a passing glance: wandering jew, *Tradescantia*. This is the same plant whose cuttings languished on a shelf in an elementary-school classroom or a crowded, sunny kitchen windowsill, roots rapidly filling a glass jar of water as new stems quested ever upward and outward. You may know more about this robust, not-especially-tidy-looking plant than you realize. The stems are juicy, almost succulent, which allows a plant to withstand the times we neglect to water it. Have you also noticed that these stems are jointed? The base of the leaves wrap like a sheath around a swollen node to produce the characteristic jointed stem. The simple leaves, too, make *Tradescantia* easily recognizable. They have smooth edges and parallel veins, and they alternate along the trailing stems.

The name Tradescantia commemorates John Tradescant the Elder, the English botanist who brought the genus from the New World into cultivation. There are several species in commerce these days, notably *T. albiflora* and *T. fluminensis*, and within these, numerous varieties. The common name, wandering Jew, references the plants' vigorous and rambling growth habit. (In an ancient legend, a man scornful to Christ on the

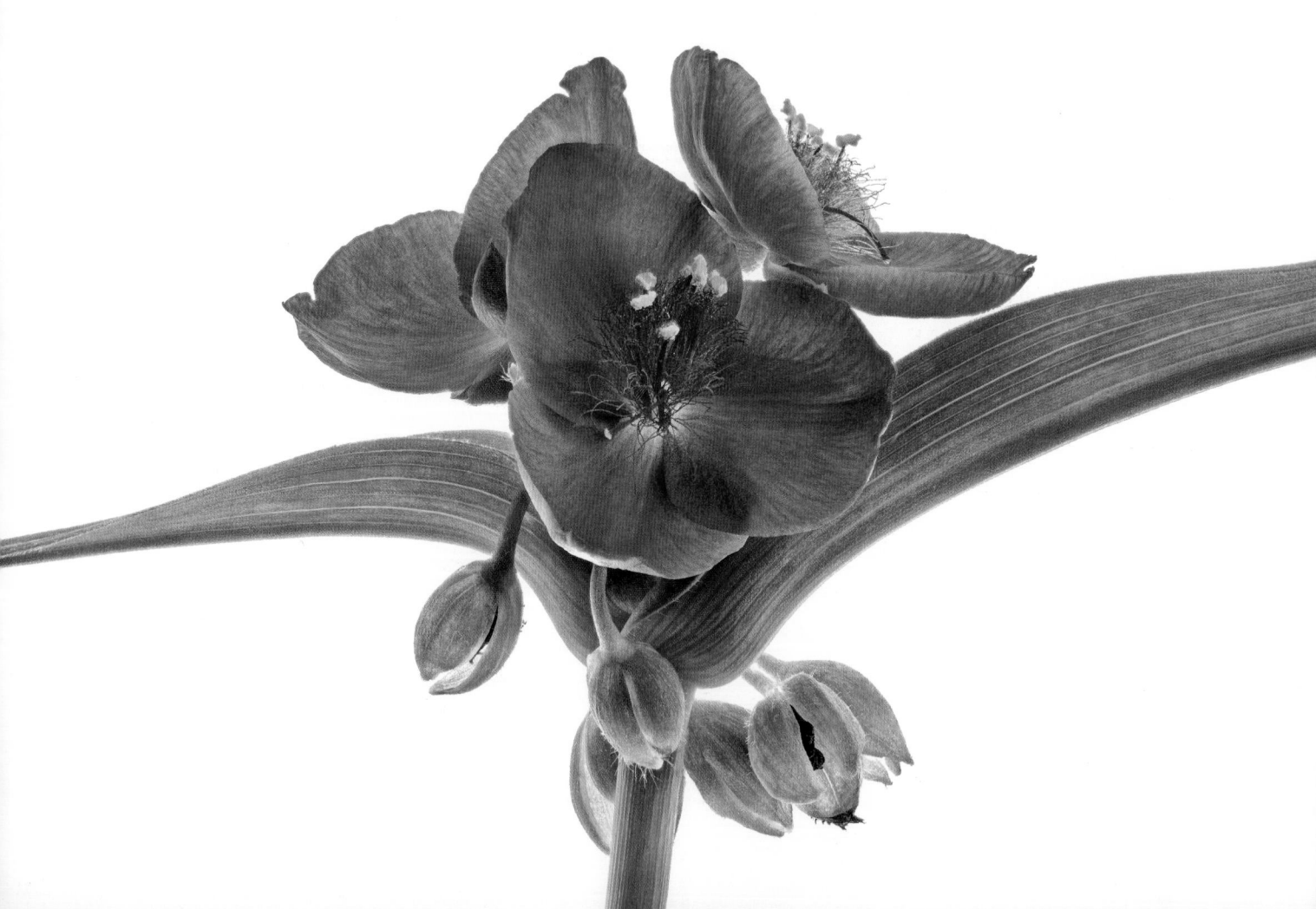

way to his execution was thence condemned to wander the earth, without rest, never dying, or at least not until the Second Coming.)

Wandering jews are in the spiderwort family, or Commelinaceae. Members traditionally grown in the ground outdoors are *Tradescantia ×andersoniana*, *T. virginiana*, and *T. pallida* (*Setcreasea pallida*), all of which answer to "spiderwort." This common name probably refers to the gooey sap, a dollop of which can be stretched like a thread of spider's silk. Foliage in the latter is usually purple, while plain, glossy green in the former two, but otherwise their kinship to the houseplants is obvious because of the sheathed leaves, fleshy, jointed stems, and rangy habit. Also included is blue spiderwort or dayflower, *Commelina coelestis*.

Here at last we meet the flowers of this family—for generally the indoor plants do not flower prolifically, if at all. And what weird flowers they are. Borne mainly in umbels, they're small, between ¼ inch to 1½ inches, and three petaled like a gently tri-cornered hat. They're backed with green, leaflike bracts and held on pedicels that may be fuzzy. Color can be white, pink, violet, magenta, bright blue, or deep purple. Six yellow anthers or stamens stand out prominently and are flanked, if you peer closely, by tiny, spidery hairs. *Tradescantia pallida* 'Purple Heart' has purple-red blooms against that rich purple foliage—undeniably gorgeous, if tricky

LEFT, PREVIOUS SPREAD, AND PAGE 86 A handsome *Tradescantia ×andersoniana* hybrid sports three-petaled flowers with a central spray of contrasting yellow stamens. Like all spiderworts, individual blossoms last only a day but are quickly replaced, including by ones in the same cluster.

to fit into a mixed flower garden. Those of *Commelina* are deep blue, although there is an 'Alba' available. In that particular species, there are two prominent petals and one tiny one.

However, as the name dayflower suggests, the blooms are not durable. Petals fall off within hours, and new flowers appear in short order to keep the show going. In this respect, the plants are reminiscent of daylilies, *Hemerocallis*, except that spiderwort flowers fall off cleanly instead of leaving you with limp, faded petals. (So, like daylilies, they do not make good additions to homegrown bouquets, nor do they dry well.)

Close observation of plants in this family has revealed that species will reliably, consistently flower at a certain time of day, and that most are bisexual (that is, each bloom has both male and female organs, positioned differently). But some plants have male *and* bisexual flowers, which open at different times, and none have nectar, only pollen. Spiderwort family flowers trick pollinators, chiefly bees and flies, into visiting anyway with adaptations such as yellow hairs, staminodes that are actually sterile, or broad anther connectives that mimic pollen.

In a garden, these plants are long blooming and require little care, and the sprawling foliage's ability to cover broad areas can be valuable. However, just like their houseplant relatives, spiderworts need occasional grooming. Indeed, cutting back hard practically any member of this family will usually inspire rejuvenation and more blooms. How interesting that a group of plants we usually take for granted has such intriguing qualities and laudable resilience. Perhaps we should be more appreciative.

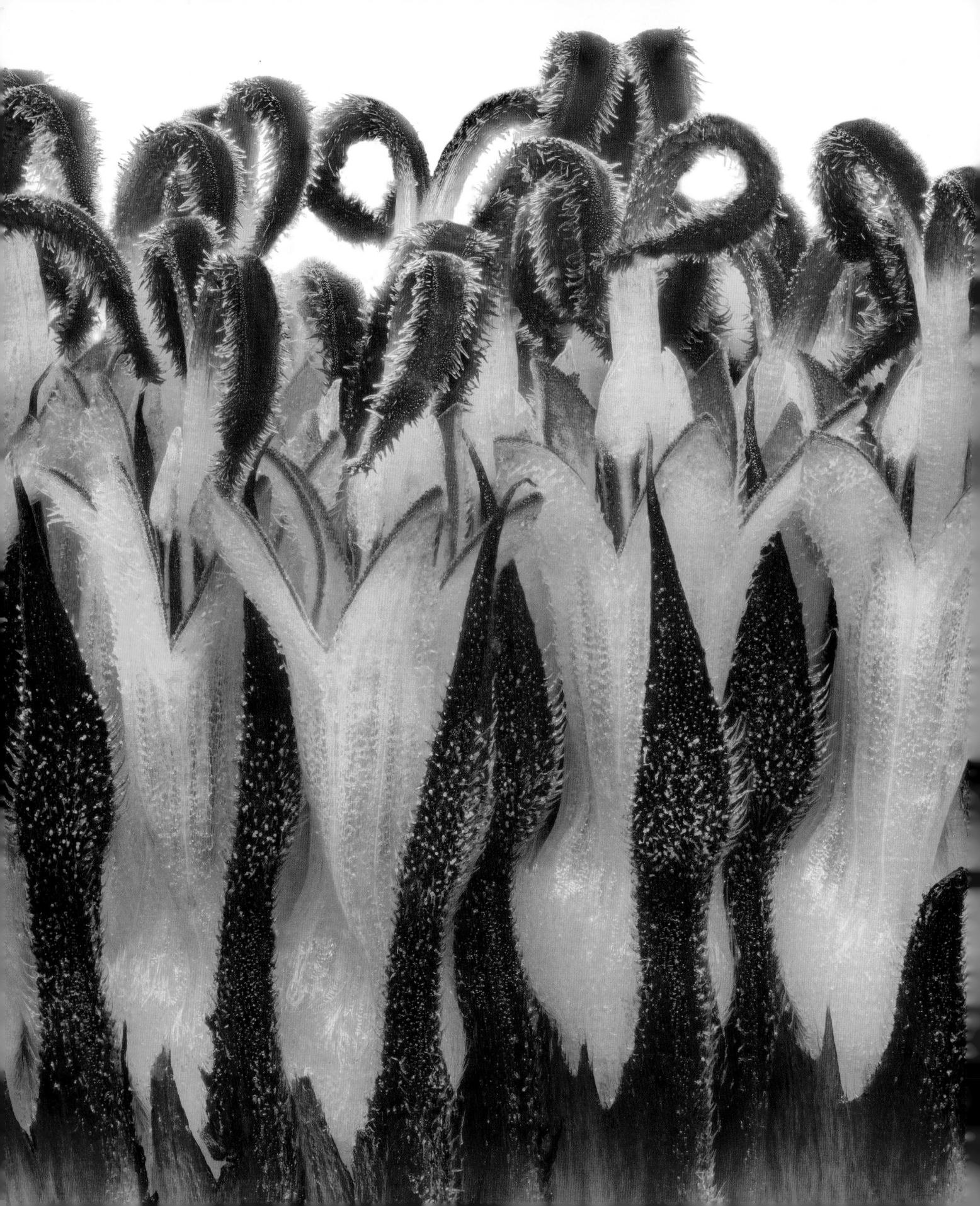

THE DAISY FAMILY

Asteraceae (or Compositae)

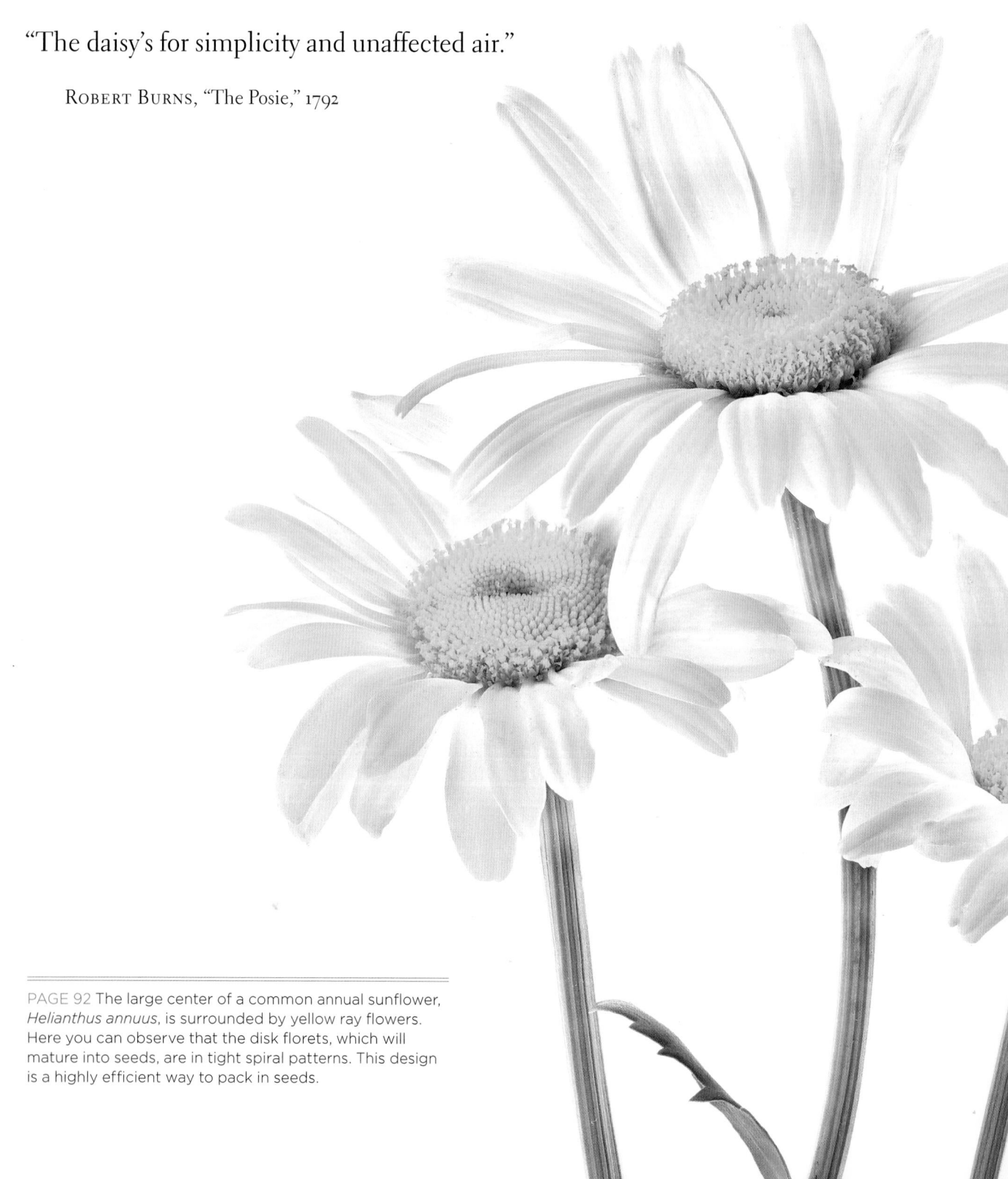

"The daisy's for simplicity and unaffected air."

ROBERT BURNS, "The Posie," 1792

PAGE 92 The large center of a common annual sunflower, *Helianthus annuus*, is surrounded by yellow ray flowers. Here you can observe that the disk florets, which will mature into seeds, are in tight spiral patterns. This design is a highly efficient way to pack in seeds.

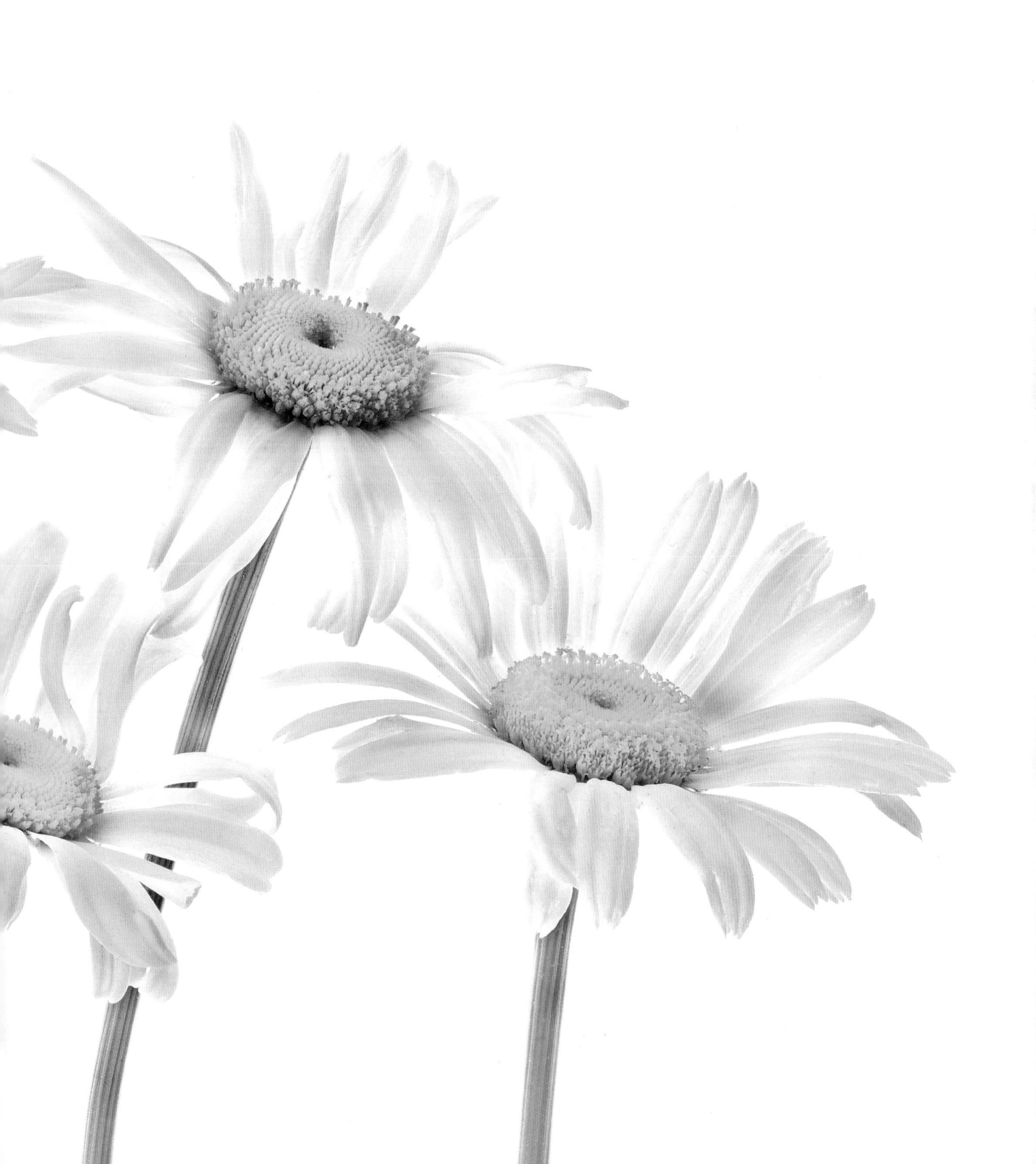

Even people who say they don't know anything about flowers always recognize a daisy. The iconic ones (there are a number of species and garden cultivars) always have the bright yellow, buttonlike center surrounded by a wheel of jaunty white petals. The flowers of other members of this huge and diverse family, called either Asteraceae or Compositae, may sport other colors and be larger or smaller, but are certainly recognizable as kin. Some of these are asters, chamomile, chrysanthemums, coreopsis, cosmos, dahlia, dandelion, echinacea, feverfew, goldenrod, hawkweed, marguerite, marigold, sunflowers, tansy, thistles, and zinnia.

Daisies hail from the temperate areas of England and Asia. The common name, however, comes from Old English. "Day's eye"—in original Anglo-Saxon, *daeges-eage*—refers to the bloom opening in the morning and closing at nightfall. Not all members of this family follow this pattern, but the ubiquitous little English daisy does.

The familiar form is deceptively simple. A daisy flower is a complicated affair, as it actually comprises two different kinds of flowers. The yellow center is made up of scores or even hundreds of tightly packed, tubular florets, each of which contains tiny pistils and stamens; these are called disk flowers. The white petals are individual ray flowers. Flies, bees, butterflies, and

BELOW Purple coneflower, *Echinacea purpurea*, has seen a lot of interest from gardeners and plant breeders in recent years. But the original species has an appealing shuttlecock look. It is thought that the petals (ray flowers) droop in order to make the plush array of central disk flowers all the more accessible to pollinators.

RIGHT Often disdained by gardeners and landscapers because it is so very variable, *Spirea japonica* was originally imported from Japan and Korea. Unlike native spireas, it sports pink or mauve flower clusters (up to 12 inches across, but often smaller) and rather fuzzy young twigs. It blooms on current season's growth, so if you are trying to control or shape a plant, prune it or cut it back in late winter or early spring.

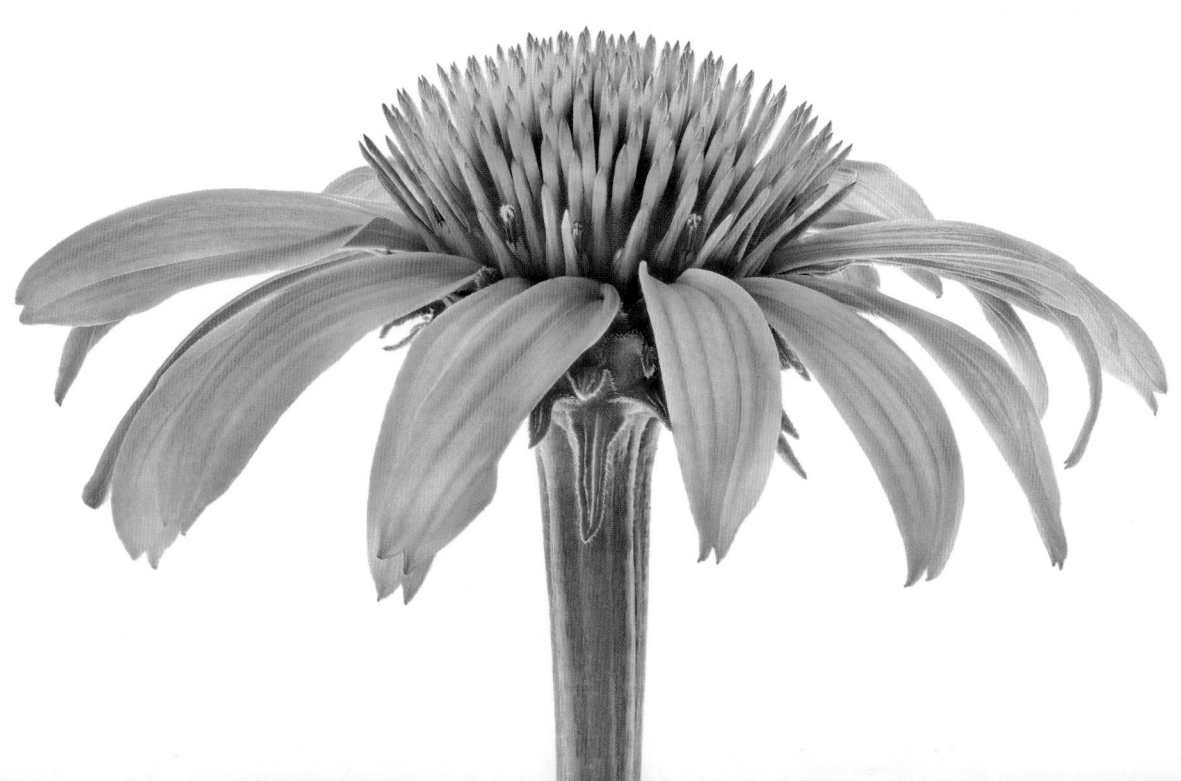

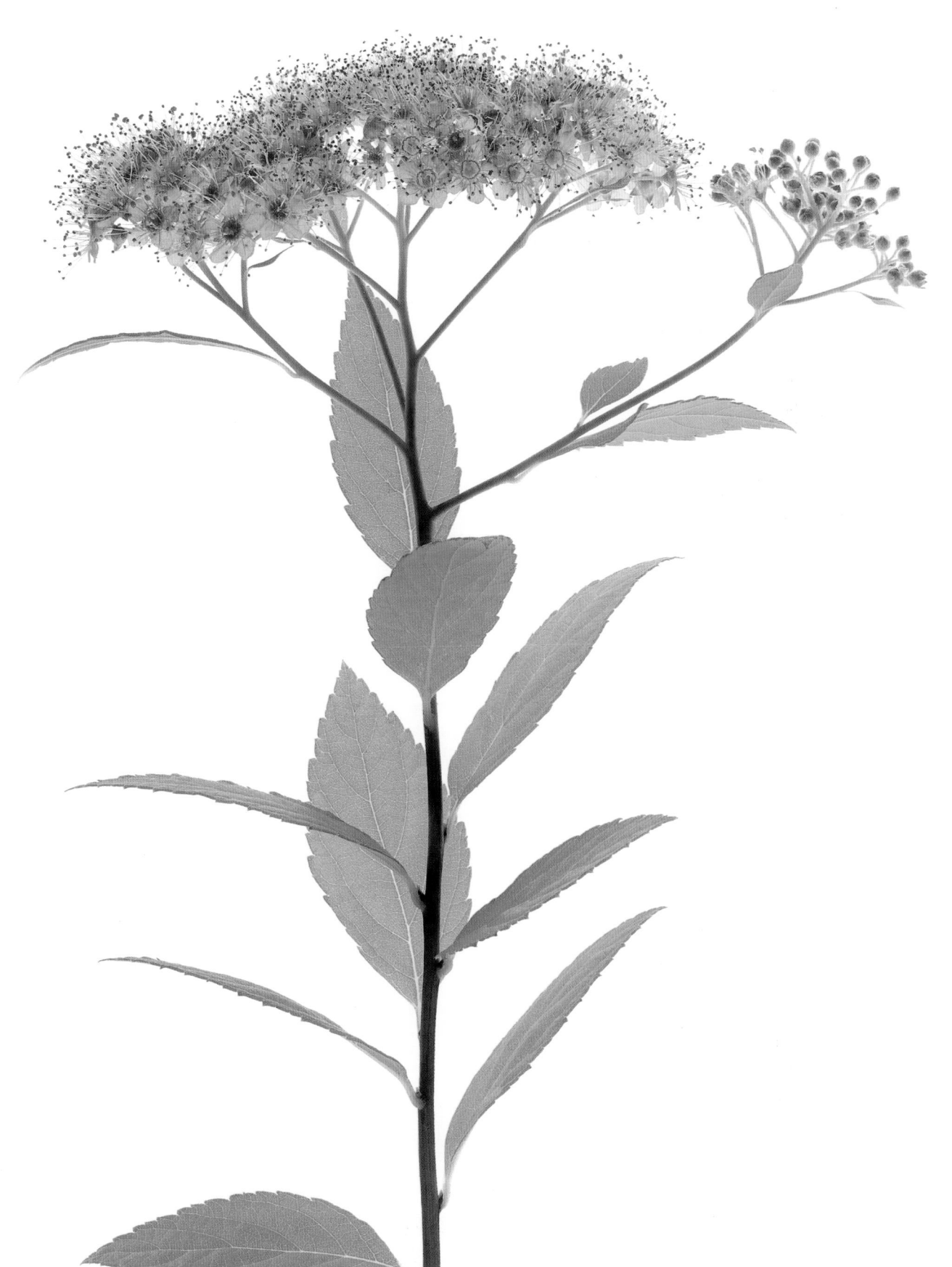

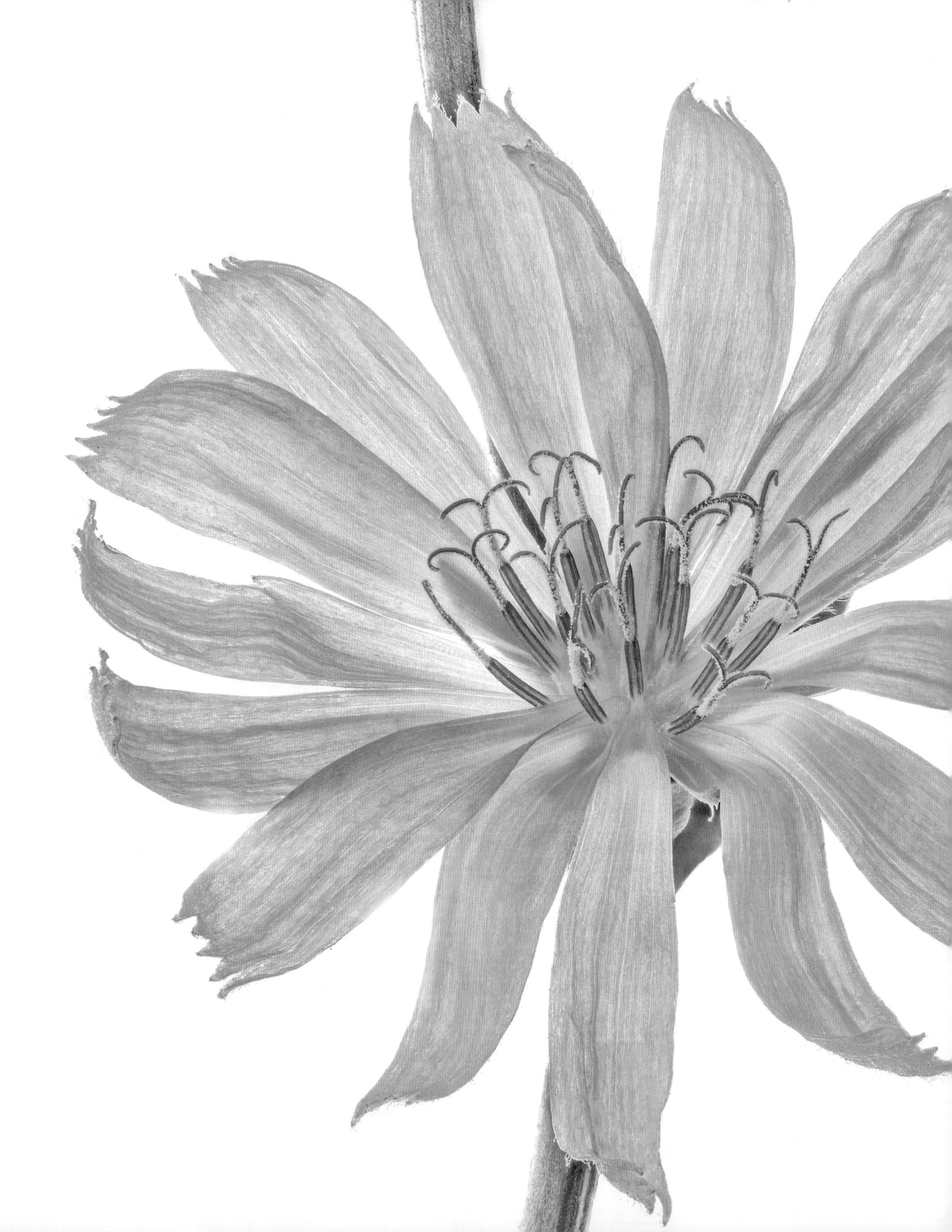

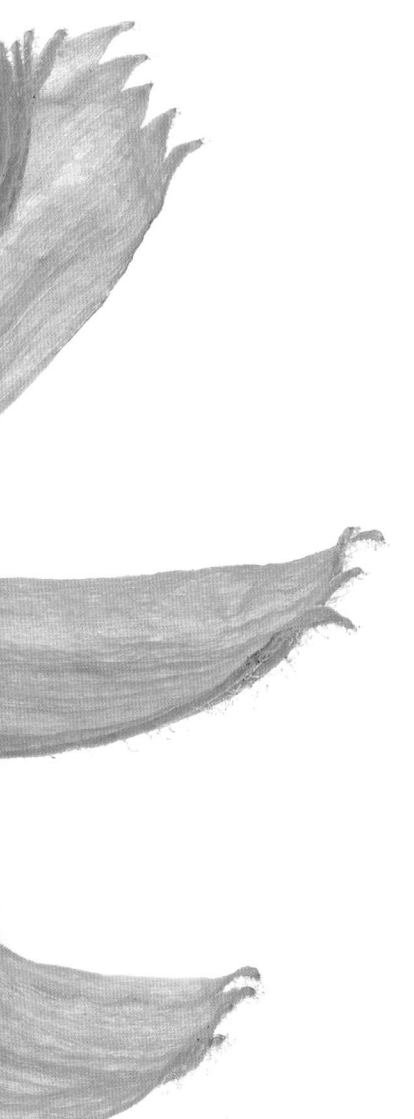

other pollinators can land on these in order to access the sweet nectar in the grouped blooms in the middle. This aggregate form is how the family's original name, composite, came about.

Knowing this makes the quaint ritual of yanking off the petals (or ray flowers) one by one while chanting "he loves me, he loves me not" seem a bit heartless. You're not dismembering a simple blossom, but an entire conglomerate. The origin of this activity is lost in history, but it is interesting to note that because daisies usually have an uneven number of petals, if you start with "he loves me," that's probably where you'll end up.

At any rate, all members of the family hearken back to this flower form, although it may not be immediately obvious. Some, such as goldenrod and hawkweed, sport flowerheads that are a mass of miniature daisies. The disk flowers may be proportionally the same size or even larger than the ray flowers. The disk flowers may, in fact, steal the show, such as in the case of sunflowers grown for their edible seeds (technically, in this family, the season-ending fruits are called achenes). In others—dandelion being the most dramatic example—the yellow ray flowers not only dominate, they are the blossom. And still other members are composed entirely of disk flowers, such as ageratum, boneset, and the blooms of the common weeds burdock and thistle.

Composite flowers have it easy when it comes to attracting pollinators. Ray petals offer ample real estate for landing, while the disk flowers provide a bounty of pollen and nectar. But where some flowers entice pollinators with an appealing fragrance, these do not. Many are scentless, and in some cases—chamomile, daisies, marigold, and tansy are the most salient examples—it would be fair to say that the scent is off-putting or bitter. (Cows will not eat daisies, which is why farmers resent

Jaunty blooms of chicory, or *Cichorium intybus*, open for only a few hours a day. Then the color of the ray flowers rapidly drains away, fading to white. Its dried, ground roots can be used as a coffee substitute and as a diuretic.

incursions of ox-eyes into their pastures.) Close study has shown that acrid juice permeates the entire plant but concentrates particularly under the blooms, in the bracts. Experts theorize that the bracts not only offer the flowers above physical support, but also prevent insects from gaining access to the nectar from below.

By this reckoning, one wonders whether daisy juice might be used as an insect repellent. Apparently it is possible. Old herbals provide recipes for topical decoctions as well as teas that are reputed to treat inflammation, bruises, and rashes. And organic gardeners who use companion planting to ward off insect pests in the flower and vegetable garden tout marigolds, zinnias, and certain daisies.

Now, back to those bracts. If you peer very closely at the underside of most flowers in this family, you ought to be able to make out scalelike bracts, which are actually modified leaves. Usually green and overlapping, they clasp together the base of the flowerhead (turn over any daisy or mum to see this). To complicate matters, in some species they are not green, but pale or whitish, and they do not overlap. Where present, observe how the disk flower slots more or less neatly inside. It's an admirable arrangement.

BELOW Rich blue perennial cornflower, *Centaurea montana*, has fringed "petals" that are actually ray flowers, while the reddish interior is made up of disk flowers. The arrangement makes a colorful, practical enticement for pollinators, including butterflies. Note also the overlapping bracts, so characteristic of this family.

RIGHT Perhaps no other fall bloomer is as magnificent as the dahlia. Enthusiasts recognize up to twenty different types. The appearance of the ray florets—flat, twisted, incurved, pointy, split at the tip, and so forth—dictate which subgroup each one belongs to. This is a waterlily type, so classified due to its double, flattened form, which allows the cupped, translucent petals to capture more light.

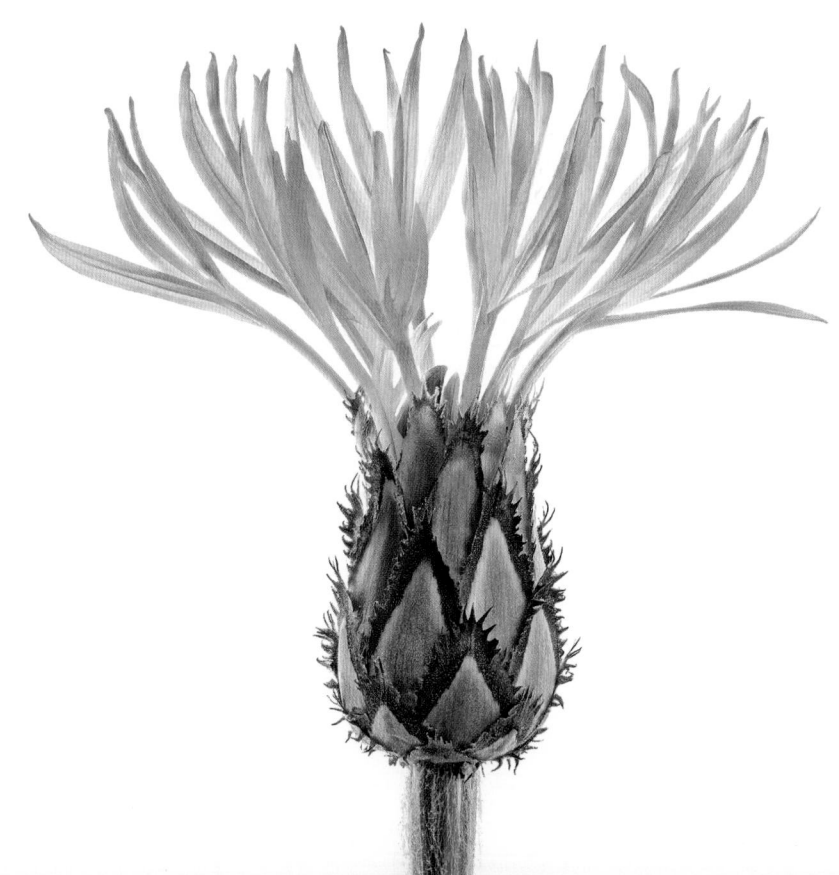

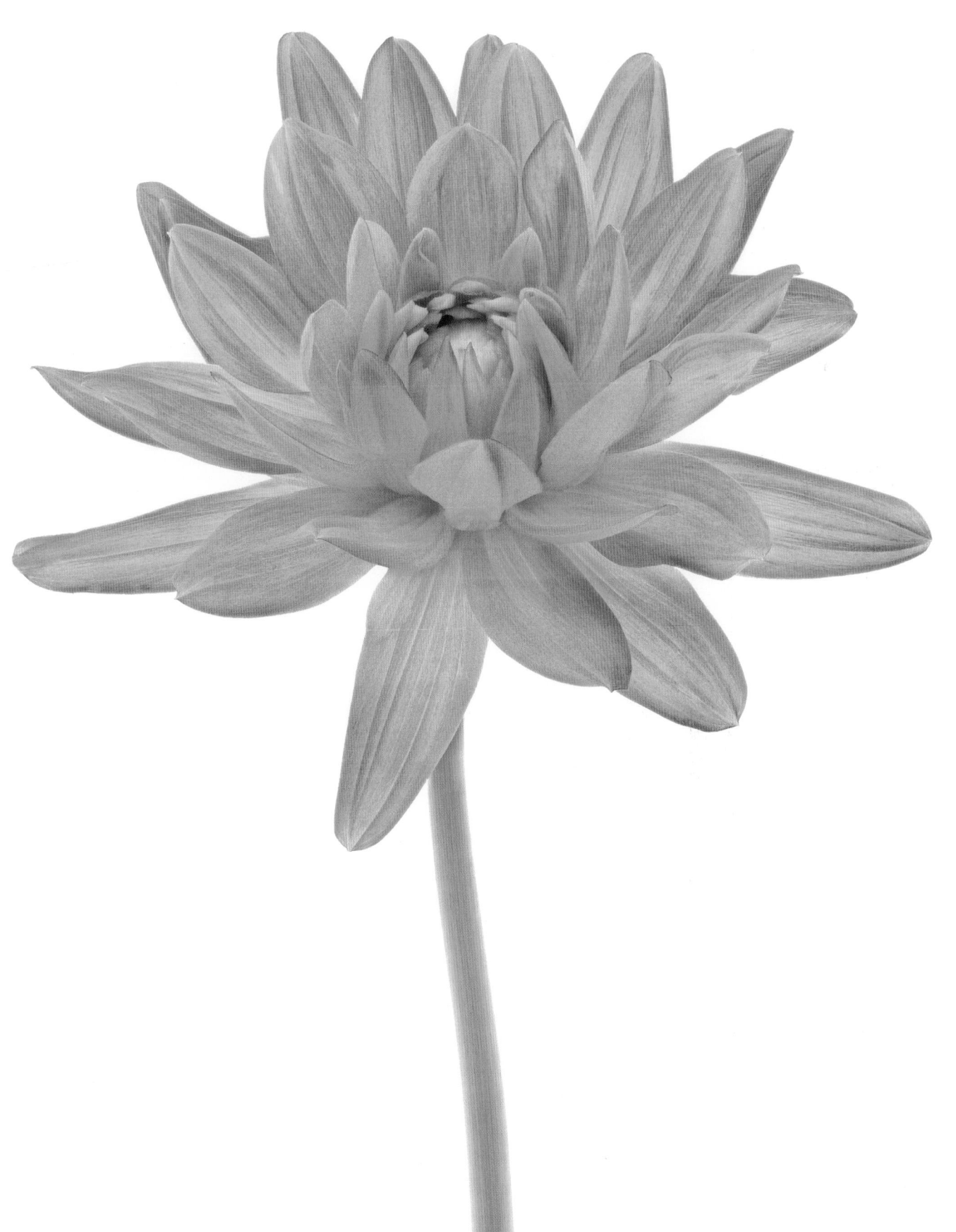

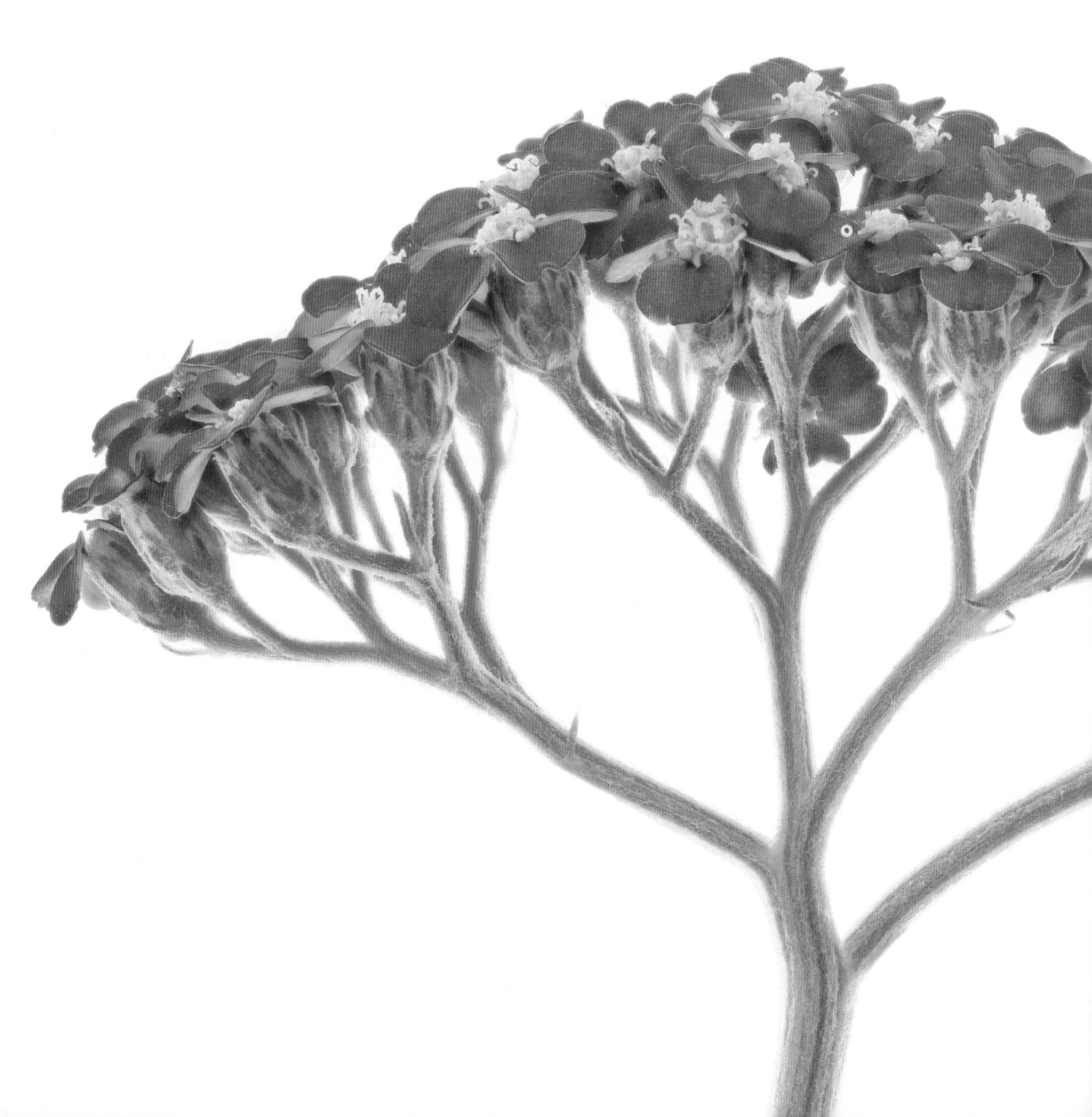

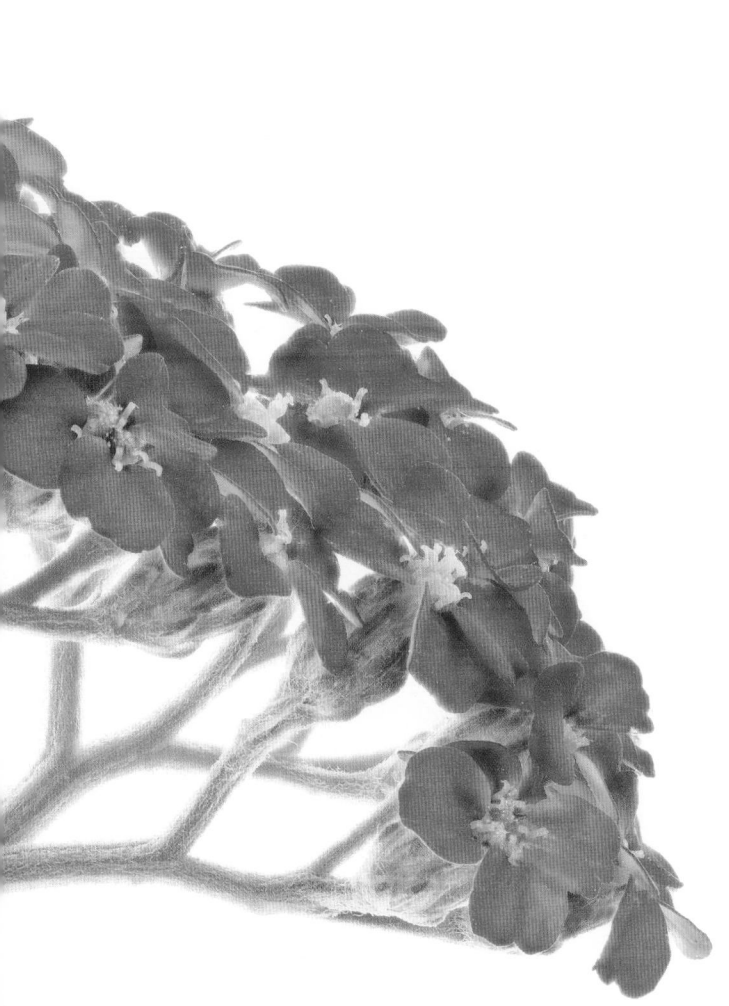

Another characteristic you can observe on many members of this family, after seeds form, is structure called a pappus. The one-fruited seeds, or achenes (think sunflower seeds), develop a means to travel—for, like all plants, they aim to make more of themselves. In some flowers, green sepals clasp a flowerbud until it opens, at which time they peel back. In the daisy family, the sepals are modified into small projections, collectively known as the pappus. This often ends up looking like a bristle or a barb that attaches to animal fur or socks. Or it can be much wispier, like the white fluff of dandelions that help the achenes parachute away on the wind, bound for new territory.

Because flowers in this family come in a variety of colors beyond the traditional white and yellow, plant breeders have had ample material to work with in their quest to extend more options to gardeners. Purple coneflower (*Echinacea*), in particular, has seen a renaissance in recent years. Purple ray flowers are passé. You can now get these stalwart perennials in rich glowing orange ('Orange Meadowbrite', 'Tangerine Dream'), fiery orange-red ('Firebird', 'Sundown'), pale yellow ('Sunrise'), and bright pink ('After Midnight'). The ones with white ray flowers (such as 'White Swan') are less exciting, as they resemble good old Shasta daisies (*Leucanthemum ×superbum*), which don't need improving with the practically flawless, bulletproof 'Becky' and 'Alaska' varieties so widely available. Some people find the stranger innovations, such as the frizzled, curled petals of the 'Crazy Daisy' Shasta daisy or the pom-pom form of the 'Razzmatazz' coneflower, a bit, well, strange.

Lacy topped, acrid-scented *Achillea*, or yarrow, hybrids come in white as well as shades of pink, red, purple, yellow, and mixes. Tiny individual flowers, true to their daisy family membership, have a composite disk surrounded by five ray florets. Interestingly, this genus has an impressive list of medicinal uses, notably for easing wounds and as an anti-inflammatory.

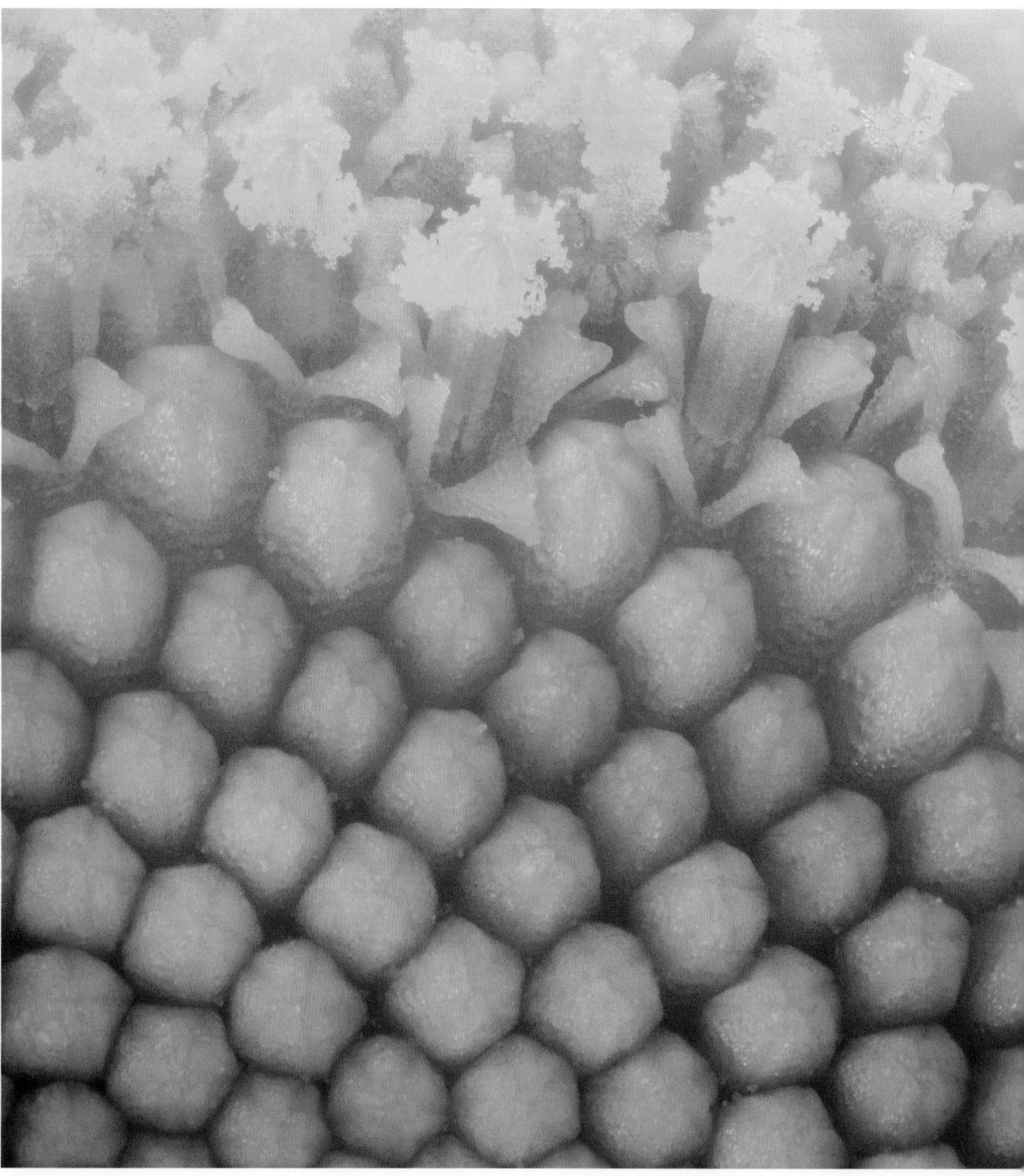

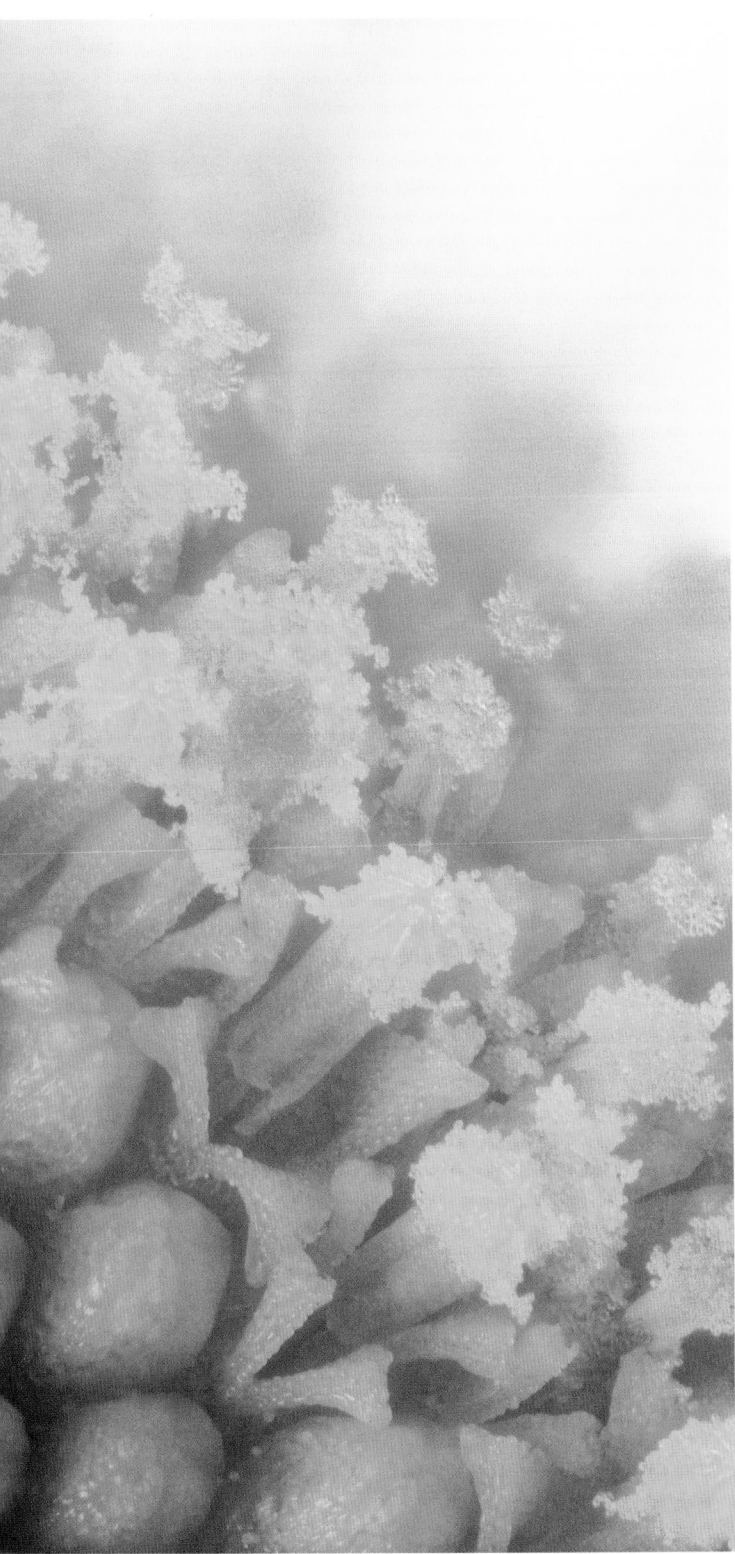

On the other hand, when plant breeders created a whole range of sunflowers for bouquet use, they did a great service to florists, market gardeners, and gardeners who like to raise their own cut flowers. These lack extensive disk flowers and do not produce a crop of edible seeds. But they retain all the best qualities of the species—namely robust, beautiful blooms atop strong stems, profuse production, and a long period of bloom. Plants shorter than a person, including numerous dwarf varieties, make it easier to appreciate these at eye level and to harvest them. You also have a choice of single-stem and branching types.

Despite their unique complexity and various quirks, both natural and horticulturally induced, members of the daisy family all share an undeniable appeal. They tend to bloom easily and abundantly, and they hold up well outdoors as well as in a vase. "Fresh as a daisy" may be a cliché, but most of these plants earn our easy admiration.

Ox-eye daisy, *Leucanthemum vulgare*, is a good classic daisy to study. Among the central yellow disk florets there are both male and female parts in stages. The outermost, fuzziest ones are female and look like little Ys, while just inside them, closer to the middle, is the male stage, a thin circle of yellow pollen. This configuration discourages a bee from self-pollinating a given daisy. If it lands on a white ray flower and walks inward, pollen attaches last; then it departs to the next daisy to inadvertently deliver pollen to the waiting outer layer of female (yellow) disk flowers.

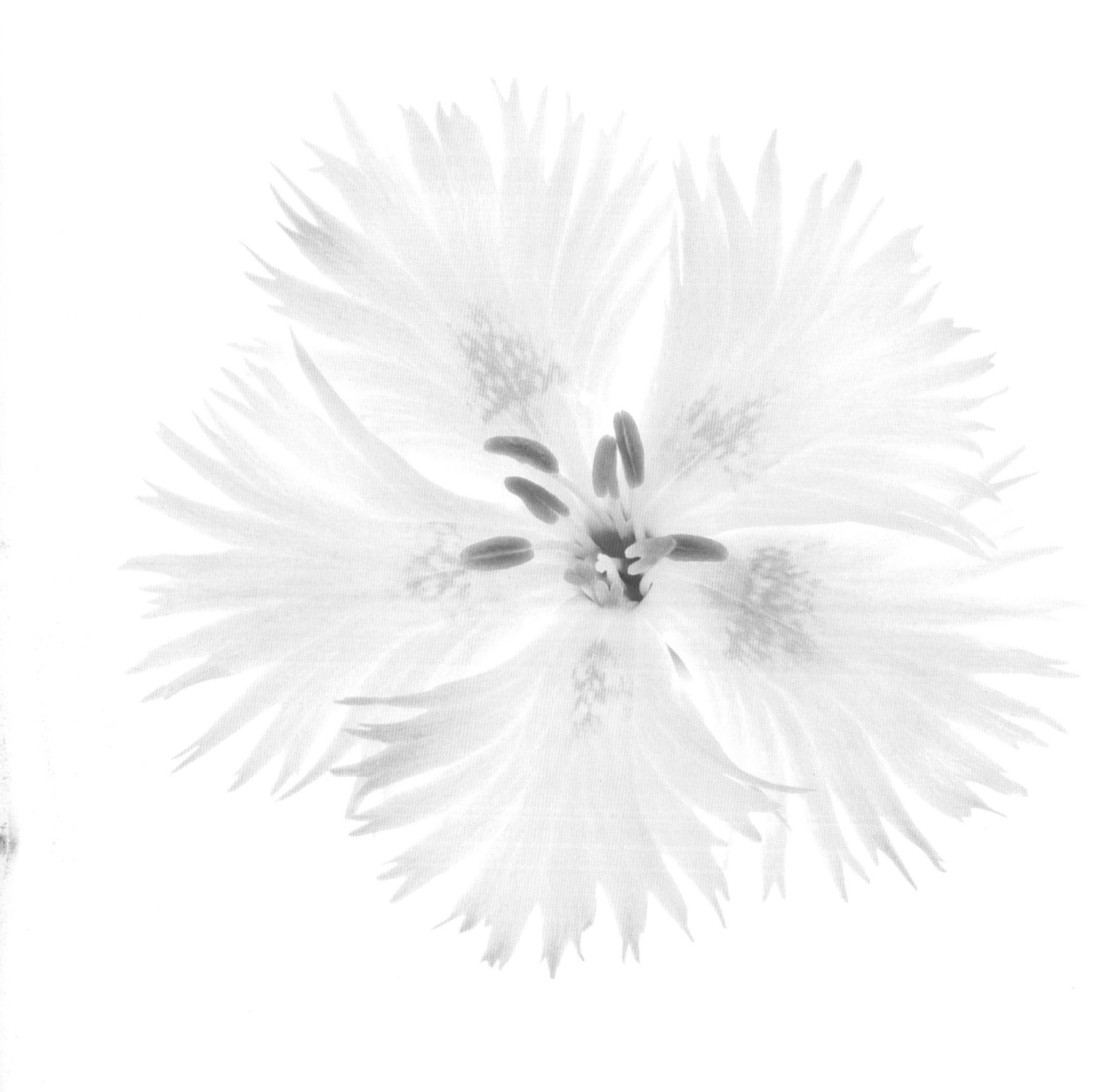

THE DIANTHUS FAMILY

Caryophyllaceae

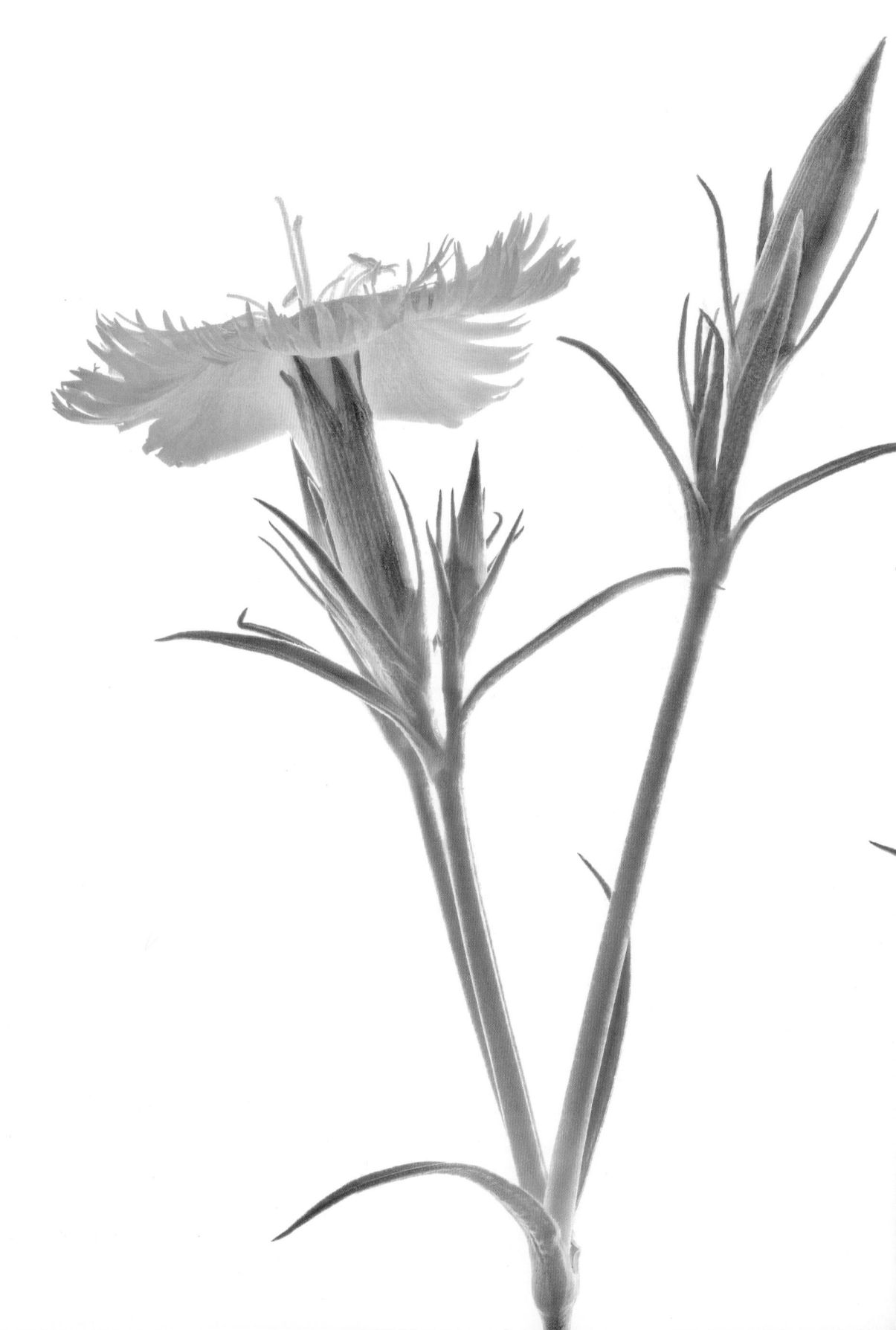

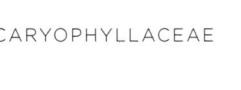

"Floral fragrance is a kind of olfactory
come-on that proclaims to a potential flower
fertilizer, 'Come hither, honey, 'cause there's
scrumptious pollen and sweet nectar hidden
inside these pretty petals.'"

JANET MARINELLI, "Planting Perfume for Pollinators," 2010

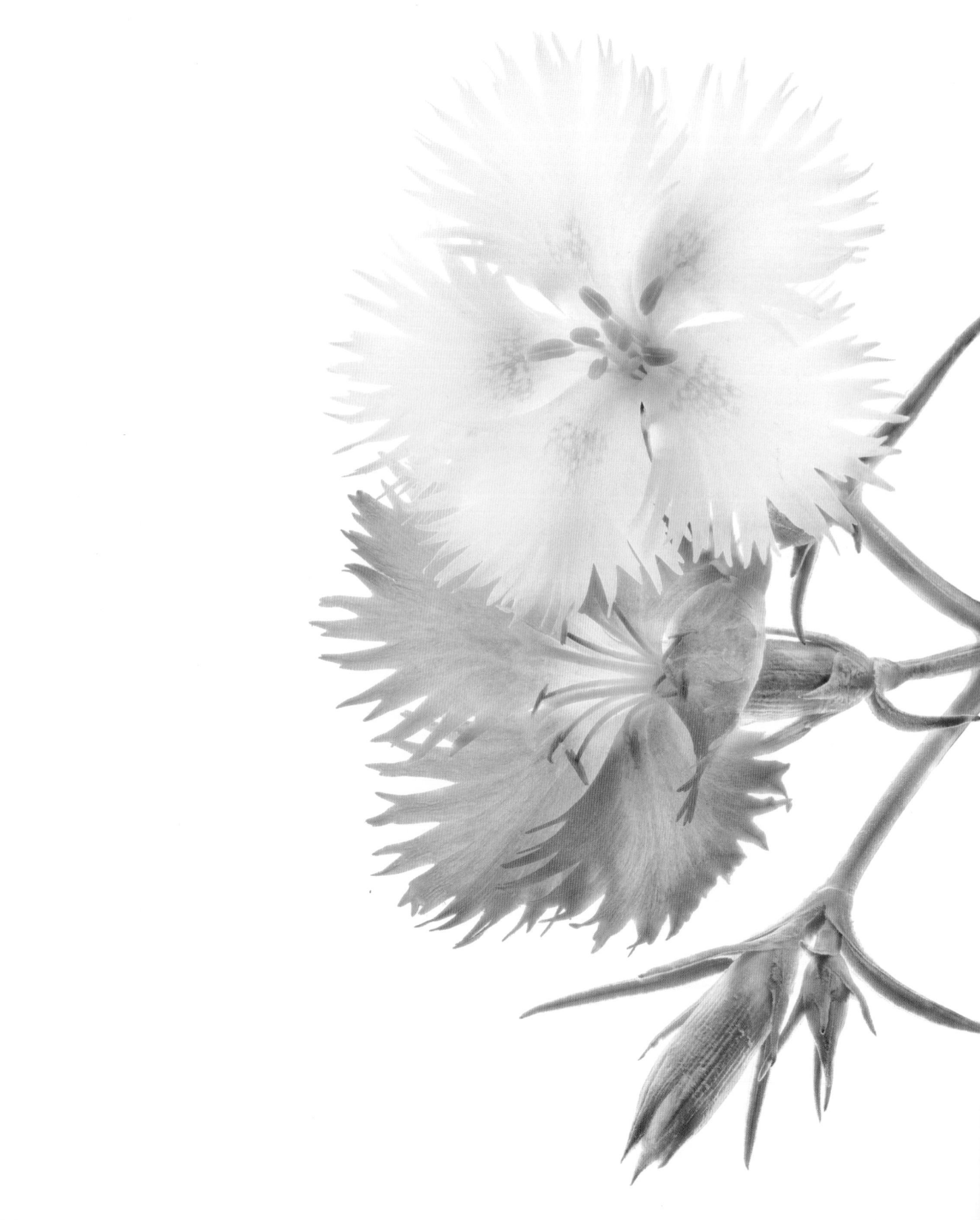

Often the very first thing you notice about pinks is their delicious, spicy-sweet scent, so reminiscent of freshly ground cloves. One whiff might send you back to your grandma's kitchen during holiday baking time, but when you open your eyes again, you're in the summer sunshine. Pinks, carnations, Sweet Williams, and all manner of dianthus flowers are grouped into the family Caryophyllaceae, which embraces more than 2000 species, including chickweed (*Stellaria*), snow-in-summer (*Cerastium*), baby's breath (*Gypsophila*), and campions (*Lychnis* and *Silene*). Gardeners and florists prize some of them, while others are considered weeds.

Not all have this splendid fragrance, but for those that do, the purpose is to entice pollinators. In the case of dianthus species, these are bees, although the white-flowered ones use the twin beacons of luminous blossoms and heady scent to draw evening moth traffic. Some species turn out to be a primary food source for the larvae of various moths.

Despite the family's large size, flower form in Caryophyllaceae is fairly uniform and easily distinguished. Look for terminal blooms, often displayed singly (which is why, for example, carnations make such good cut flowers). If the stems fork or branch, usually you can still identify one as the top flower. Most have petals and sepals in groups of five. Stamens within number five to ten. Where present, a tubular calyx consistently has five lobes and is accompanied by one or more pairs of bracts. Flowers harbor both male and female parts and ultimately produce a capsule with a seed or seeds inside.

Petal edges may be smooth, but are frequently frilly, fringed, or even deeply incised. So while it could be that the common name pink refers to flower color—so many of these are pink or rosy, as well as red or white—it is also possible that the uneven edges lead to this name. The verb *pink* dates from the fourteenth century and means "to decorate with a perforated or punched pattern." Hence pinking shears, which give fabric or craft paper a fringed edge.

Foliage in this family is also an identification aid. Rather grassy and smooth, it is often gray-green or blue-green, which complements the flowers. Often stems are swollen at the nodes, which may also be sheathed—think of a carnation flower's sturdy stem. Lower-growing types form mounds and tufts that endear them to people who have modestly sized mixed flower borders or a rock garden. A further asset is that nibbling deer seem to avoid this group.

Pinks and their relatives have been popular in gardens for centuries. Because they are so variable and interbreed freely, the upshot is confused nomenclature. You may find named cultivars are attributed to one species or another, while others have unknown or complex parentage. Nonetheless, there are certainly plenty of splendid plants for gardeners to choose from, no matter what are called.

There is an undeniable appeal to Caryophyllaceae. Whether well mannered or weedy, with small or bigger blossoms, they always seem to carry themselves with endearing pluck, as if they are confident we will love them.

LEFT, PREVIOUS SPREAD, AND PAGE 106 Modern border or cottage pinks, *Dianthus plumarius*, come in white and shades of rose and pink. The common name, however, comes from the frilly, fringe-edged petals, which are pinked, or serrated. All are spicily fragrant, making them a favorite with people as well as bewitched pollinators like bees, butterflies, and moths.

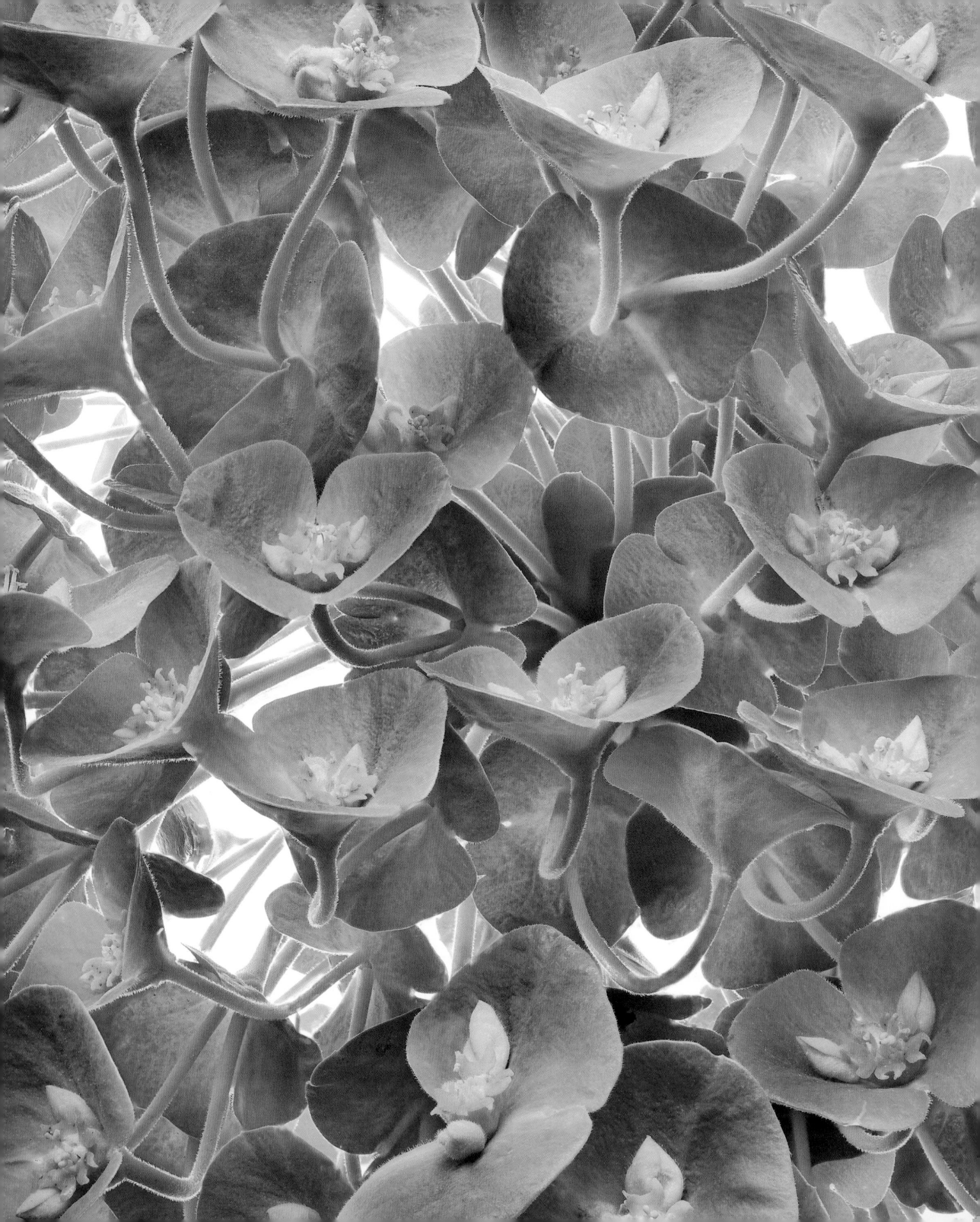

THE EUPHORBIA FAMILY

Euphorbiaceae

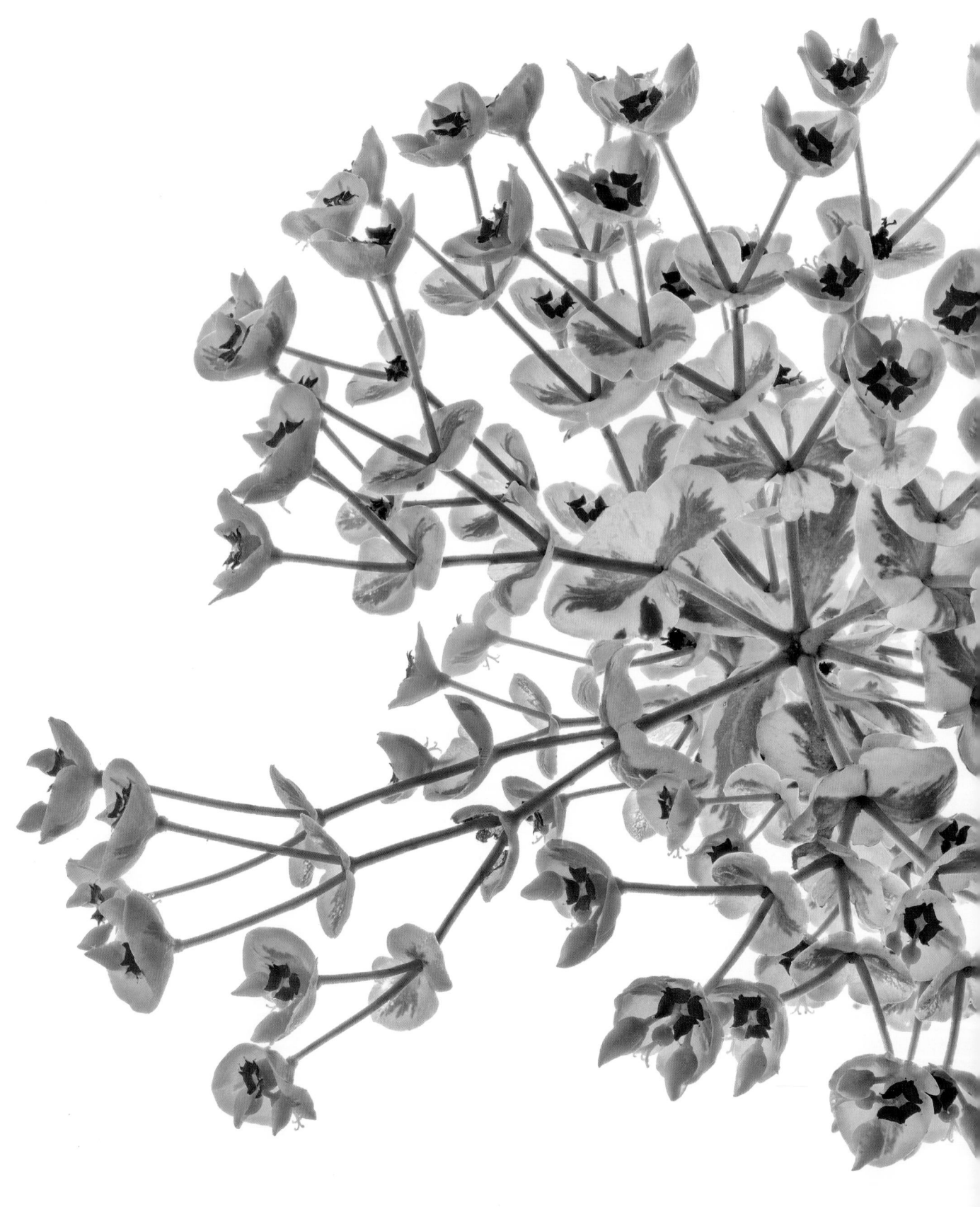

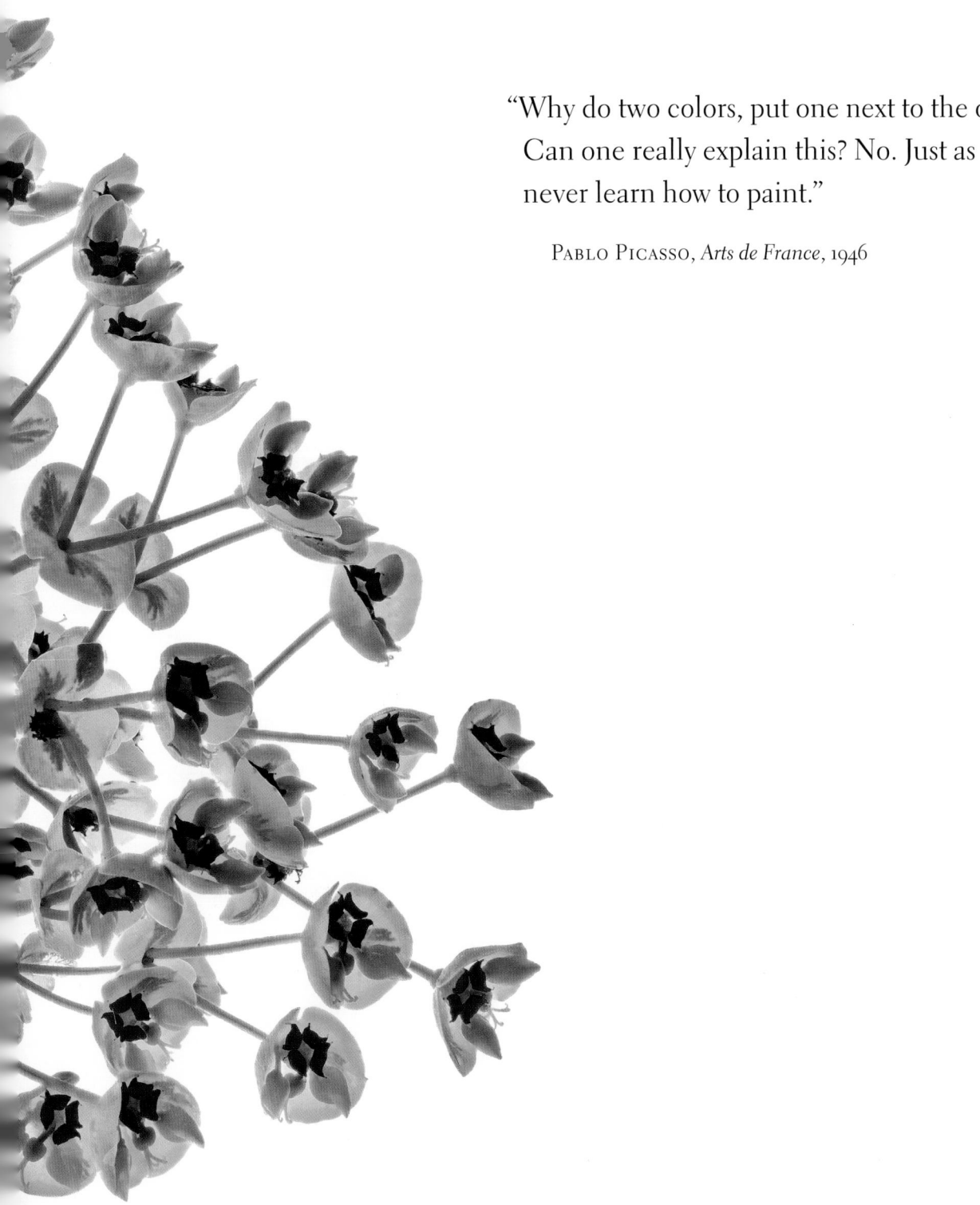

"Why do two colors, put one next to the other, sing? Can one really explain this? No. Just as one can never learn how to paint."

PABLO PICASSO, *Arts de France*, 1946

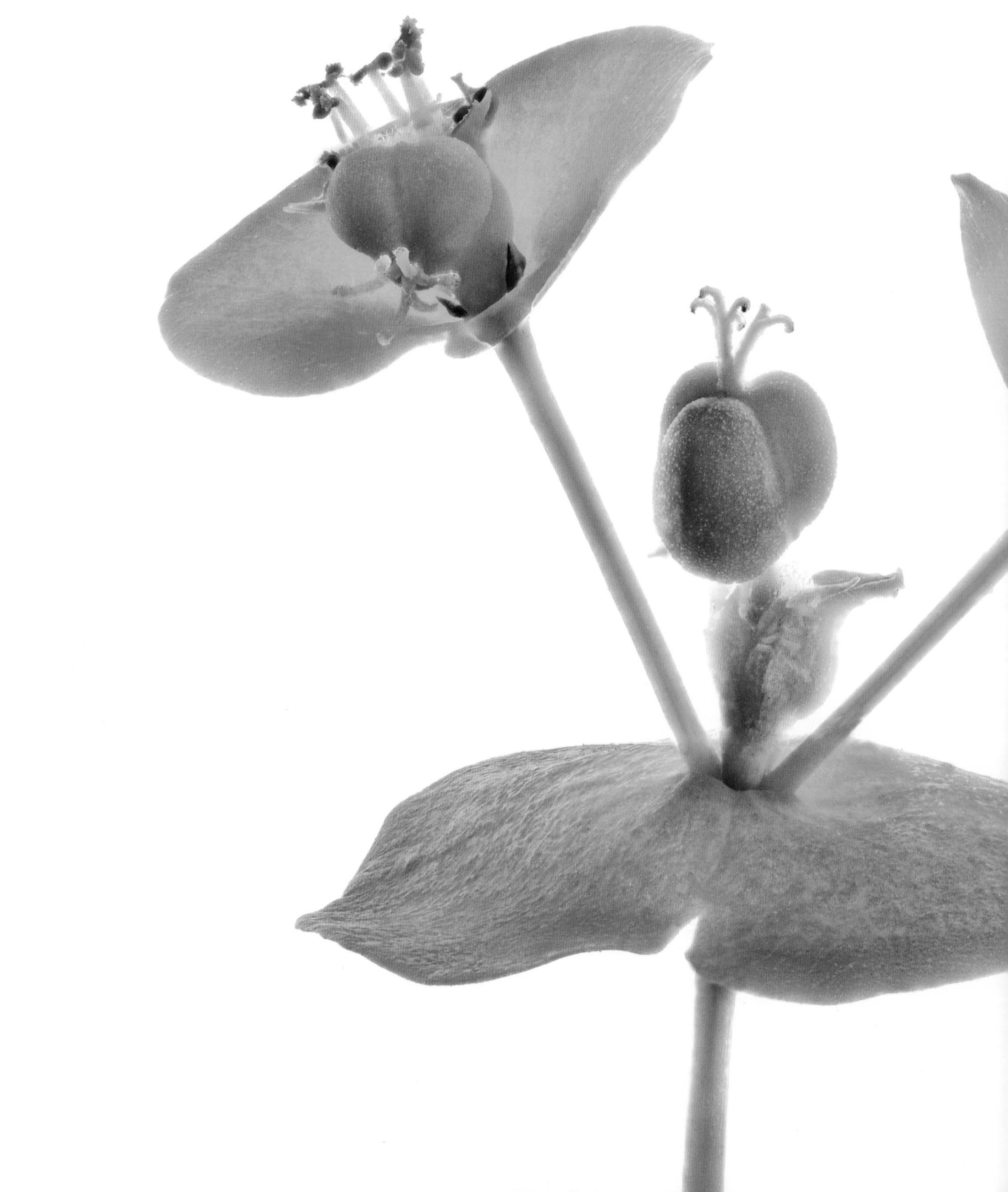

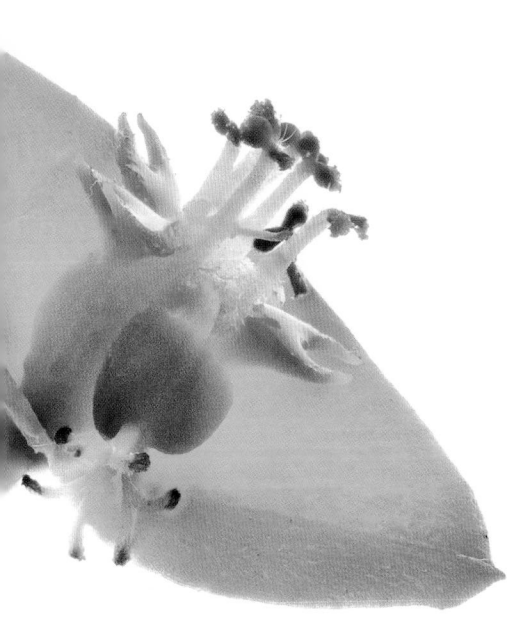

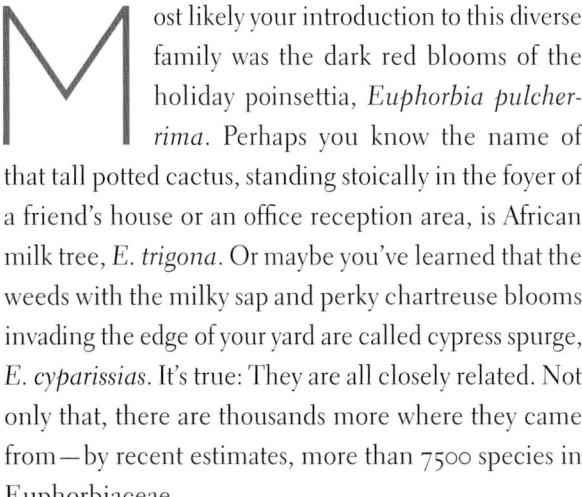

Most likely your introduction to this diverse family was the dark red blooms of the holiday poinsettia, *Euphorbia pulcherrima*. Perhaps you know the name of that tall potted cactus, standing stoically in the foyer of a friend's house or an office reception area, is African milk tree, *E. trigona*. Or maybe you've learned that the weeds with the milky sap and perky chartreuse blooms invading the edge of your yard are called cypress spurge, *E. cyparissias*. It's true: They are all closely related. Not only that, there are thousands more where they came from—by recent estimates, more than 7500 species in Euphorbiaceae.

Those brightly hued poinsettia bracts—the red, white, or pale green "flowers" are technically not—appear in autumn or so. Shorter days inspire these to color, which is how the plant became such a wintertime favorite (it is native to the Southern Hemisphere, specifically Mexico and Central America). In peak color, they certainly make a sharp, festive contrast with the dark green foliage. You can try to inspire a repeat performance by keeping a plant alive all the way through the following fall, then confining it to a darkened closet for two months, an exacting process that may be more trouble than it is worth.

There are dozens of varieties. The major improvement, though, has not been with the bract colors, but with the leaves, which used to fall off all too easily, turn-

Strange and strangely appealing flower forms characterize the herbaceous euphorbias. The actual flowers of *Euphorbia* ×*martinii* 'Ascot Rainbow' (PREVIOUS SPREAD) are tiny and red, but set off dramatically by green- and yellow-striped bracts. Those of *Euphorbia characias* subsp. *wulfenii* (PAGE 112) are soft yellow accented with minute dark purple glands and nestled in light green bracts, for a "flower" that almost looks edible. Wood spurge or "Mrs. Robb's bonnet," *Euphorbia amygdaloides* var. *robbiae* (LEFT), has yellow-green flowers and fleshy green bracts; between one and three seeds develop inside diminutive pods. When they are ripe, pods will explode if touched and thus disperse the seeds.

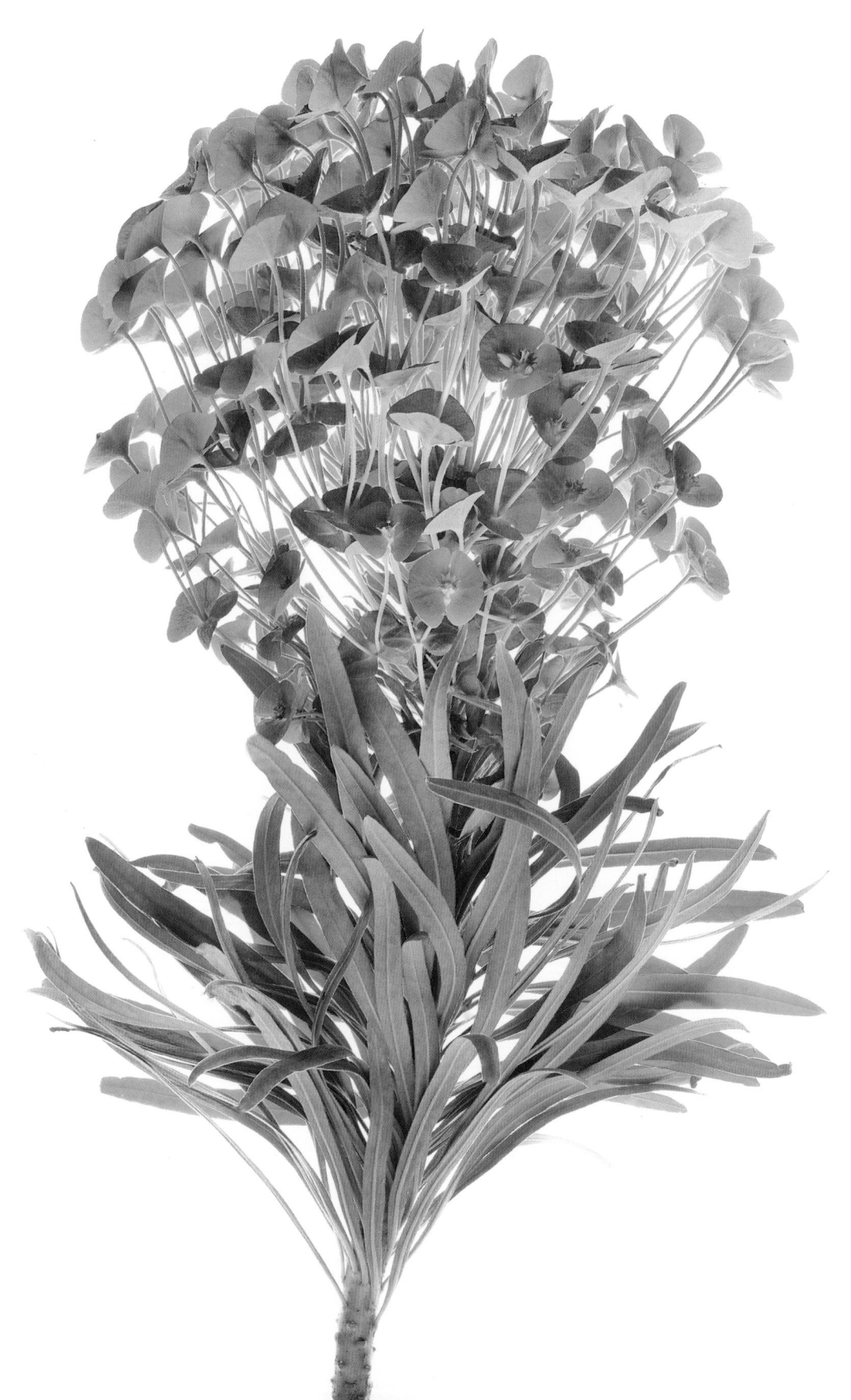

ing—to quote tart-tongued garden author Thalassa Cruso—"almost at once into ridiculous, bare-legged caricatures." The nursery industry selected for those that hang on to their foliage better, and forged on.

Incidentally, all euphorbias, including poinsettia, contain a milky white latex sap that can be mildly toxic. It can cause a skin rash or burning eyes (if you rub your eyes after exposure) and, if ingested, an upset stomach to the point of vomiting. But, as a recent edition of the Plant Delights mail-order nursery catalog points out, we ought to "consider this a bonus rather than a problem because deer don't like to eat euphorbia plants."

Which brings us at last to the garden-variety and wild herbaceous euphorbias, known as spurges and placed in the Chamaesyce subgroup. This encompasses the above-mentioned cypress spurge, *Euphorbia cyparissias*, as well as wood spurge, *E. amygdaloides*; leafy spurge, *E. esula*; myrtle spurge, *E. myrsinites*; and myriad other species and crosses, plus all their many cultivars. Snap off a stem of any of these and, sure enough, the white sap will run.

Generally speaking, foliage in this subgroup is slender and crowded along the stems. Color varies with the species, cultivar, and season. Blue-green and gray-green leaves are common and can be quite lovely, especially

Like a bountiful and busy bouquet, the inflorescence of Mediterranean spurge, *Euphorbia characias* subsp. *wulfenii*, shows off its green bracts. The true flowers are tiny and lodged in the centers.

when the plants are massed. And here again, the "flowers" are not flowers. Spurge bracts, for the most part, are green or greenish-yellow, though they also come in brown, dusky red, dark red, and purple.

Peer into the middle of the inflorescence of one of these herbaceous spurges (or the poinsettia flower or cacti imposters), and you will observe the true euphorbia flower. Tiny, cup-shaped structures, termed *cyathia* by botanists, are in fact flower clusters. The male ones are so small as to be barely detectable; they may even be reduced to one stamen (red in poinsettia, generally green or yellow in the other types). The female flower is often little more than a wee stalked pistil. The nectar is usually embedded, as glands, in the cup's rim or in super small petal-like bracts, very much smaller than any showy, colorful ones above or surrounding it.

Truly, the reason to grow the spurges is for the full (and deer-resistant) color show, often featuring jazzy or unusual but eye-catching contrasts. Among the tempting hybrids now on the market are dramatic 'Charam' (or 'Redwing'), with bristly blue-gray foliage, crowned with red spikes that open to reveal a glittery golden interior, and flashy 'Ascot Rainbow' (*Euphorbia ×martinii*) with yellow-striped leaves and bracts and tiny but glowing red flowers. Like their cactuslike cousins, these tolerate sun and prolonged heat in style. Overall, nobody really needs to justify their affection for euphorbias and their "flowers." If you find the plants beautiful, by all means grow and savor them.

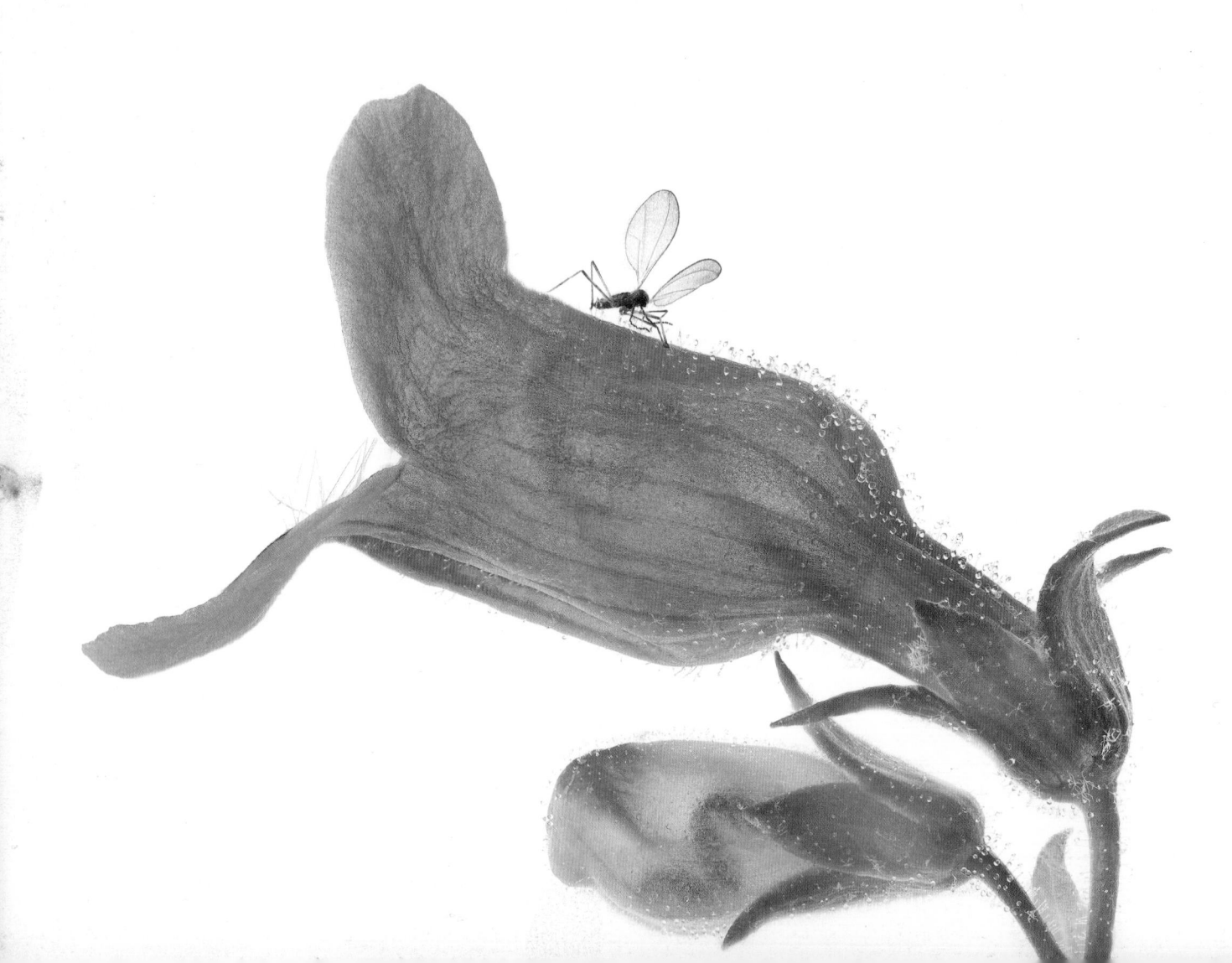

THE FIGWORT FAMILY

Scrophulariaceae

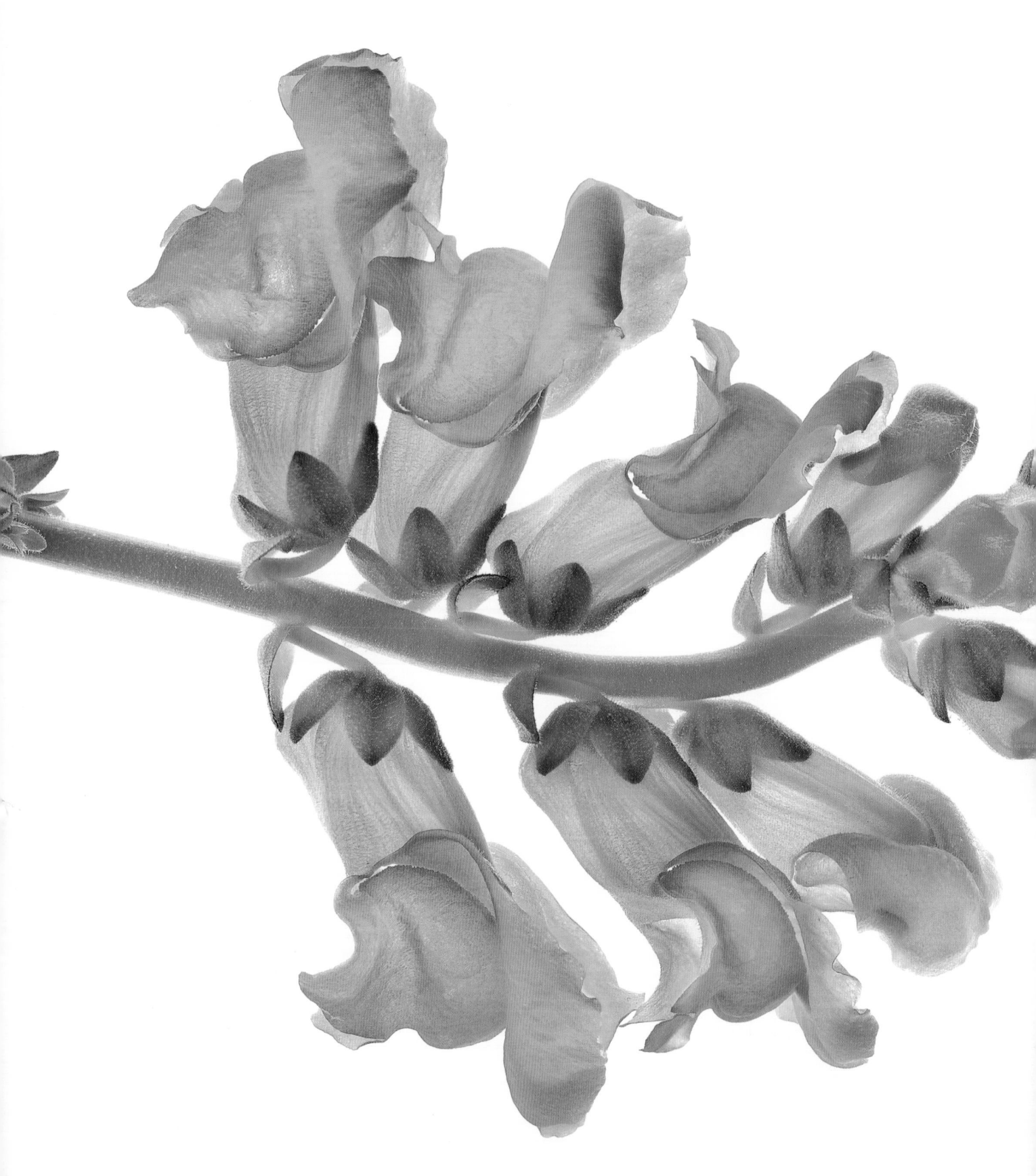

"Also in this rich garden pass we
gathered many fine grasses and carices,
and brilliant penstemons, azure and
scarlet, and mints and lilies, and scores
of others, strangers to us, but beautiful
and pure as ever enjoyed the sun or
shade of a mountain home."

JOHN MUIR, "Rambles of a Botanist Among the
Plants and Climates of California," 1872

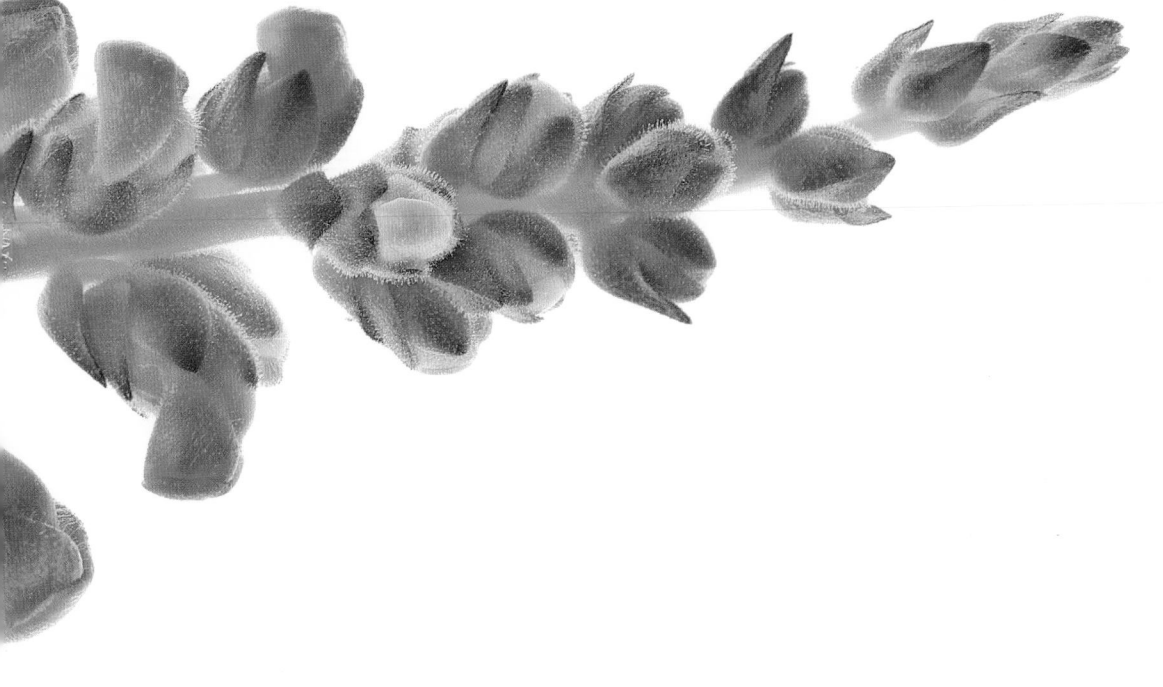

It's great fun to pinch open the closed jaws of snapdragon flowers, *Antirrhinum*, but did you ever wonder how such a flower gets pollinated? Bumblebees arrive, attracted by color and sweet scent (and perhaps, where applicable, also guided by colorful stripes that follow the veins in the petals). They are big and strong enough to shoulder past the closed flower to access the pollen and nectar within.

PAGE 120 Penstemons get their other common name, beardstongue, from a weird little flower feature. Like other members of the figwort family, they have five stamens, but one of these, which is sterile, has a small, hairy tuft. Shown here is richly colored 'Midnight Blue'.

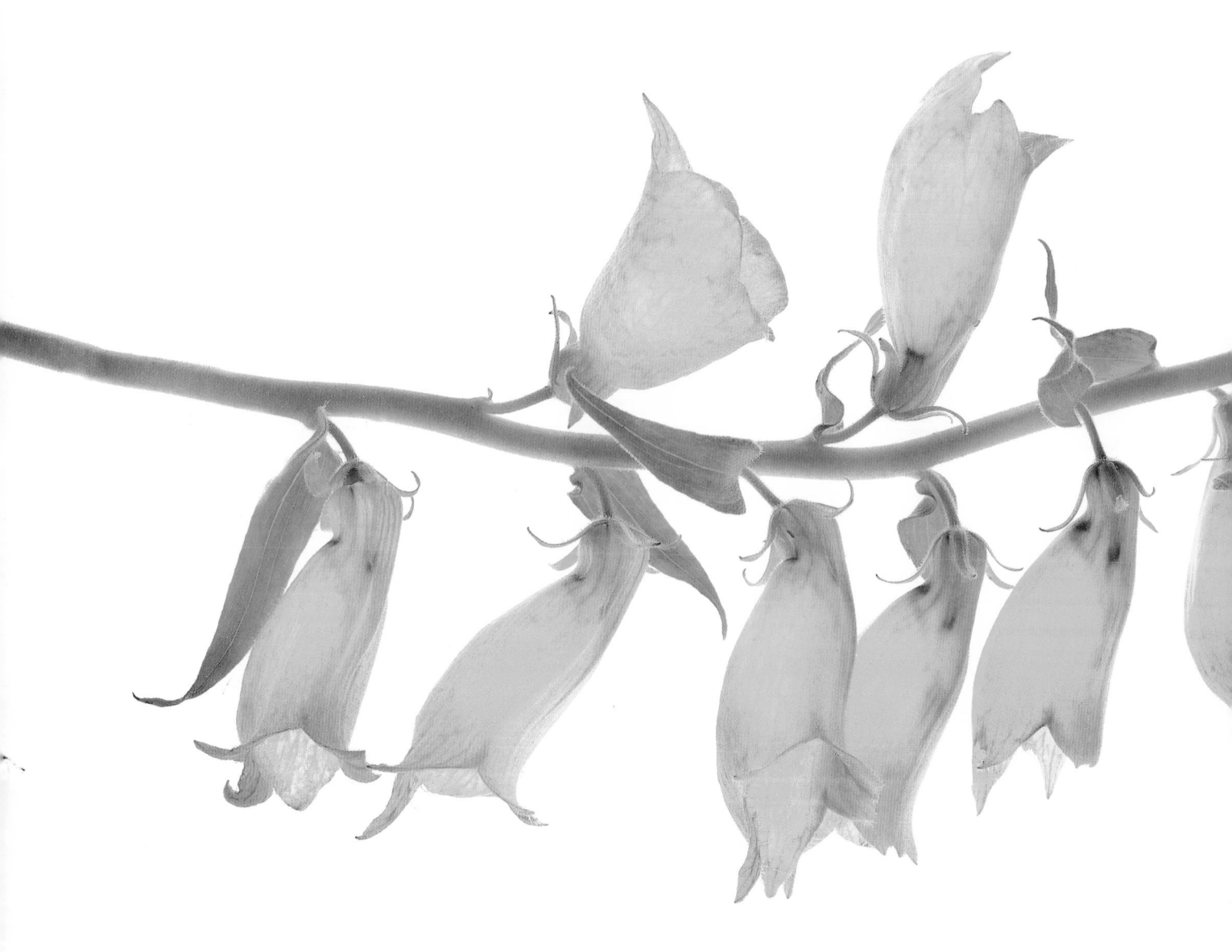

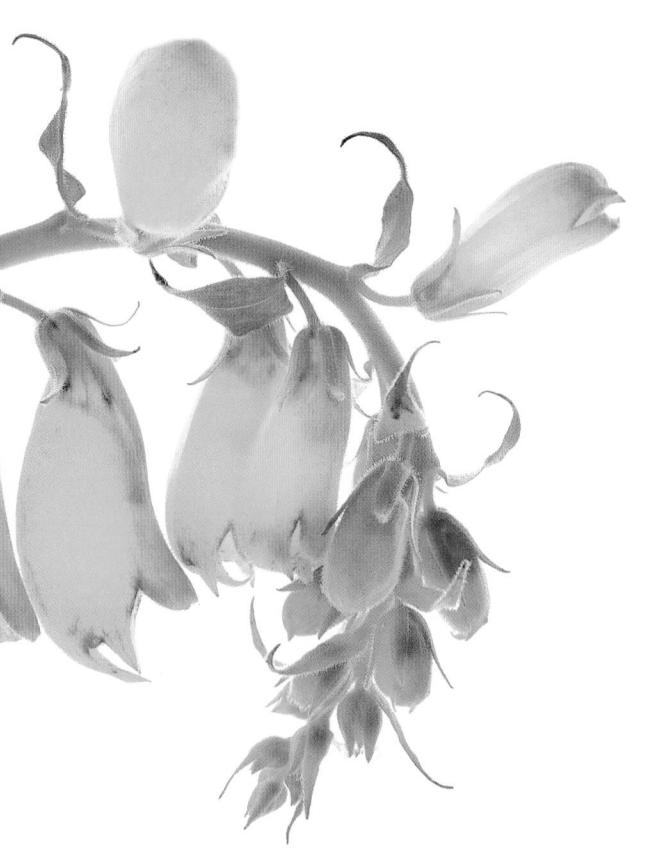

When you first hear mention of the figwort family, or Scrophulariaceae, you may be pardoned for thinking you are not familiar with it. The common figwort, *Scrophularia nodosa*, is not often seen in the United States, and while it has historically enjoyed some popularity in its native British Isles as a medicinal plant, it is not a garden beauty (*wort* is certainly a clue). It was once used to treat scrofula, a disease of the lymph nodes related to tuberculosis. A poultice made from the leaves was also used to treat hemorrhoids, or "figs." A sprawling plant with little green and purple flowers, it is not especially handsome, and its scent is described as fetid. So, not a sparkling résumé, at least from the point of view of ornamental gardening.

And yet quite a few commonly grown and widely enjoyed garden plants and pretty wildflowers reside in this family: penstemon (*Penstemon*), monkey flower (*Mimulus*), snapdragon (*Antirrhinum*), foxglove (*Digitalis*), mullein (*Verbascum*), Indian paintbrush (*Castilleja*), Cape fuchsia (*Phygelius*), and the ubiquitous weed butter-and-eggs (*Linaria*). Immediately you can make some amateur generalizations: They all have spikes or clusters made up of numerous small, two-lipped flowers, and they tend to be durable long bloomers. Most thrive in or tolerate hot, dry growing conditions.

Whether a mullein or a common snapdragon, this family's two-lipped flowers are very distinctive. They have four or five petals joined to the corolla, accompanied by four or five sepals joined to a calyx. Usually the form is tubelike. At the top, the petals flare outward, while the lower ones form a turned-down lip. Interestingly, the "four or five" characteristic appears within as well. Peer in, and you can observe two pairs of stamens,

An individual tubular flower from a yellow foxglove, *Digitalis grandiflora*, can be up to 2 inches long. These occur in spikes only on one side of the 2- to 3-foot stems. The brownish netting or spots on the interior guide pollinators inward.

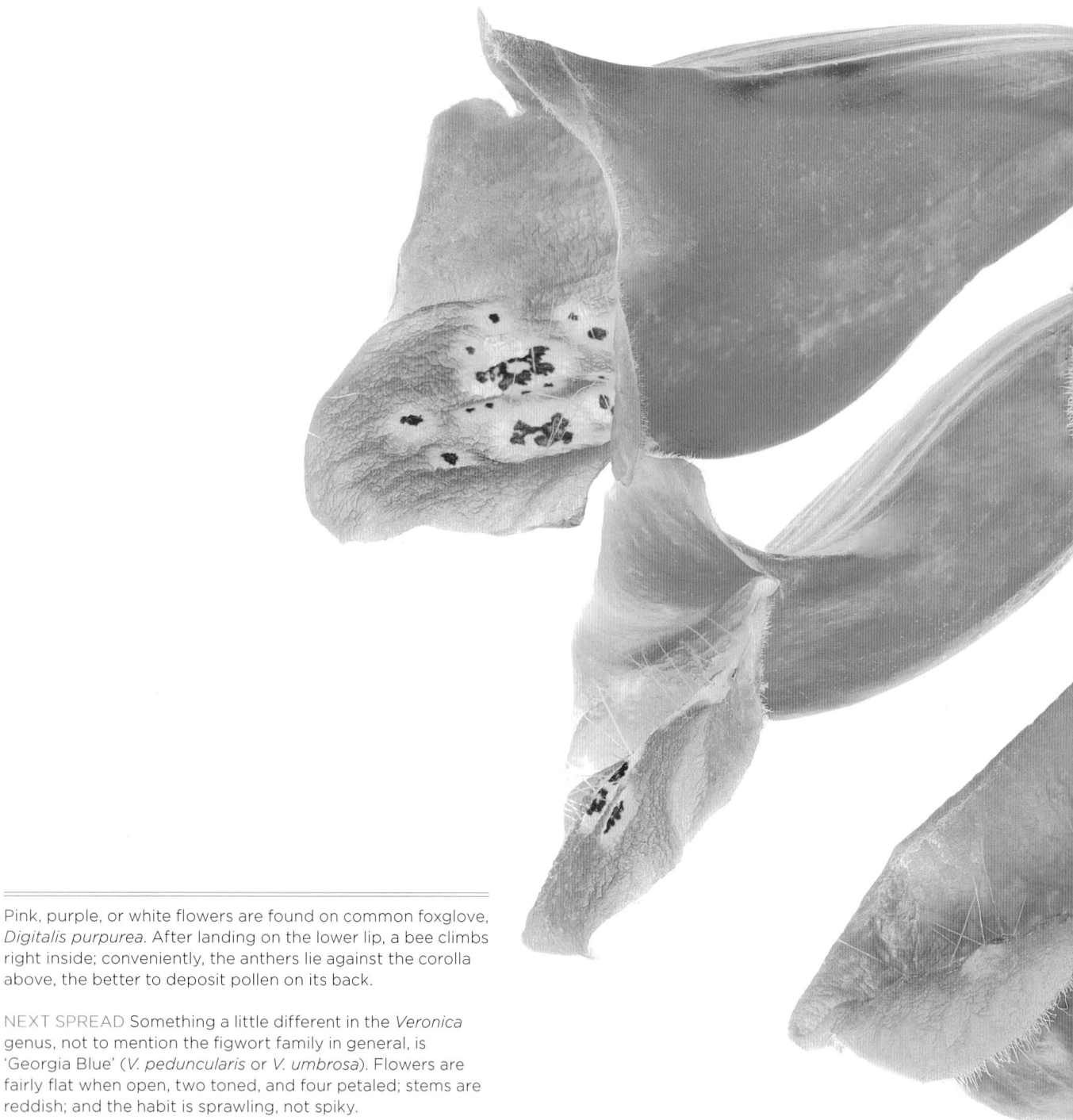

Pink, purple, or white flowers are found on common foxglove, *Digitalis purpurea*. After landing on the lower lip, a bee climbs right inside; conveniently, the anthers lie against the corolla above, the better to deposit pollen on its back.

NEXT SPREAD Something a little different in the *Veronica* genus, not to mention the figwort family in general, is 'Georgia Blue' (*V. peduncularis* or *V. umbrosa*). Flowers are fairly flat when open, two toned, and four petaled; stems are reddish; and the habit is sprawling, not spiky.

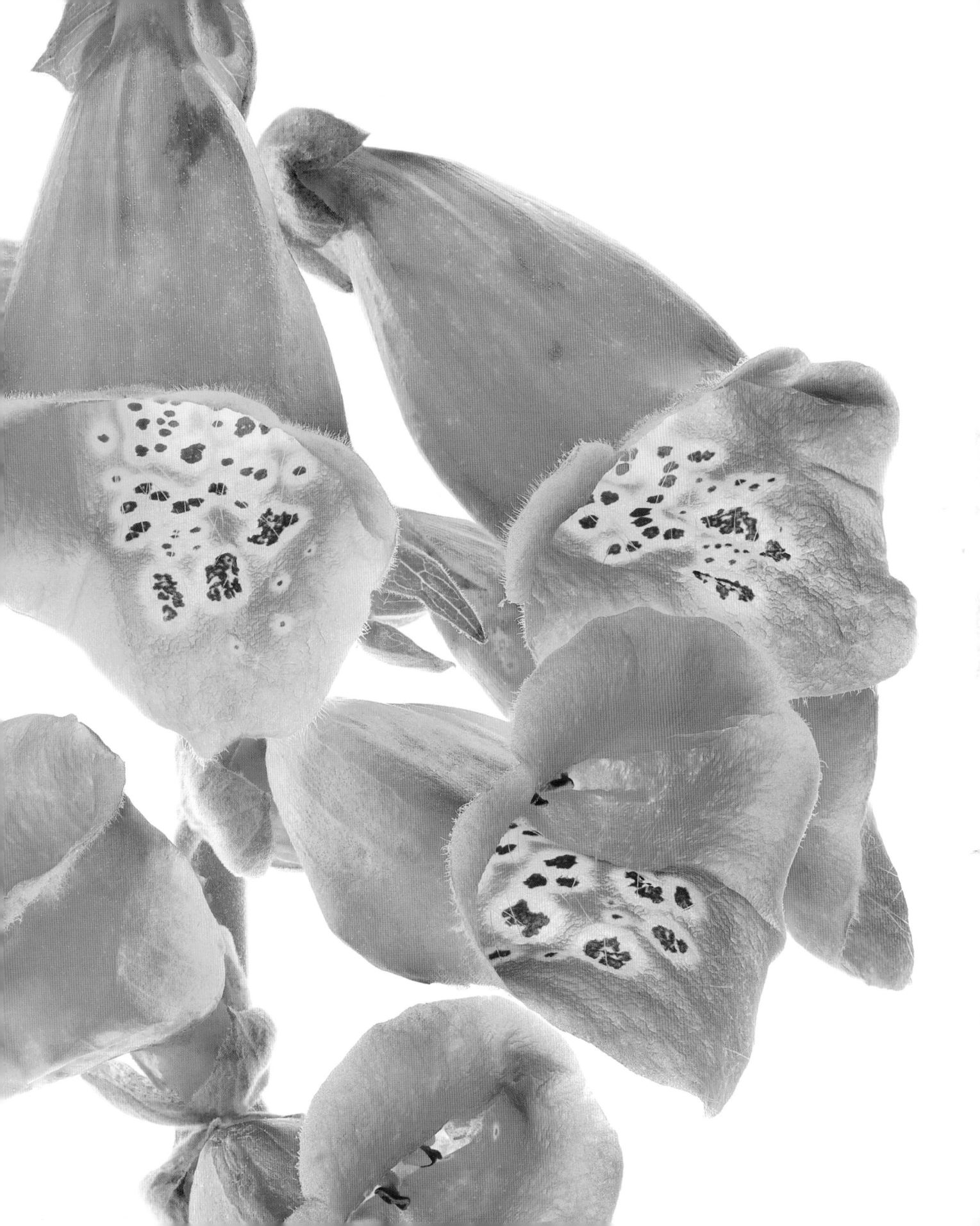

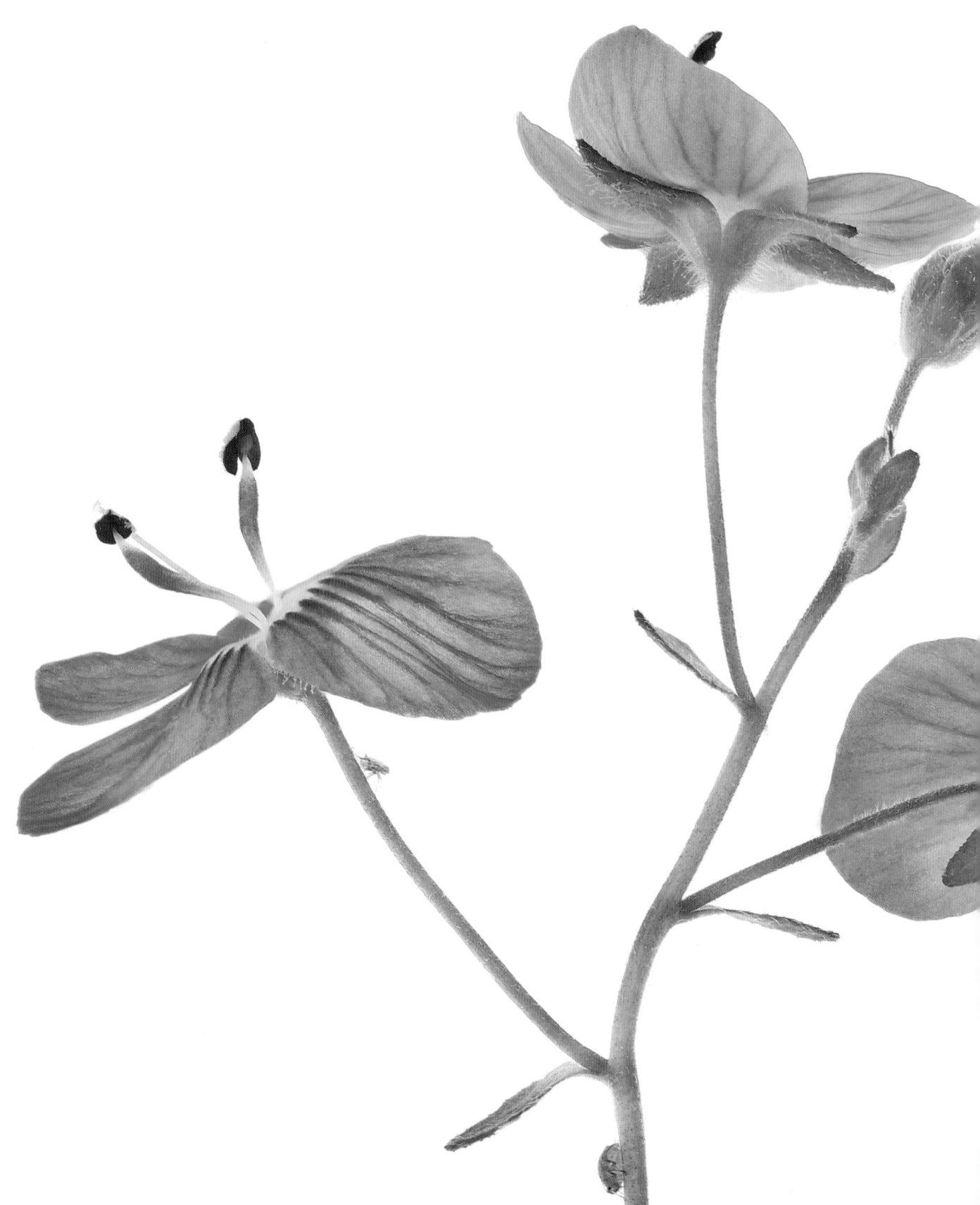

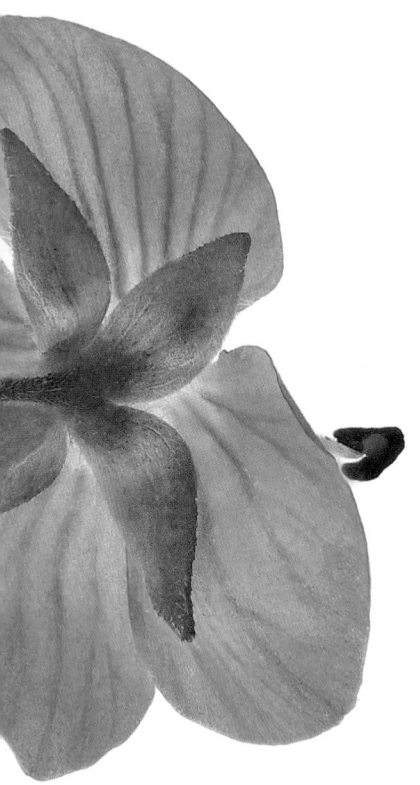

topped with anthers; the fifth stamen, when present, is sterile. In penstemons, this fifth one sports a small tuft of hairs, which accounts for the common name, beardstongue. At any rate, after the flowers fade, the fruit develops into a two-chambered capsule that holds tiny, waferlike seeds. Left to their own devices, most Scrophulariaceae will slowly but surely form small colonies.

If you come upon a member of this family in full, recent bloom, whether on a hike or in your garden, it's fun to carefully pluck off an individual bloom and hold it up to the light. Many times, you can actually see the nectar level at the bottom of the spur. (If pollinating insects have been by, the level will be lower.) This is observable even in the tiniest members, such as butter-and-eggs (*Linaria*). Generally Scrophulariaceae are pollinated by bumblebees, because they not only have the strength to push open the flower's mouth, they also have tongues long enough to reach way down inside and get at that sweet nectar. Hummingbirds and sphinx moths also forage on flowers in this family. Colorful dots or patches within the flower tubes of some are meant to both attract and guide pollinators.

For gardeners, the draw of plants in this family is their sturdiness and ample bloom period. Often the flowers need only a boost in size or color to round out a plant's appeal, and horticulturists, plant breeders, and nurseries have heeded this call. Snapdragons, which come in many colors and bicolors, are the most obvious example.

But quite a few of the wild and weedy ones have been significantly improved. While flowers of mullein species are usually a shy soft yellow or white, there is now a broader range of appealing color choices. *Verbascum ×hybrida* 'Southern Charm', for example, comes laden with pink, cream, light orange, and yellow flowers, and is truly delightful in garden and vase alike. 'Pink Pixie' (no specific parents touted) produces vivid pink spires. Biennial mulleins have been coaxed into producing

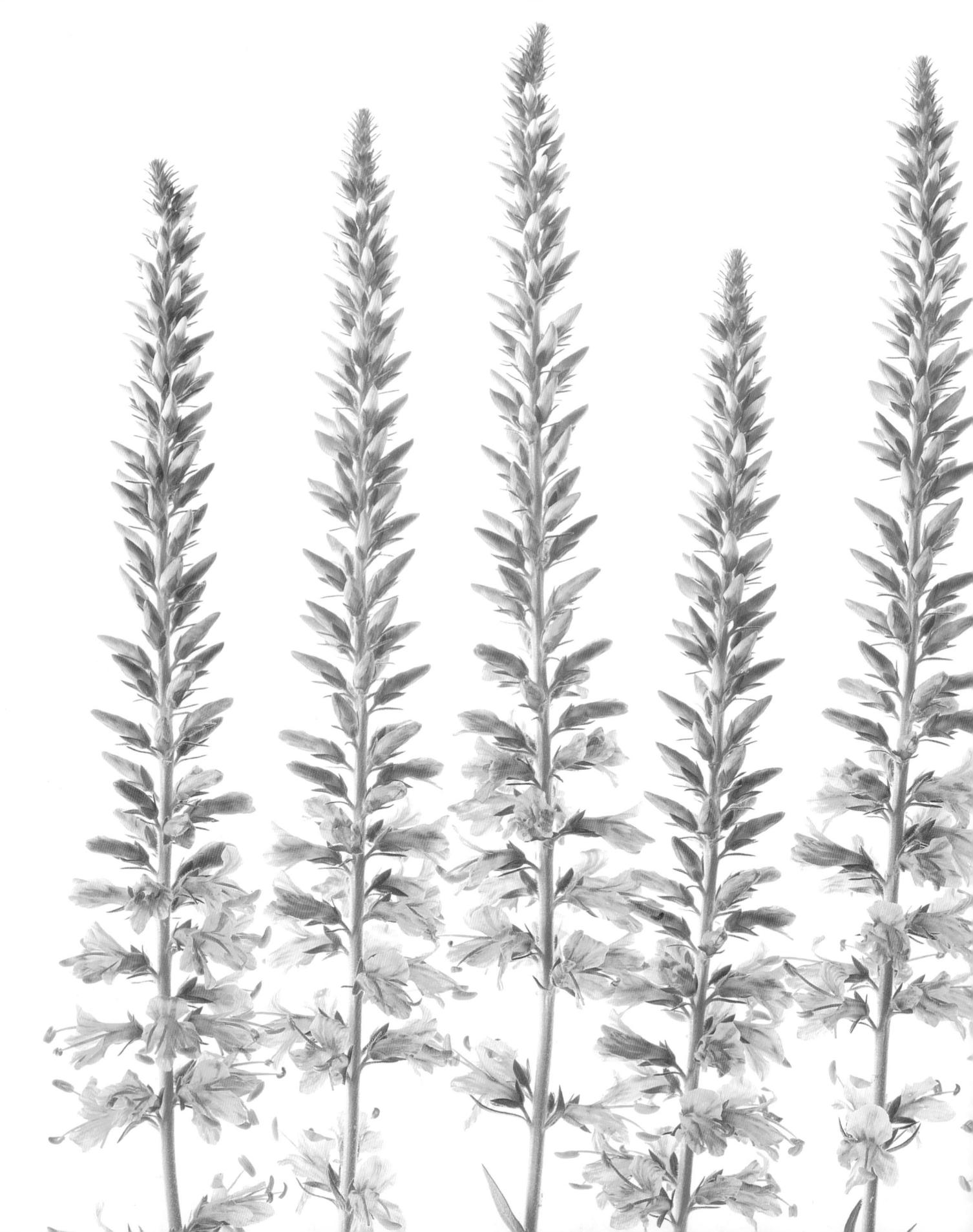

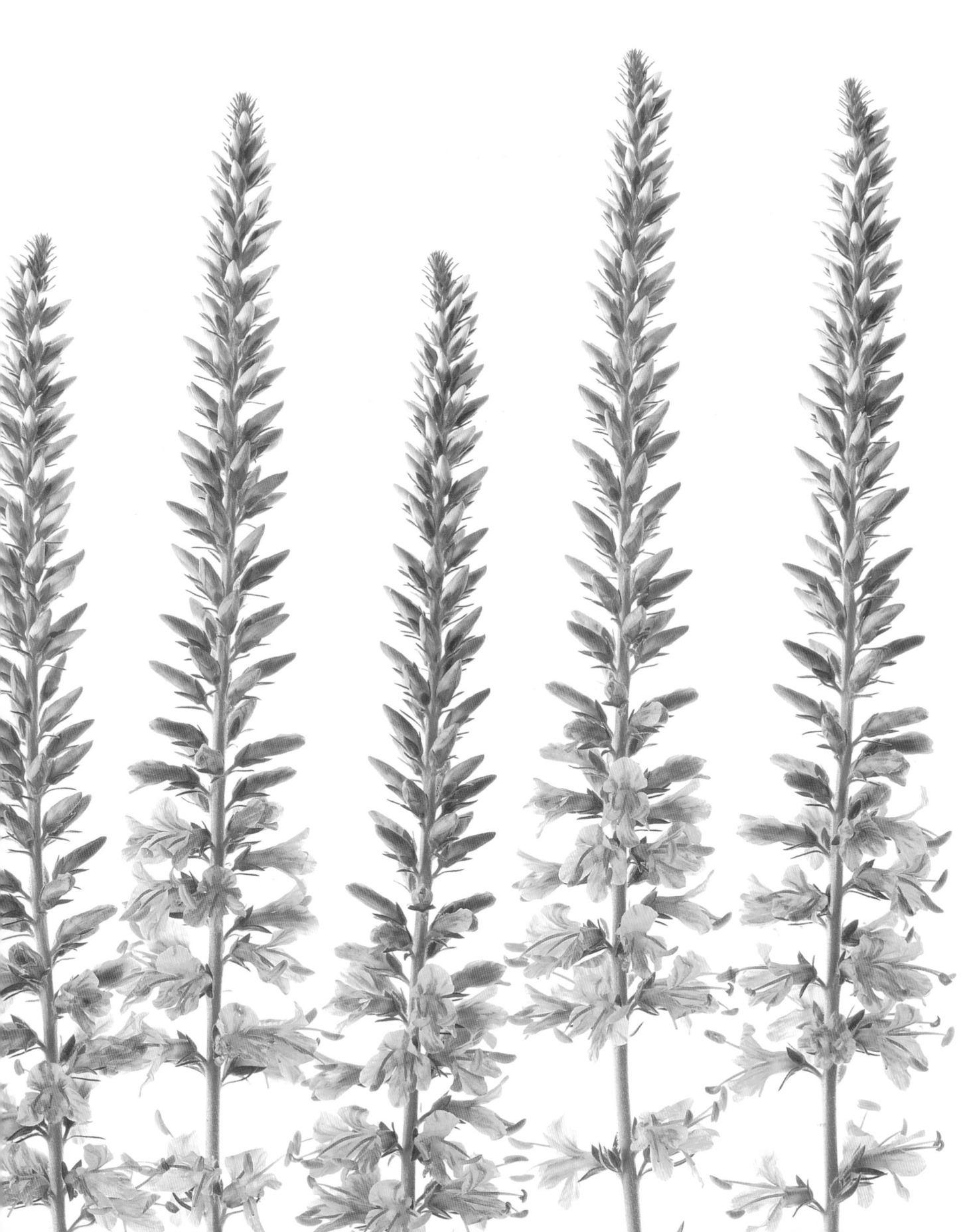

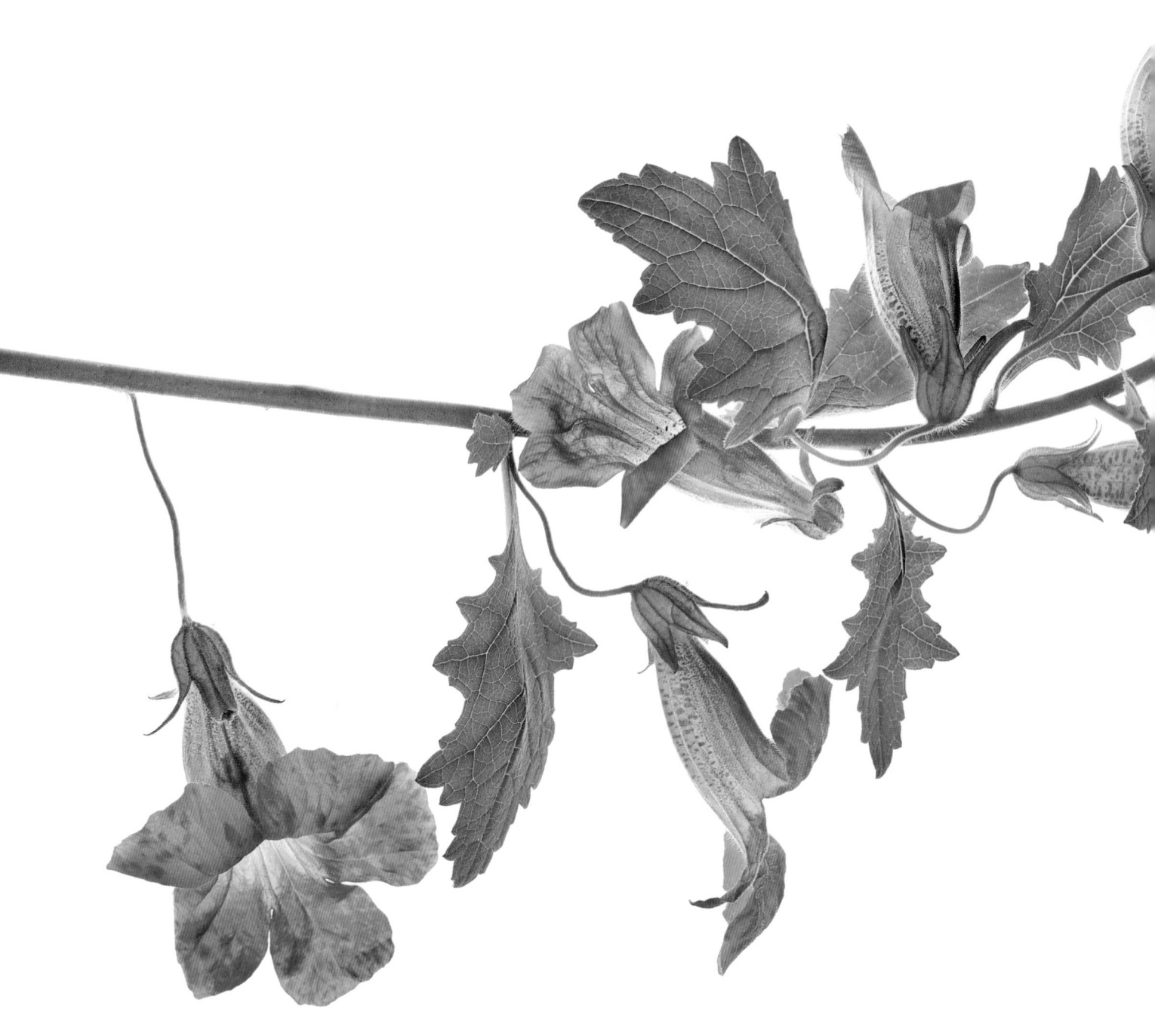

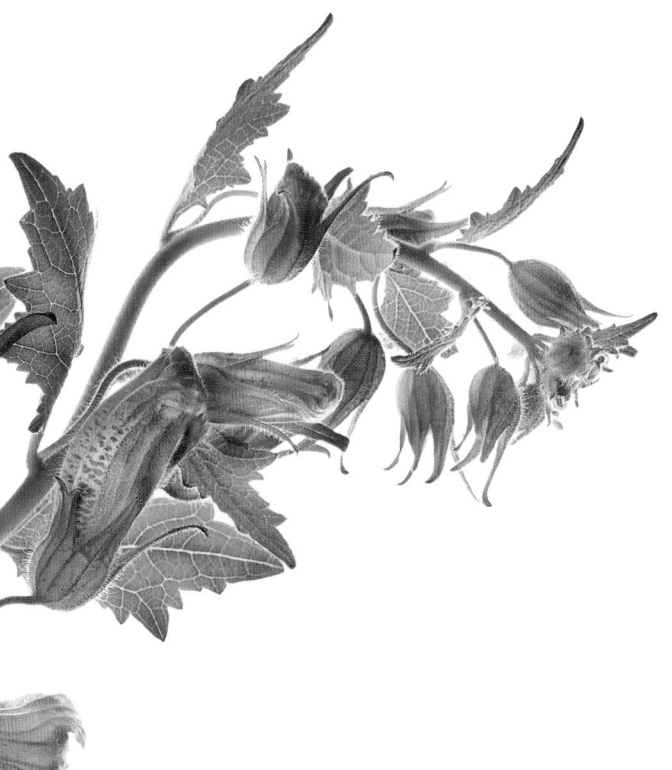

bloom their first year. The horticultural varieties also tend to be shorter than their undomesticated cousins, 2 to 3 feet tall instead of 5 or 6 feet, which endears them further because it is easier to fit them into mixed borders or cutting gardens.

Native American penstemons have also enjoyed refinement and selection, with praise and popularity following. The Perennial Plant Association named *Penstemon digitalis* 'Husker Red' Plant of the Year back in 1996, and it continues to be widely grown today. Its flowers are cream to pale pink; the name actually comes from the reddish stems and red-tinged leaves. Planted in small groups or, where space allows, en masse, this penstemon is arresting. Brilliant colors in other penstemons also abound, notably the hybrid two-tone Phoenix series, which have white throats that contrast dramatically with red, magenta, or purple outer edges. These were also bred to be of manageable size for most backyard gardens, topping out at 18 inches high.

The South African native Cape fuchsia (not at all a real fuchsia), *Phygelius*, also comes in brilliant colors and is long blooming. In mild-climate areas, it can bloom from spring all the way through November if summer is not overly soggy. And the cultivar names hint at the vivacious show you can expect. 'Devil Tears' is deep red in bud, opening to reveal glowing red tubes rimmed in fiery orange-red. 'Sunshine', of course, has bright yellow tubes. And so on.

Overall, though, members of the figwort family are hardly show-offs. Instead, they carry on from one year to the next, reliably blooming and gradually expanding their numbers. Such steadfastness ought not be taken for granted, whether we are admiring them in a meadow or along a roadside, or inviting them into our flower borders.

Adorable *Rehmannia elata*, which is sometimes known as Chinese foxglove or Beverly bells, does better than its foxglove cousins in hot and humid areas. Yellow spots within the rosy blooms seduce pollinators further onward and upward to the waiting pollen.

PREVIOUS SPREAD Summer is prime time for *Veronica* 'Red Fox' ('Rotfuchs'). It blooms in spikes and attracts bees as well as the occasional browsing hummingbird. Note how flowers in the spike open from the base upward.

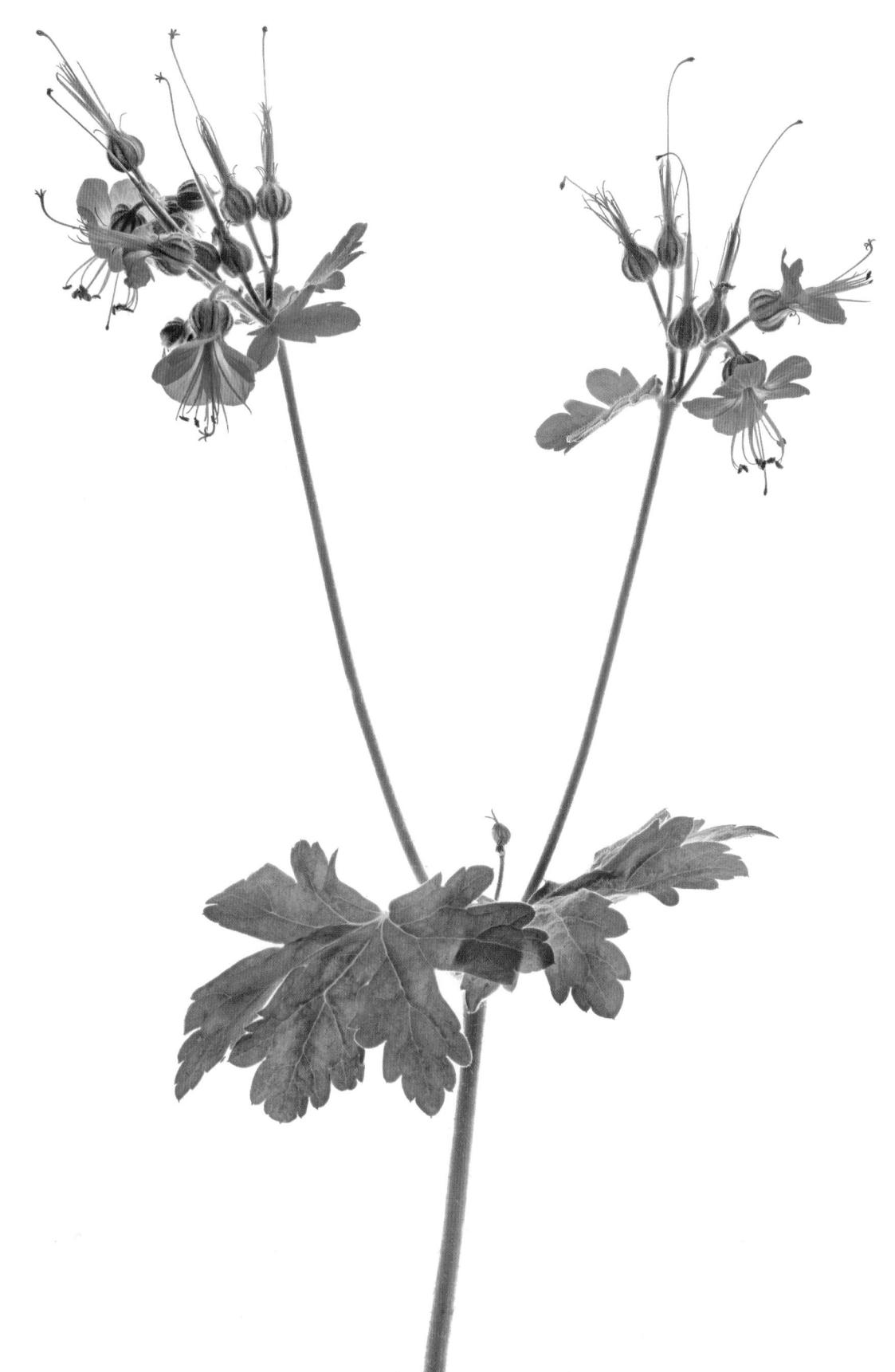

THE GERANIUM FAMILY

Geraniaceae

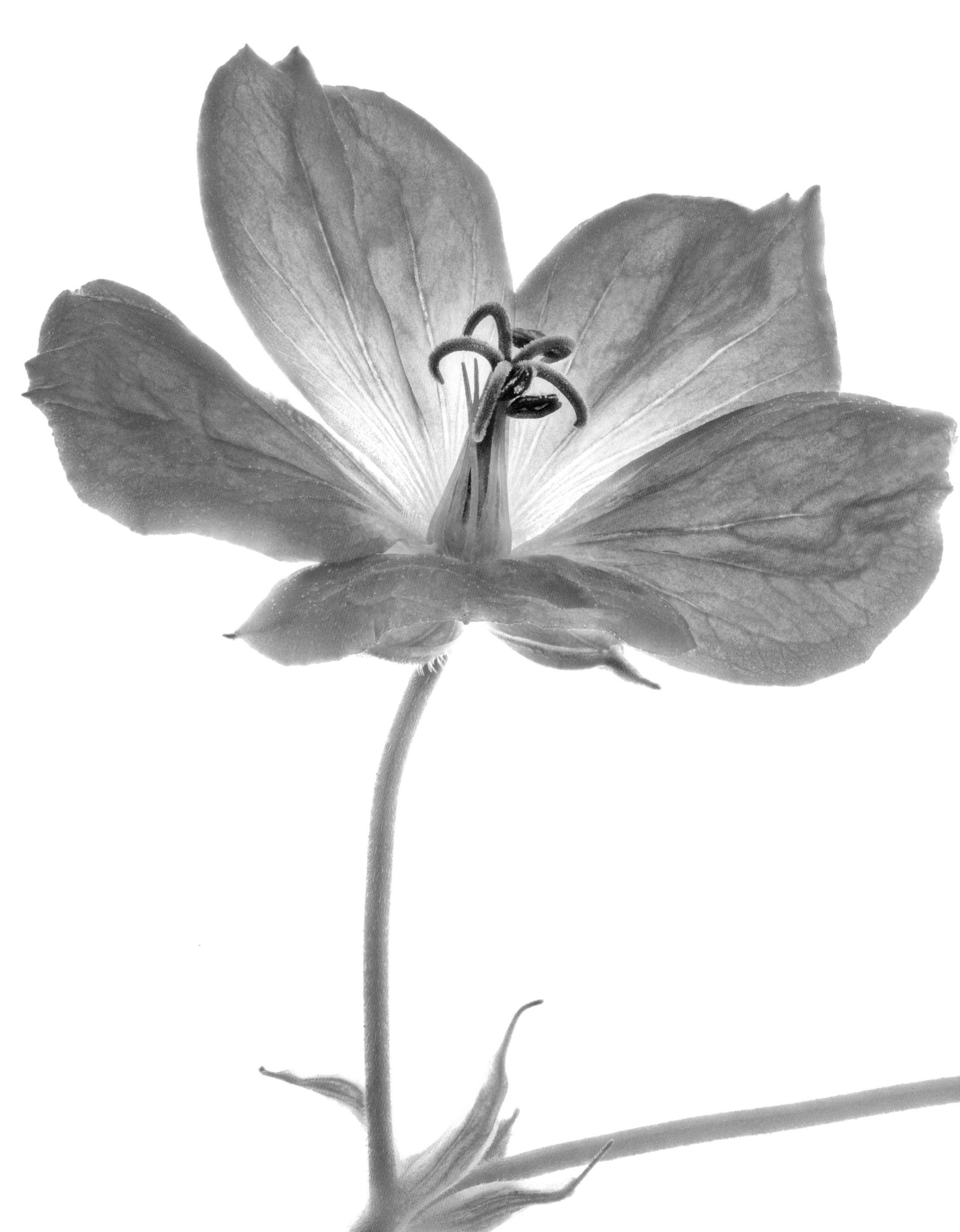

"Buttercups and geraniums cover the meadows, the latter appearing to float on the grass—of various tints. It has lasted long, this rather tender flower."

HENRY DAVID THOREAU, *The Writings of Henry David Thoreau: Journal, Volume IV*, 1852

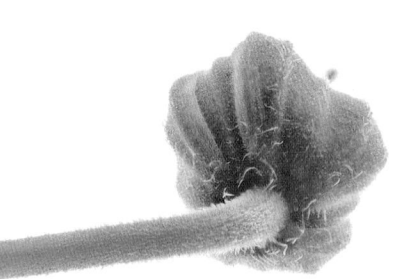

The exquisite blossom of *Geranium* 'Rozanne' displays the refined perfection of the plant family's five-part plan: five petals supported from below by five sepals, and the five-parted unfurled stigma (five "stigma branches," if you will), which eventually gives way to a five-chambered seedpod.

PAGE 134 Low-growing *Geranium macrorrhizum* is appreciated for its abundant clusters of pink-hued flowers, which are held above the aromatic leaves. Measuring an inch or smaller across, they are nonetheless striking, with inflated calyces behind and protruding stamens out front.

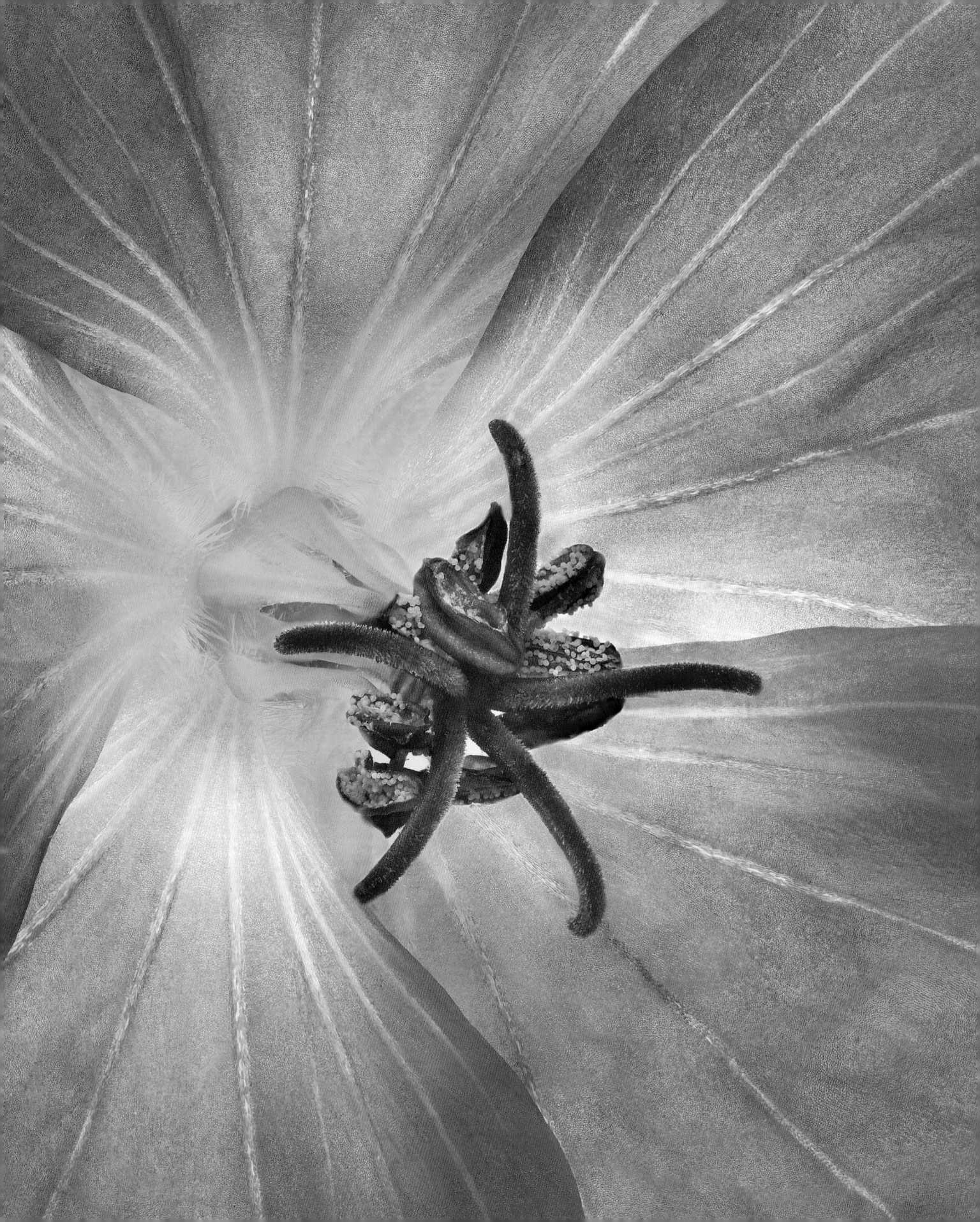

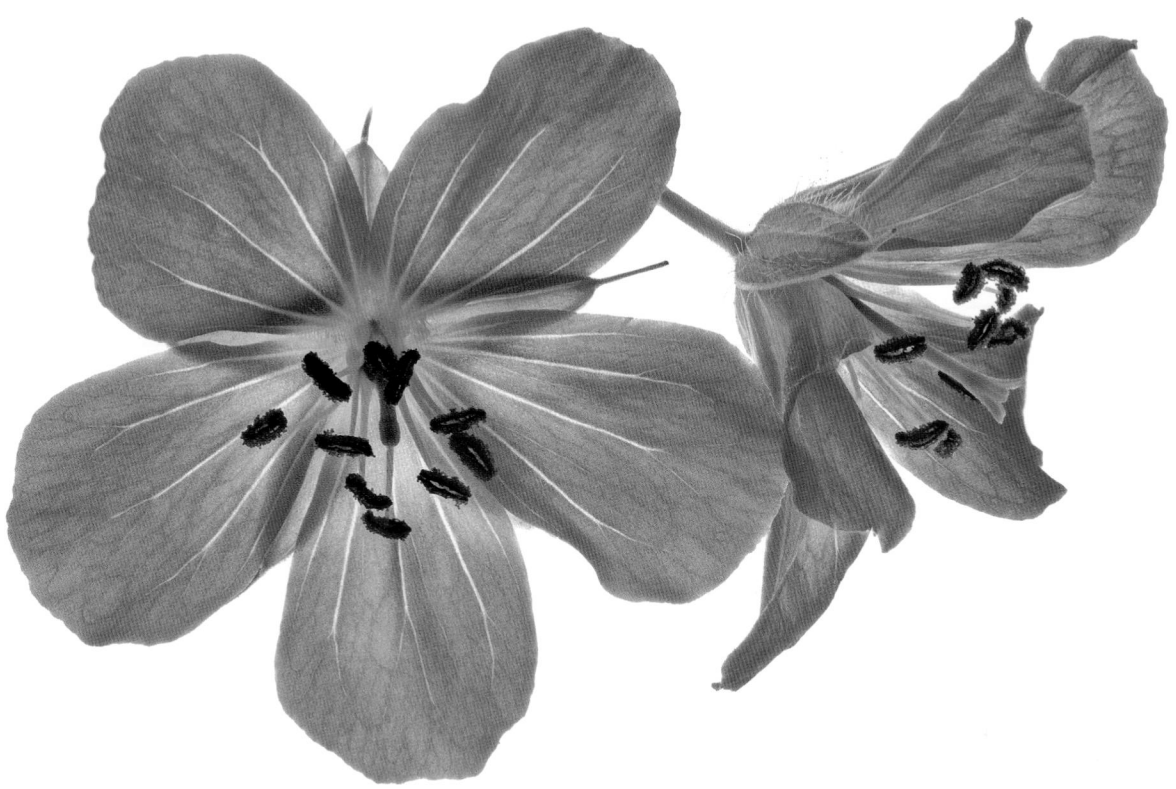

Here is a tribe of pleasant, long-blooming flowers. You may spot patches of them trailside in woods of dappled sunshine, particularly the wee purple-pink blossoms of Herb Robert, *Geranium robertianum*, or Bicknell's cranesbill, *G. bicknellii*, or the ½ inch, pink-purple ones of wild geranium, *G. maculatum*. Or, as Thoreau did, happen upon either of these in a sunny meadow, dancing above the tops of the grasses. Or notice them decorating an overgrown lot, a weedy roadside embankment, or a sidewalk crack, for they do self-seed and get around.

Herb Robert grows widely in both North America and Europe. Various other species are found in Europe as well, and some are indigenous to Asia, Africa, and even New Zealand. The genus name comes from the ancient Greek word *geranos*, which means crane. This refers to the seedpods, which are pointed like the beak of a crane. Thus the English nickname, cranesbill. Often they are collectively called hardy geraniums.

"Hardy geraniums" distinguishes the perennials from the popular bedding, pot, and window box plants that we also call geranium. Those big, bold, cluster-flowered plants are *Pelargonium*, today's splashy and colorful hybrids derived from lineage that began with a South African species. They still are not very cold hardy, which is why we tend to grow them as annuals. And yet,

A closer look at the flowers of *Geranium* 'Rozanne' shows that there are, in fact, ten fertile stamens awaiting a pollinator—so this flower ABOVE is less mature than the one at LEFT. Thus self-pollination is avoided because the male parts of the flowers develop and ripen before the female parts.

they are indeed in the family Geraniaceae. As further confirmation of what appears at first to be an unlikely kinship, you may have noticed that your pelargoniums also develop those distinctive pointy seedpods.

Despite the divergent looks and name confusion, this family demonstrates consistency of form. Cranesbill blossoms display plant parts in fives: five petals, five sepals, stamens in groups of five, and a pistil with five styles, each attached to one seed for a total of five seeds. And if you pluck off an individual pelargonium flower from one of those big, full, umbel-like clusters (pseudoumbels), you will find that it does not look too different from its cousins. Sure enough, there are five petals, stamens, sepals, and so forth.

More often than not, any geranium family petals have slim lines, or their blooms sport contrasting center eyes. Either of these acts as a navigation guide for pollinating insects, principally bumblebees, various flies, and the occasional hummingbird. Carefully slip your finger under petal lines, when present, and you may be able to observe that they are transparent windows through the petal, not an actual colored line. While peeking in, notice also little hairs at the very base of each petal. These protect the store of nectar, or nectar spot, and a pollinator must reach past this slight barrier.

When flower petals finally fall off, the flower stem (pedicel) helps hold up the developing beak. Those pointy seedpods are amazing. As the pods mature, the elongated styles come under tension, curl up as they dry, and then, like a spring, catapult the small ripe seeds up to several feet away. The remains linger on the plant, looking as curly as a reflexed wild lily flower.

Pelargonium tinkering, meanwhile, has been going for on a long time and is evidently quite complex.

Rather than trying to create or cull out your own backyard hybrids, your best bet is to acquire excellent new varieties at a well-stocked garden center or via a mail-order seedhouse. As you know, red, maroon, purple, pink, and white colors and bicolors abound (with or without "zonally" marked accompanying foliage); the holy grail remains orange. 'Orange Appeal', which debuted some time ago and even won a European Fleuroselect medal in 1991, is not easy to find—and, by some reports, not a genuine or satisfying pure orange.

An odd side note about pelargoniums: their foliage and flowers contain a compound that is harmful to Japanese beetles. Evidently the pests become paralyzed after nibbling on the plants, although they do not die. You might watch for this phenomenon or even try these as companion plants to the beetles' common victims, such as your rosebushes, hollyhocks, or dahlias. Other creatures may eat the temporarily prone pests, or you can swoop in and dispose of them before they have a chance to revive.

Whether you prefer the graceful perennial cranesbills or the big-flowered pelargoniums; whether you prefer the wildflowers, love the old tried-and-true hybrids, or like to hunt for the best of the new introductions, there's no denying that Geraniaceae is an outstanding flower family. There are no serious pest or disease problems, they are easy to grow, and they pump out blooms over a long period. Generous and pretty—that's a winning combination.

Every spring, graceful stalks of wild geranium, *Geranium maculatum*, display pinkish-purple flowers. Each one remains open for three days, then sheds its petals. The female part then develops into a fruit, elongating into a thin beaked point that reminds some people of a crane's bill, hence the other common name.

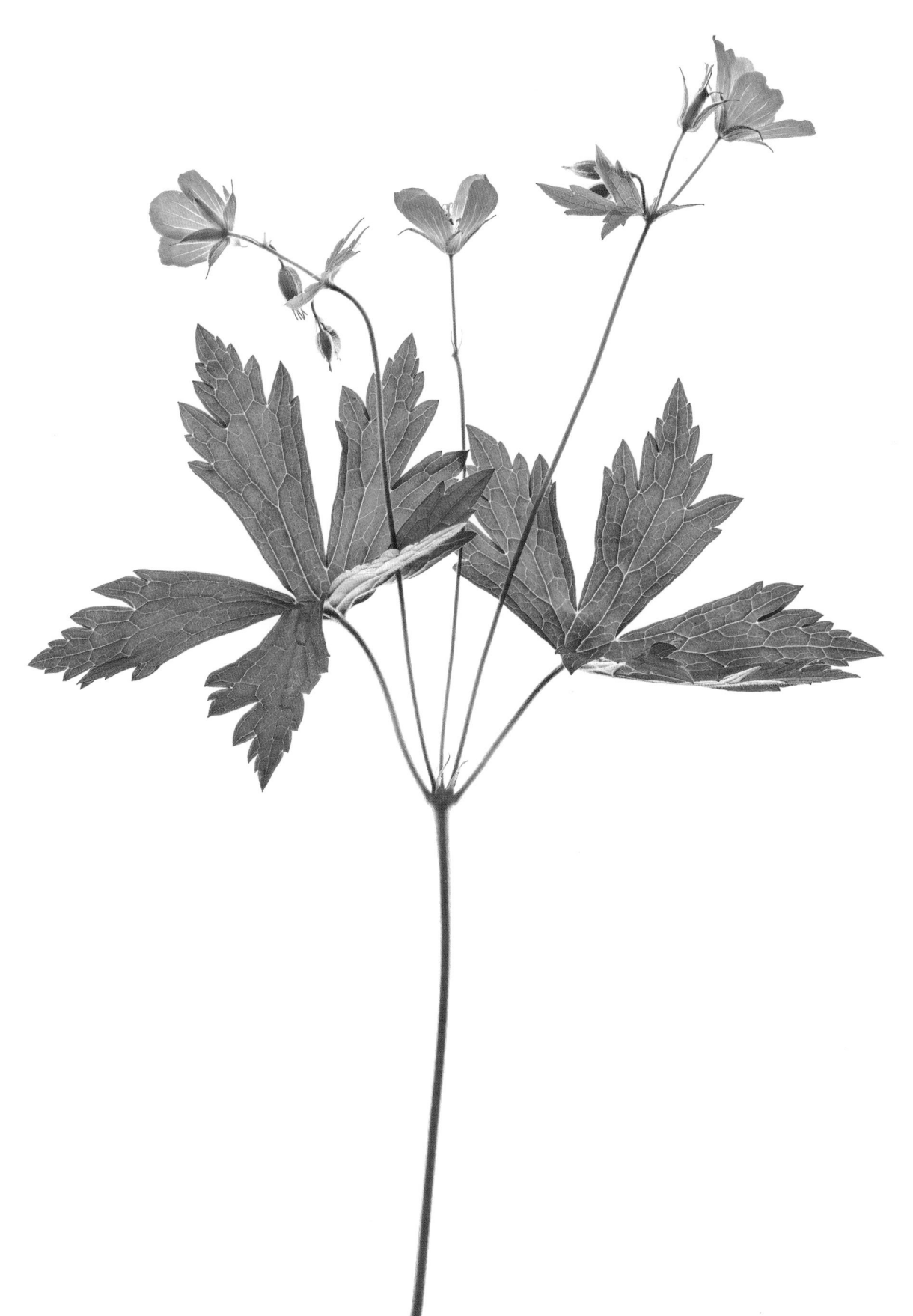

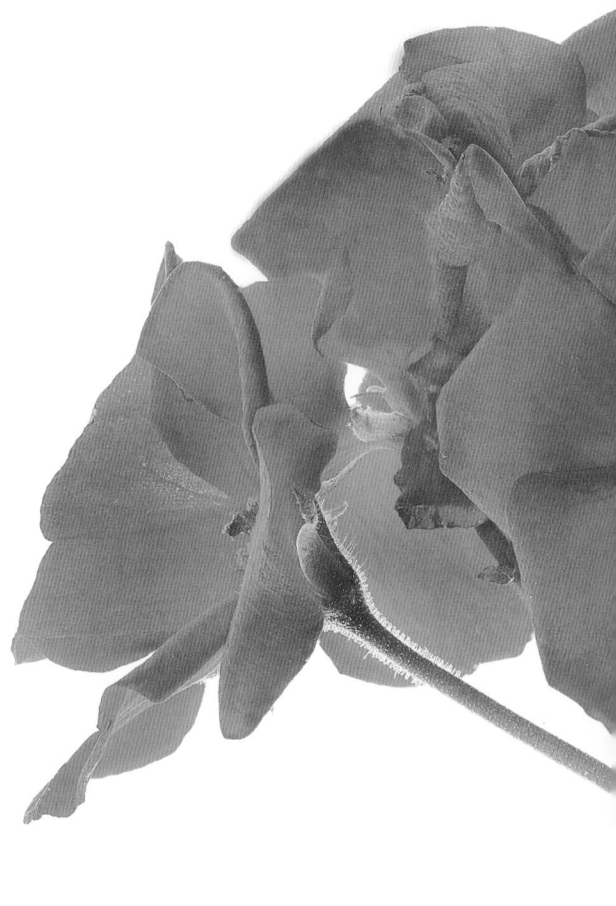

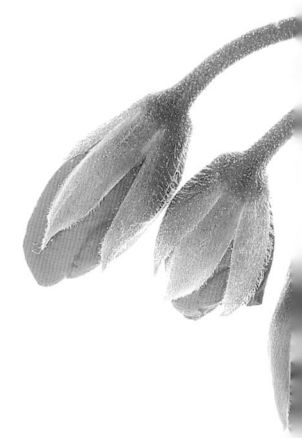

Most people associate the name geranium with these big cluster-bloomers, which do belong to this family but are in the genus *Pelargonium*. Like their relatives, though, individual flowers are five petaled. Shown here is 'Salsarita Dark Red'.

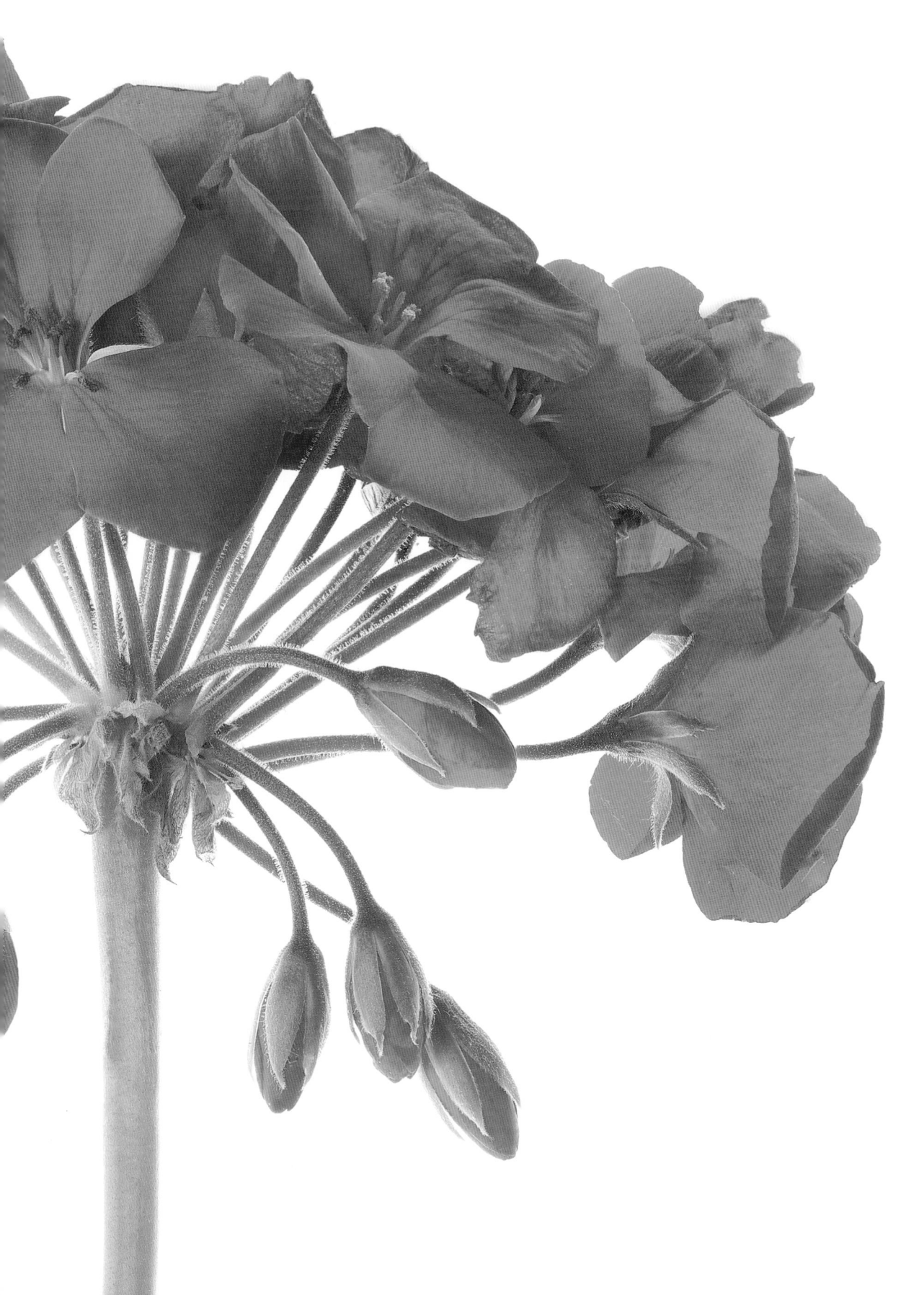

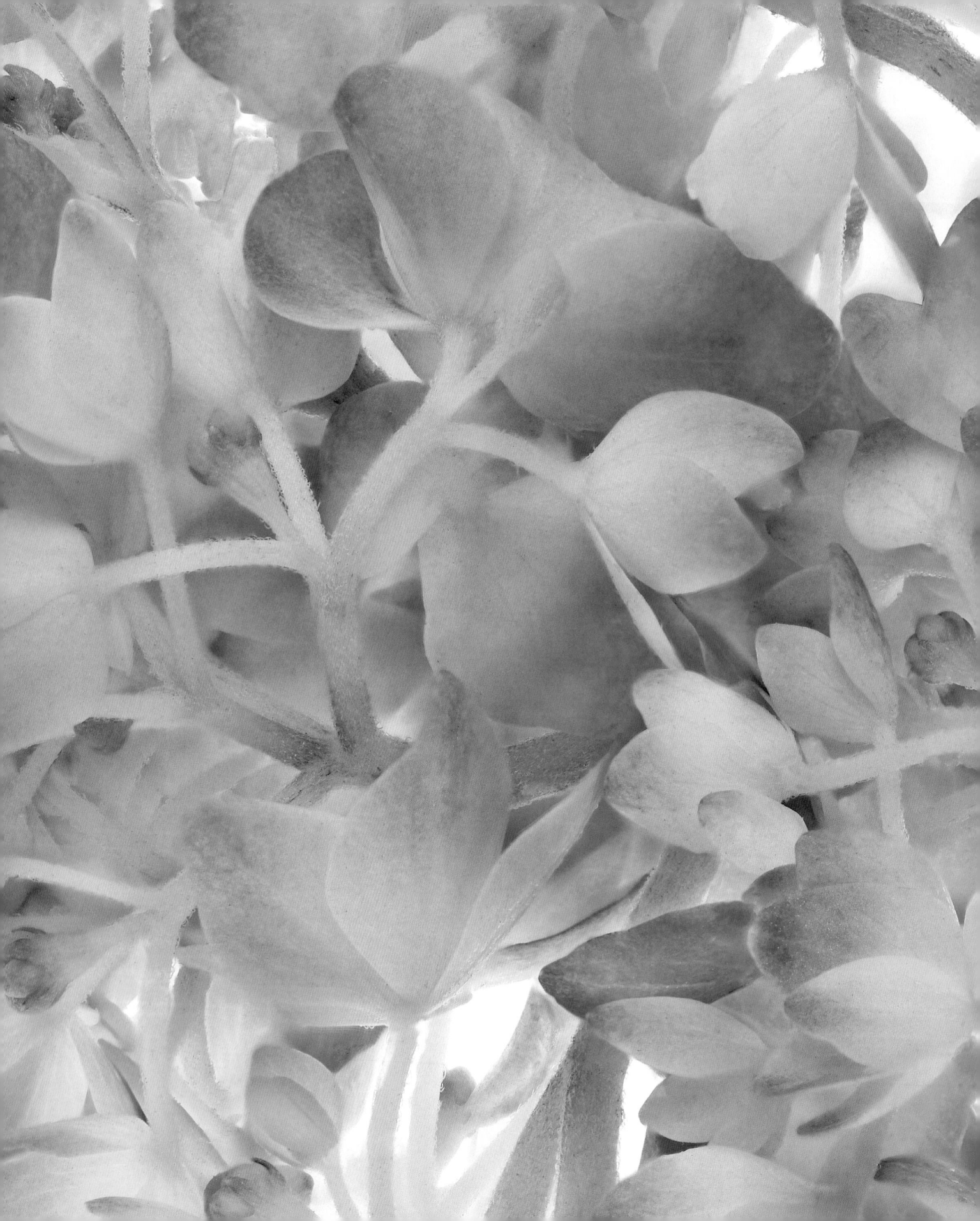

THE HYDRANGEA FAMILY

Hydrangeaceae

"Ice cream on green cones
   white hydrangeas in full bloom
cool the summer day"

Haiku by CDSinex, 2011

The creamy white blossoms of oakleaf hydrangea, *Hydrangea quercifolia*, have the same mix of sterile and fertile florets as their relatives. The sterile ones may be an inch or more across and have four petals; abundant, tiny fertile ones hide behind them.

PAGE 144 Mophead-type hydrangeas, like this pretty 'Endless Summer', feature big, ball-like clusters of mostly sterile flowers. The tiny fertile ones are sheltered among and under them.

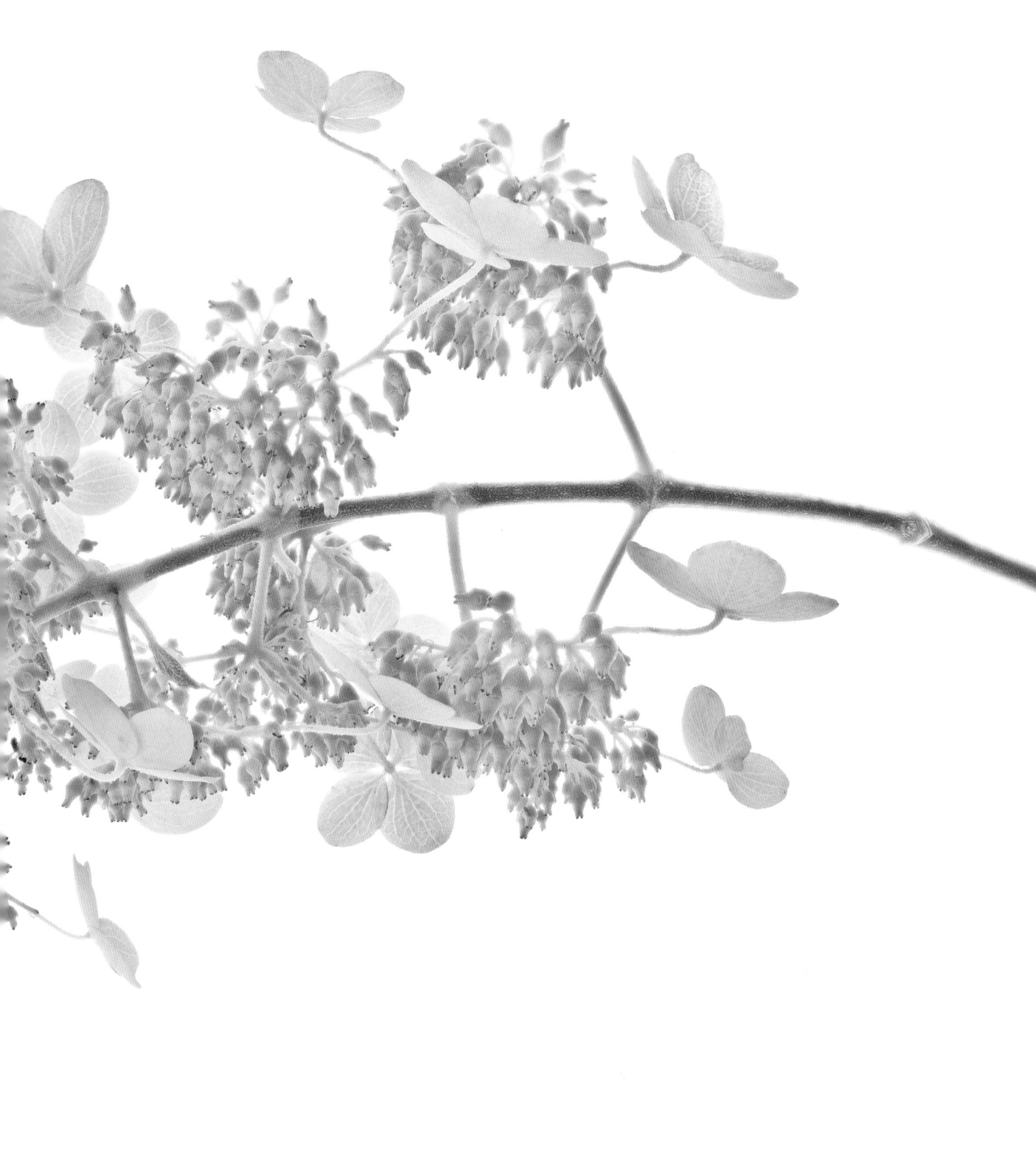

Drive down the residential streets of many small American towns, and you are sure to spot the PeeGee hydrangea, the nickname for *Hydrangea paniculata* 'Grandiflora'. Usually big, and often unruly in habit, the familiar shrubs display easily recognizable "grand" flowers in pyramidal spikes, sometimes measuring up to a full fluffy foot high. These emerge white in midsummer, but soon moderate to pale green and deep shades of pink. By the time cooler fall weather arrives, they darken to bronze, and as winter approaches, they fade to parchmentlike shades. And yet, everyone from regular folks to landscape professionals are often disdainful of PeeGees, calling them sloppy, coarse, overplanted, and even "a granny plant."

But if you take care of a PeeGee, water it well in dry spells, groom it, and perhaps trim off the lower branches, a specimen can be quite elegant. And those blossoms are undeniably splendid for bouquets at any stage in their color evolution, including dried. Plus it's lovely that they flower when they do, for the spring-blooming shrub show is long since over and, aside from the roses, garden shrubs are pretty much in plain green foliage mode. Plus, hydrangea flowers last extra long, carrying on for weeks and even months. Top that, forsythias and rhododendrons.

You no doubt also know the mophead-type hydrangeas, which are *Hydrangea macrophylla*, or the Hortensia group. Not quite as cold hardy, they're still widely grown. Acidic soil boosts blue, while alkaline soil turns the flowers pink or mauve (the key is how much aluminum sulfate is present). Some gardeners have fun tampering with soil amendments and fertilizer to see what color changes they can provoke, although this is easier when you grow one in a large container rather than in the ground. However, the white hydrangeas, including the gorgeous snowball blossoms of *H. arborescens* 'Annabelle', will not respond in this fashion.

When the pastel-hued *Hydrangea macrophylla* Endless Summer series debuted a few years ago, the snazzy newcomers delivered a popularity jolt to this species. They were a breeding breakthrough because they also bloom on new (this year's) growth. The result is a bush with a lot more flowers over practically the entire summer—or, as the marketers crow, "Blooms like crazy! All season long! Year after year!" They've been followed by the similar Forever & Ever series in all sorts of great colors, so what at first seemed like a novelty appears to be turning into the wave of the future for this sort of hydrangea.

Yet another variation is the lacecaps, *Hydrangea macrophylla* var. *normalis*. These probably also qualify as "granny plants," as the blue-flowered 'Mariesii Perfecta' has been around since 1904. Nowadays, it (or a version) can be found as 'Blue Wave'. Lacecaps also come in pink, purple, white, and bicolors, and tend to be more graceful and circumspect in growth habit than some of their kin.

The more flattened clusters of the lacecaps make hydrangea flower morphology easy to see. The bigger, showier florets that tend to dominate or stay to the outer edges are always sterile. The smaller, inner ones are fertile, and with pollination they will produce fat little fruits that lead to chambered seeds. It's typical to have four or five petals, and between four and ten sepals. Peer closely—the true petals are clearly separate from one another, while the sepals are fused.

Thus the pollinators, various flies and bees, may have to root around a bit to find the true flowers once they arrive. Hydrangea flowerhead form also explains

Aptly named lacecap hydrangeas are distinguished by large sterile flowers around the outside and a mound of little fertile flowers (which look like buds at a glance) inside. Notice how on this hybrid, *Hydrangea serrata* 'Bluebird', as is typical, the colors of the two kinds of flowers are in contrasting hues.

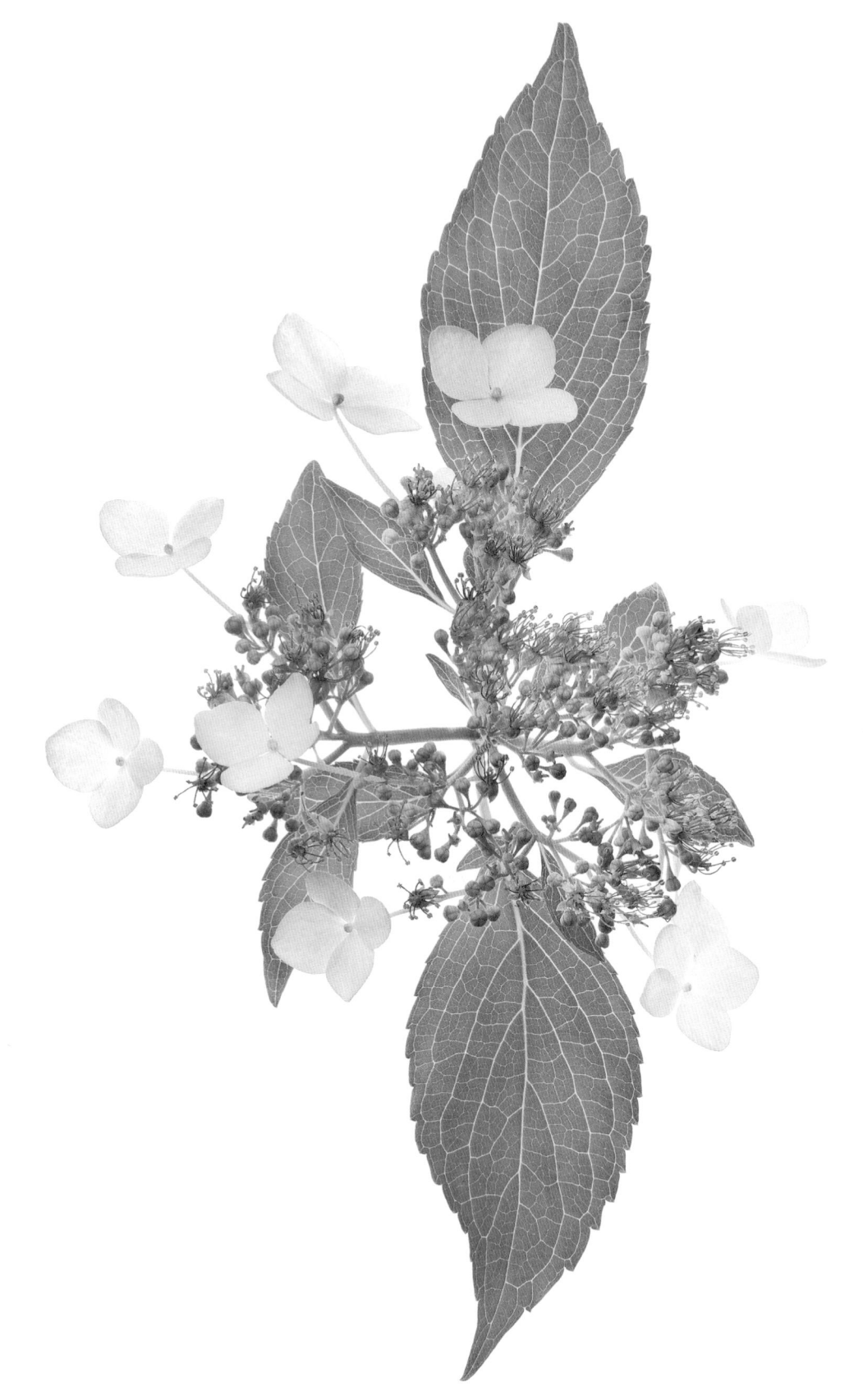

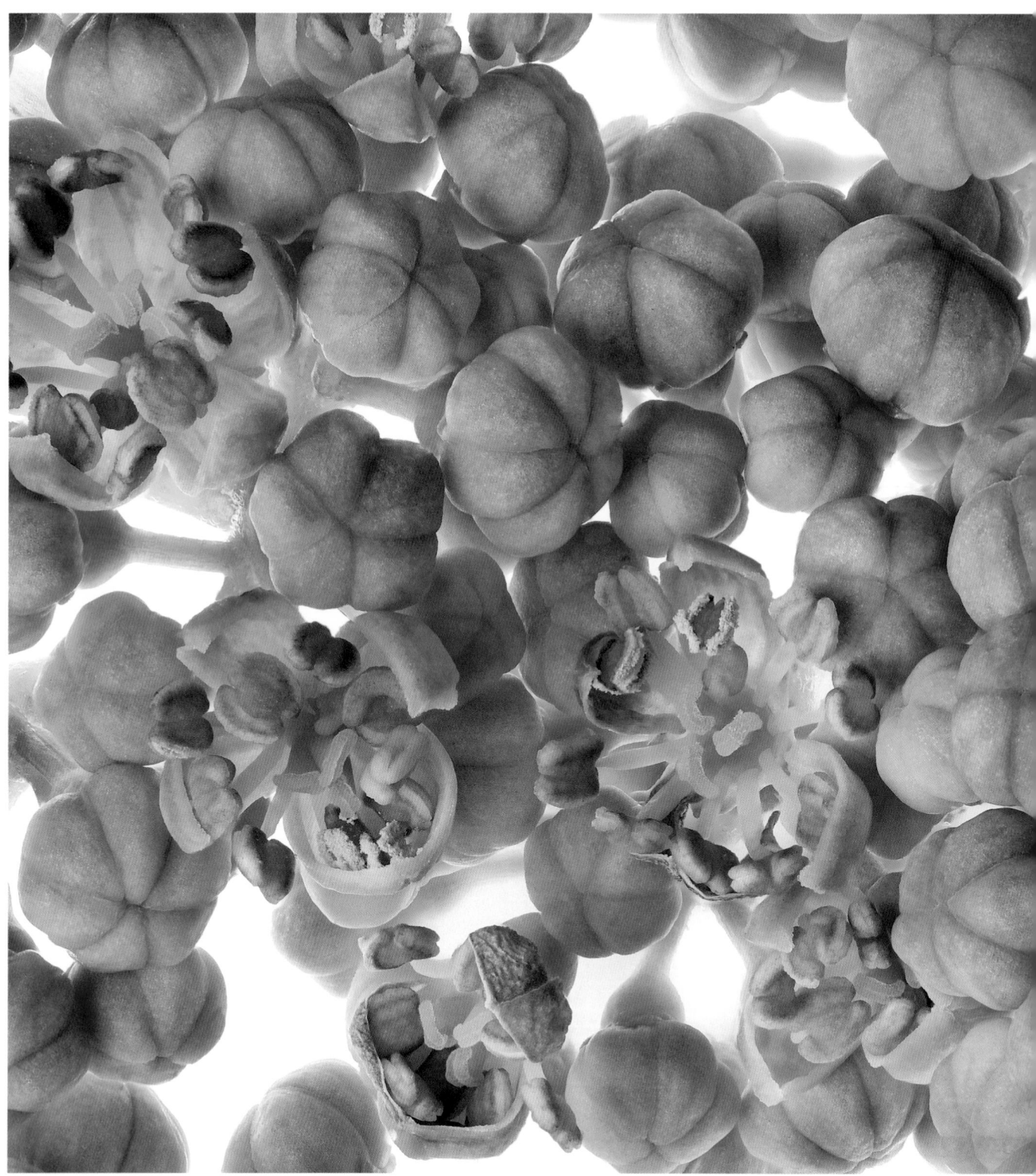

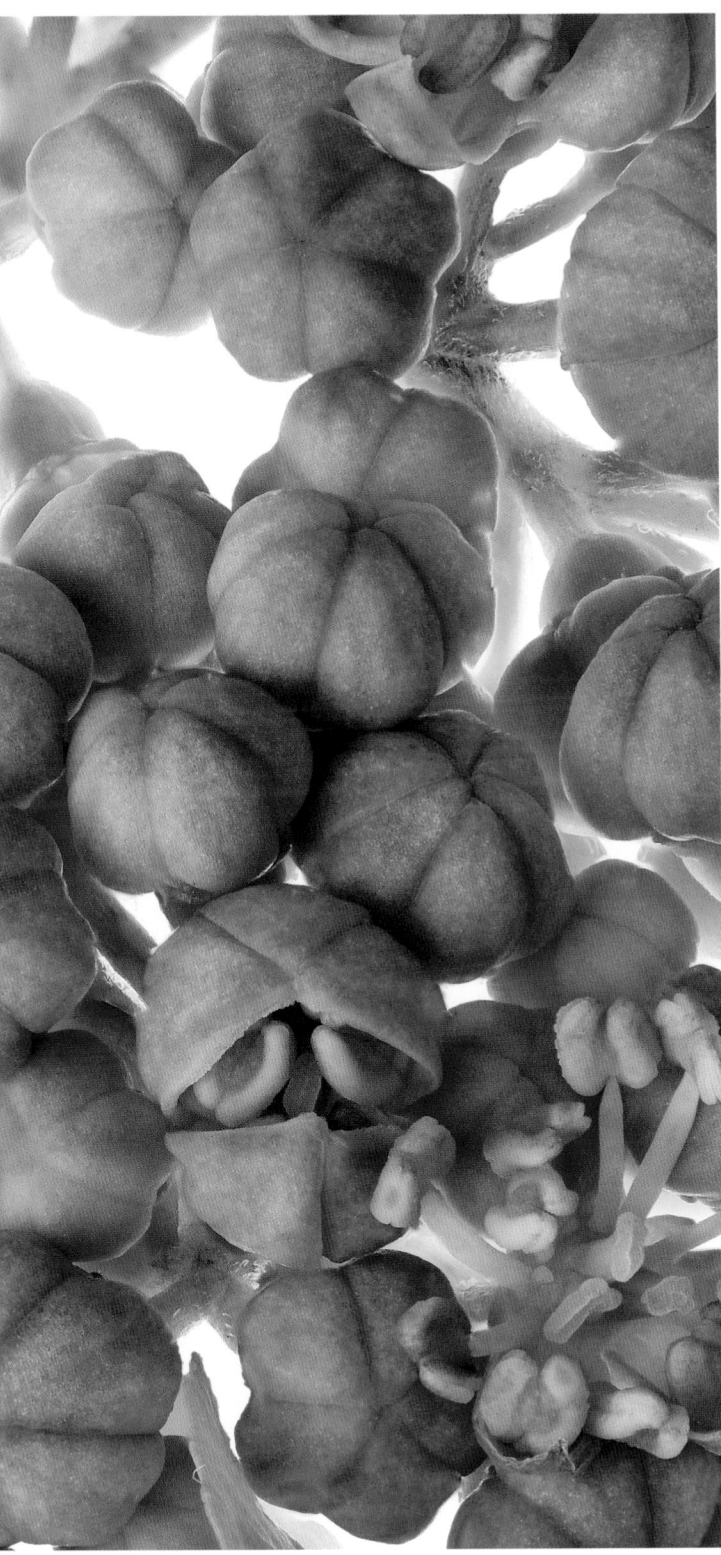

why the plants bloom for so long. The sterile parts (sepals included) don't have to produce fruit or seeds, so they can hang on for quite a while.

So widely grown are the various hydrangeas that the other members of the family seem almost shoved aside. But Hydrangeaceae does have other members worthy of attention and praise, however. One is *Deutzia* (there are several species in cultivation), a shrub that enchants with scads of fragrant white flowers every spring. But the sentimental favorite is arguably mock orange, *Philadelphus*. Its leaves are plain and ordinary looking, too, but when it blooms in early summer in a cloud of yellow-centered white flowers, you may lose your heart, as their spicy-sweet orange blossom fragrance is so wonderful. Like its cousins, though, this plant can become large and rangy, so if your yard is small or crowded, consider 'Dwarf Snowflake', which gets only about 3 feet tall and wide.

And thus our tour closes, approximately where it started, with an old-fashioned plant that you may never have chosen to grow or paused to examine. Perhaps the takeaway is that even the most familiar plants not only genuinely deserve their reputations as stalwarts, but also have dimension and charm.

Fertile hydrangea flowers ripen slowly over a week or so. First they break open to reveal many two-lobed anthers, coated with yellow pollen; the stigma, in their center, has three lobes.

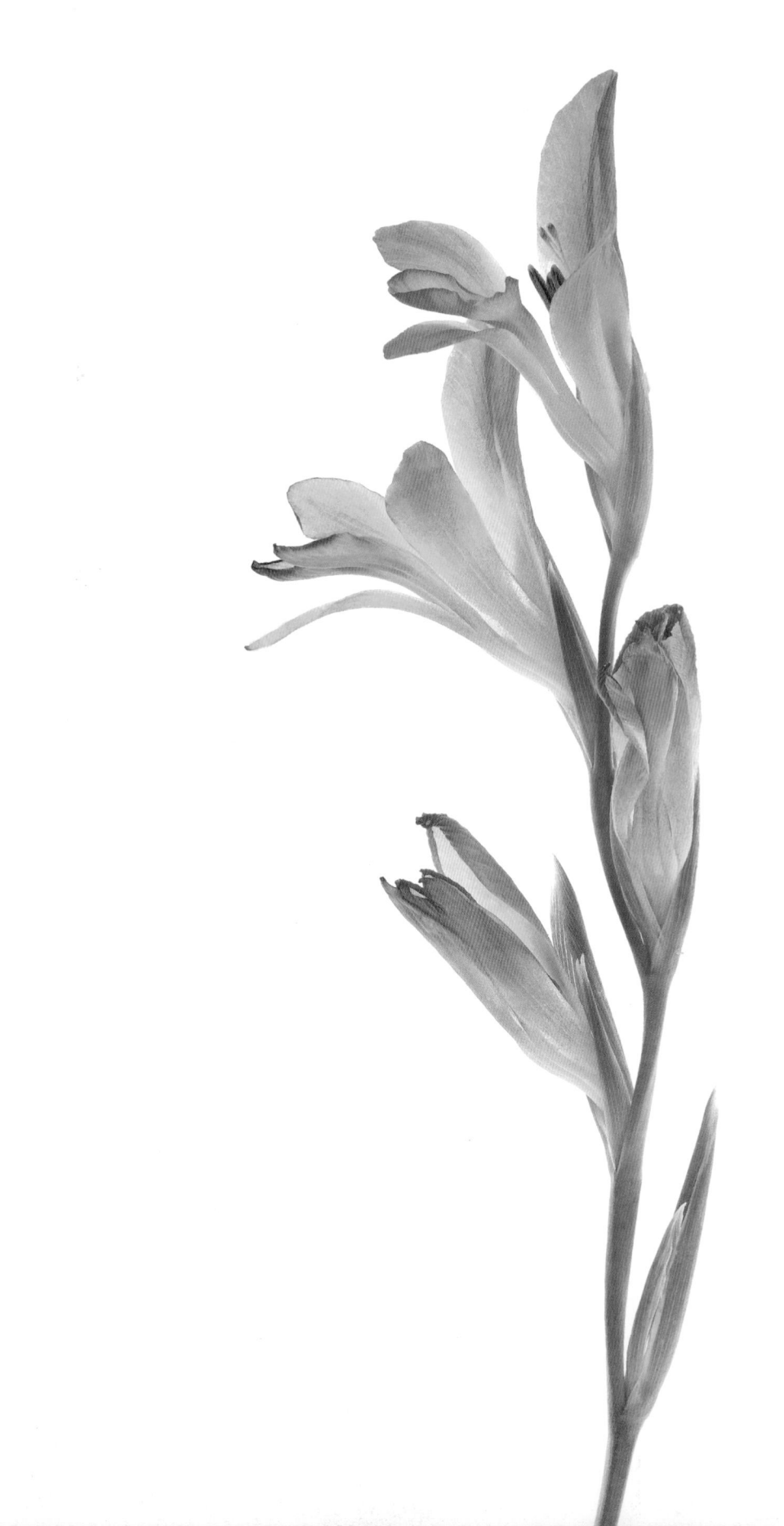

THE IRIS FAMILY

Iridaceae

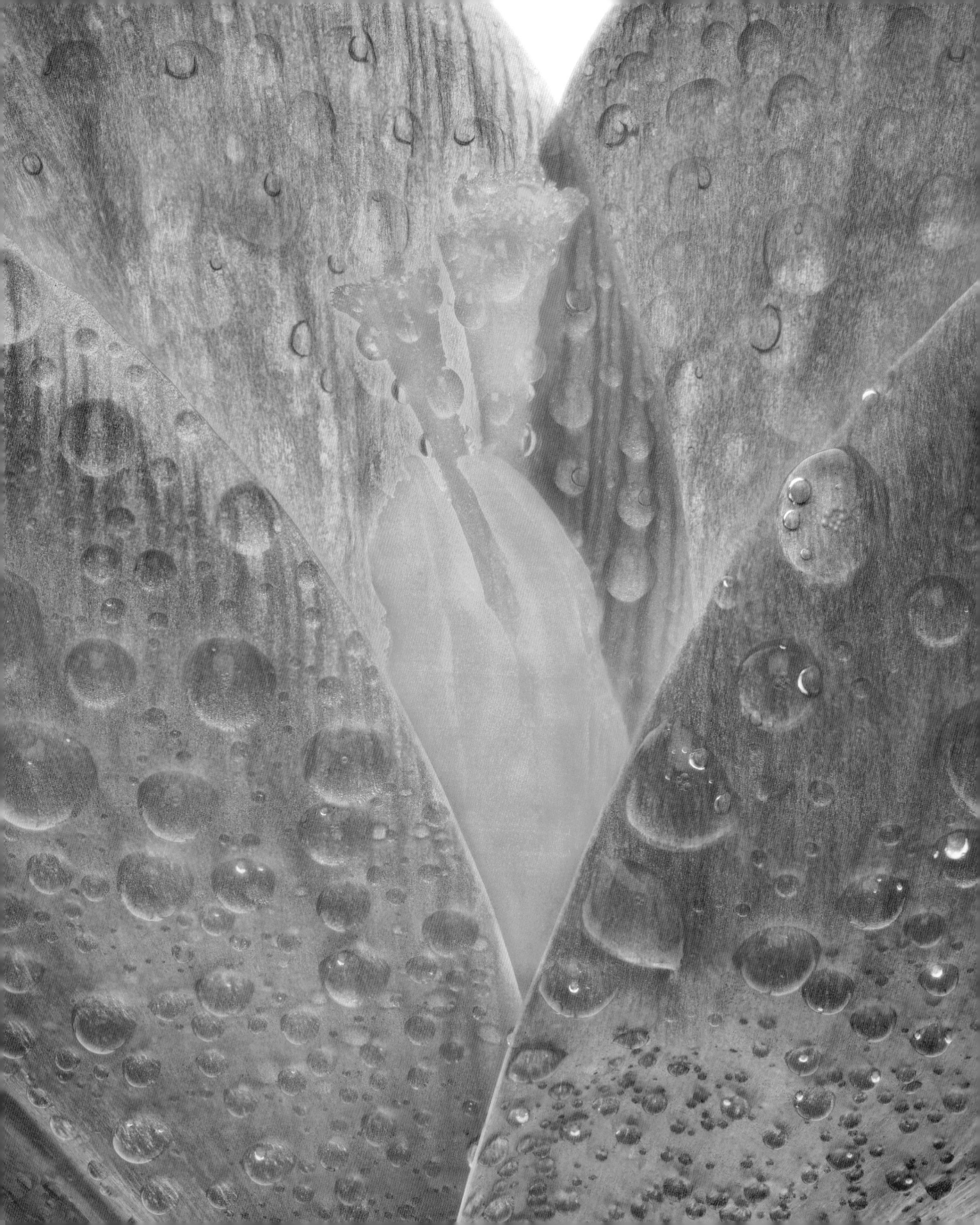

"Then we had the irises, rising beautiful and cool on their tall stalks, like blown glass, like pastel water momentarily frozen in a splash, light blue, light mauve, and the darker ones, velvet and purple, black cat's ears in the sun, indigo shadows. . ."

MARGARET ATWOOD, *The Handmaid's Tale*, 1985

So small and common are *Crocus* hybrids that we rarely kneel to view their unique details. They come by their short stature honestly—the stem emerging from the ground is actually the tubular base of the corolla. Their cupped form is actually made up of six clasped petals or, rather, petal-like tepals, often purple, pale purple, yellow, sometimes white. Showy stamens within—yellow, orange, red, or white—make a sharp-looking contrast.

PAGE 152 The so-called hardy glad, *Gladiolus communis* subsp. *byzantinus*, is cherished for its graceful flower spikes. Individual blooms are shaped like funnels, and the lower petals (tepals) are striped white to beckon and guide pollinators—tempting maybe even hummingbirds.

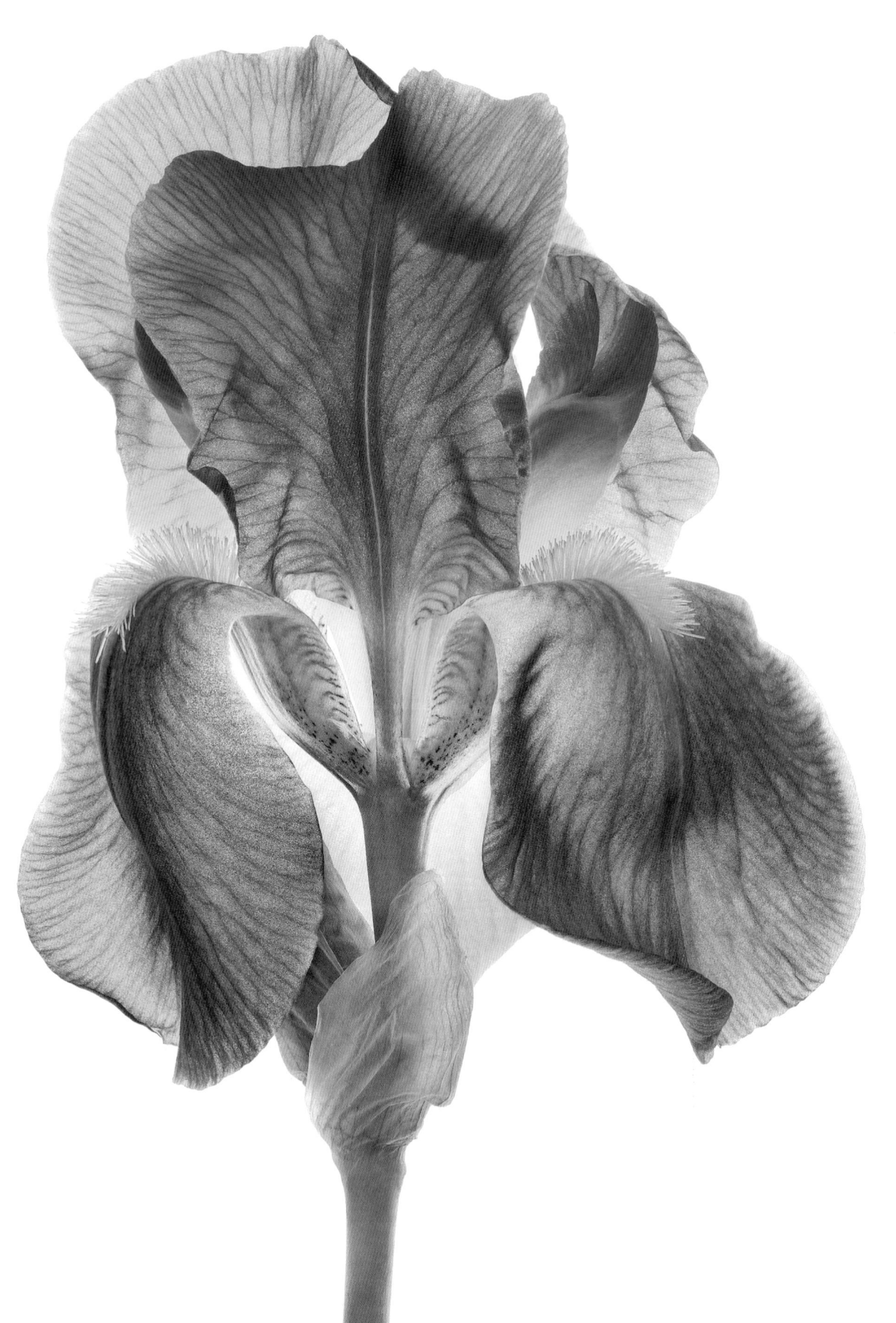

People who enjoy irises are like those who love ice cream: no plain vanilla, thank you—I think I'll try that tempting new flavor. Few other flowers have been so extensively planted and hybridized. You can have anything you want, from classic royal purple or pale blue blossoms to light yellow, bright orange, or deep pink ones. And among the tall bearded irises, the bicolors are especially creative and luscious (for example, cream-topped apricot, gold-topped bronze, raspberry red and bright white). Additionally, texture can be silky or velvety, sometimes both on the same flower.

A lot of irises thrive with their feet in the water, so to speak, such as yellow *Iris pseudacorus*, the usually purple-splashed and always sumptuous Japanese irises, *I. laevigata* and *I. ensata* (*I. kaempferi*) and their many cultivars, and the colorful Louisiana iris group (apparently derived from five different species). And let's not forget the sturdy and dependable—in cooler climates—Siberian irises, *I. sibirica*, plus the floral hyperbole of blowsy, robust tetraploids patient hybridizers developed to bring us bigger flowers on stouter stalks.

Enthusiasts seek out the low-growing, small-flowered iris species native to Mediterranean hillsides and waterside meadows. They find their plants at specialty nurseries or share seeds, and try them in small numbers in pots or mixed flowerbeds or encourage them to spread out, space permitting, to create great color swathes. The American native dwarf crested iris, *Iris cristata*, is a particularly lovely one, light purple in the original and also available as a soft white 'Alba'. Collectors lust after the sweet purple-hued roof iris, *I. tectorum*, which owes its name to the fact that it blooms on house roofs in its native Japan (it also comes in an 'Alba' form). All this and more await anyone willing and able to wade in and explore this abundant genus.

The colonizing propensity of many irises brings to mind the same habit found in their smallest relatives, crocuses. When growing in an agreeable location, they will multiply their stout rootstocks—technically, rhizomes in *Iris* and corms in *Crocus*—with each passing year, a process known as naturalizing. The spring-flowering bulb industry has mainly worked with the resilient, cold-hardy "Dutch crocus," *C. vernus*, to favor larger flowers and improve flower color. Most are white, a shade of purple, or yellow, and some are striped, but strong blue and deep pink ones have been developed and are worth seeking out if you are tired of the same old thing.

It is interesting, but perhaps not surprising, that Iridaceae also includes gladioluses. Here, too, you find myriad pretty pastels, bold solids, and dazzling color duos—literally hundreds of choices. And species glads, notably ones hailing originally from South Africa and their horticultural derivatives (bred to extend color range and bloom period), are exotic looking and graceful enough to be compared favorably with orchids.

Unlike their cousins, though, glads make excellent, long-lasting cut flowers. You may wonder why are many irises are poor cut flowers. They open and fade away too quickly, florists say, and their rather stiff stems are not always an asset. Your best bet is to harvest or buy them completely closed, when their buds are wrapped up neatly like a tiny umbrella. A vase of lukewarm water will start the show.

Complicated, gorgeous, and colorful, tall bearded irises are true garden aristocrats. From the moment a blossom emerges from its papery sheath like a gift being unwrapped, the show is on. First the falls, three sepals, flare downward and outward, while the three inner petals, the standard, arch upward. On the falls is the beard, a dense strip of delicate colored filaments. Colors reveal and deepen until the full glory of a flower is before you. It will last about three days, but if your plants are happy and healthy, more are waiting in the wings. With these beauties there are literally thousands of cultivars to choose from—it seems hybridizers have truly found a rainbow palette.

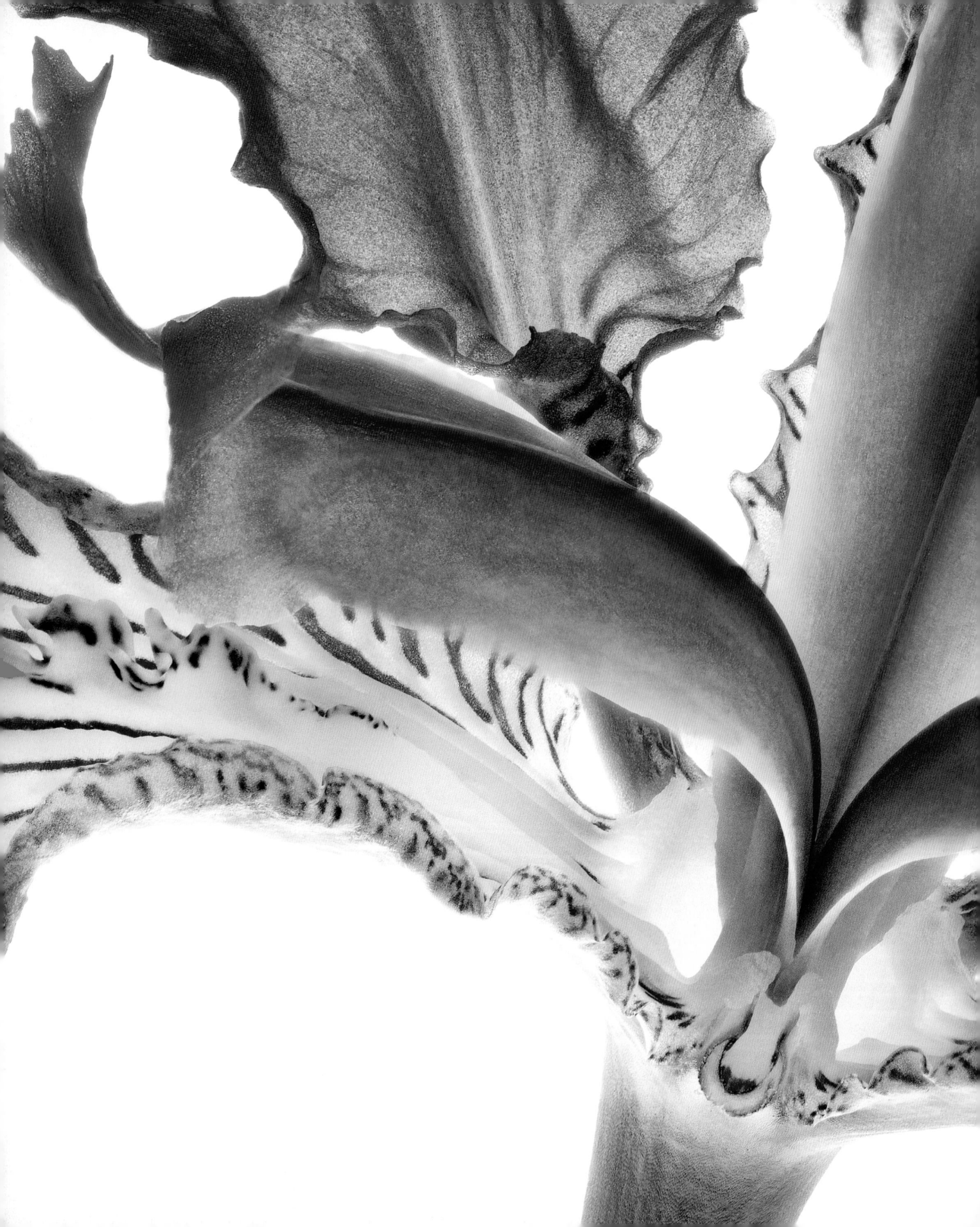

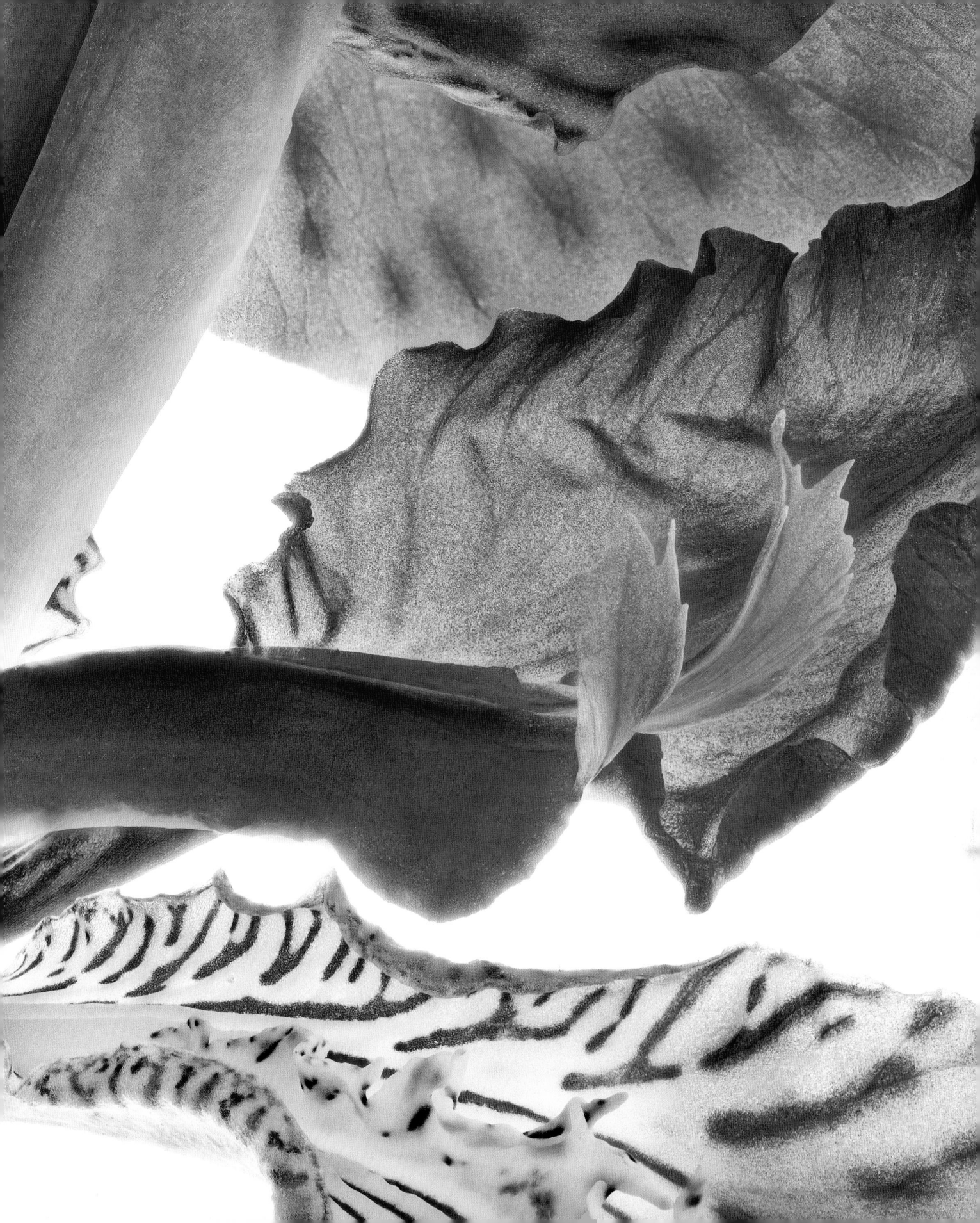

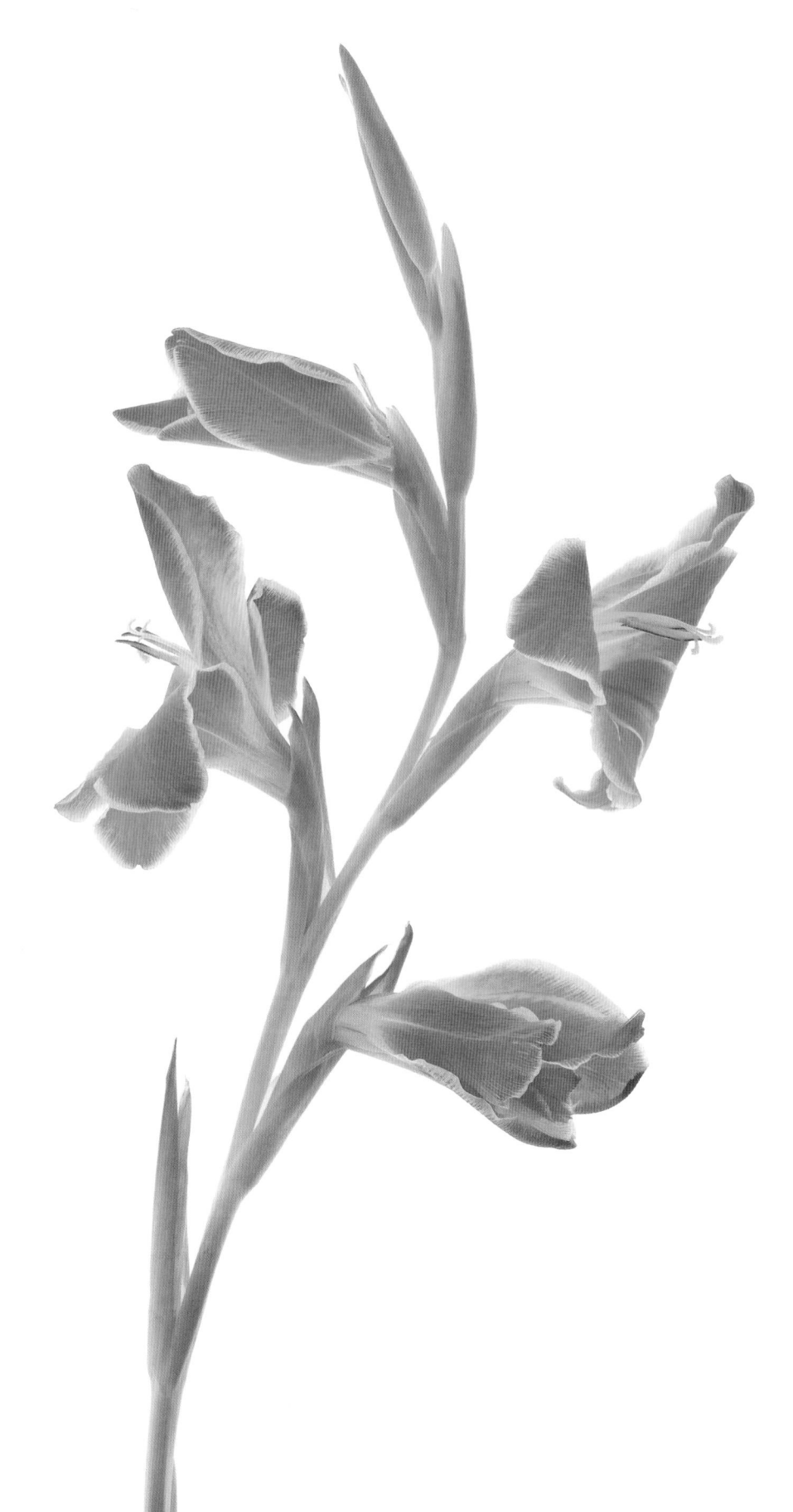

In any event, distinguishing characteristics of this family, which actually encompasses around 2000 species, are fairly easy to spot. They all arise from a bulb, rhizome, corm, or tuber, and they all sport slender, pointy, swordlike leaves with parallel venation, as is typical of monocots (other monocots, such as lily family members and grasses, do too). The flowers all share a basic three-part design: six petals often presented in two whorls, joined by three stamens at the bloom's center. Iris family flowers also have an "inferior ovary," which means that it is situated below the other floral parts, a feature that is particularly obvious if you look at an iris that has gone to seed or a glad whose petals have dropped off.

The structure of bearded irises conforms to this template, even if the blooms appear dauntingly complicated. The three outer petals, technically sepals, wrap around an emerging bud. As the bud swells and unfurls, these flare outward and downward to become the falls. Meanwhile, the three inner petals, called the standard, arch upward. The style arms then branch outward to be more accessible to bees and other pollinating insects.

Glads, including this old favorite cultivar 'Atom', best reveal their kinship with irises when still in bud, resembling an unfurled but surely colorful umbrella. Once they open, though, you can confirm that their petals and stamens are indeed in sets of three.

PREVIOUS SPREAD Splashy, strange-looking markings accent an *Iris tectorum* flower down near the base of the petals, guiding pollinators. But layout plays a role too. Each stamen has pollen on only one side. When an insect arrives, the receptive stigmatic lip above receives pollen from the last flower it visited, rather than from its own flower.

The small but prominent beards adorning the falls are actually a compact strip of colorful filaments. The falls meet the rest of the flower at the haft, which is often made more conspicuous with contrasting colors or patterns.

Individual gladiolus flowers look somewhat less complex, and conform to the sets-of-three format. Long-tongued flies and moths tend to take on the gladiolus pollination job, and markings on the petals and sepals help guide them. Chalice-shaped crocus flowers are, of course, simpler and more uniform in design, tempting bees and others with pollen-dusted orange stigmas—unless the absence of active pollinators or heavy spring rains thwarts the natural process.

Plant breeders who have intervened at what they hope is just the right moment with little paintbrushes and their own crossing goals will attest that this painstaking process can yield many more sterile or unattractive new plants than horticulturally worthwhile ones. Even when iris breeders avail themselves of the "trick" boost of treating ripe seeds with colchicine (a substance derived from the autumn crocus, *Colchicum autumnale*), viable and attractive seedlings of potentially good new and improved varieties are proportionally few.

One might speculate that iris family plants hedge their own reproductive bets with their propensity to expand their numbers vegetatively, increasing via their own root systems over the years. Gardeners can allow this natural tendency, hasten it along with supplemental water and fertilizer, or intervene by dividing and replanting burgeoning clumps throughout their yards. You can't have too much of a good thing, can you?

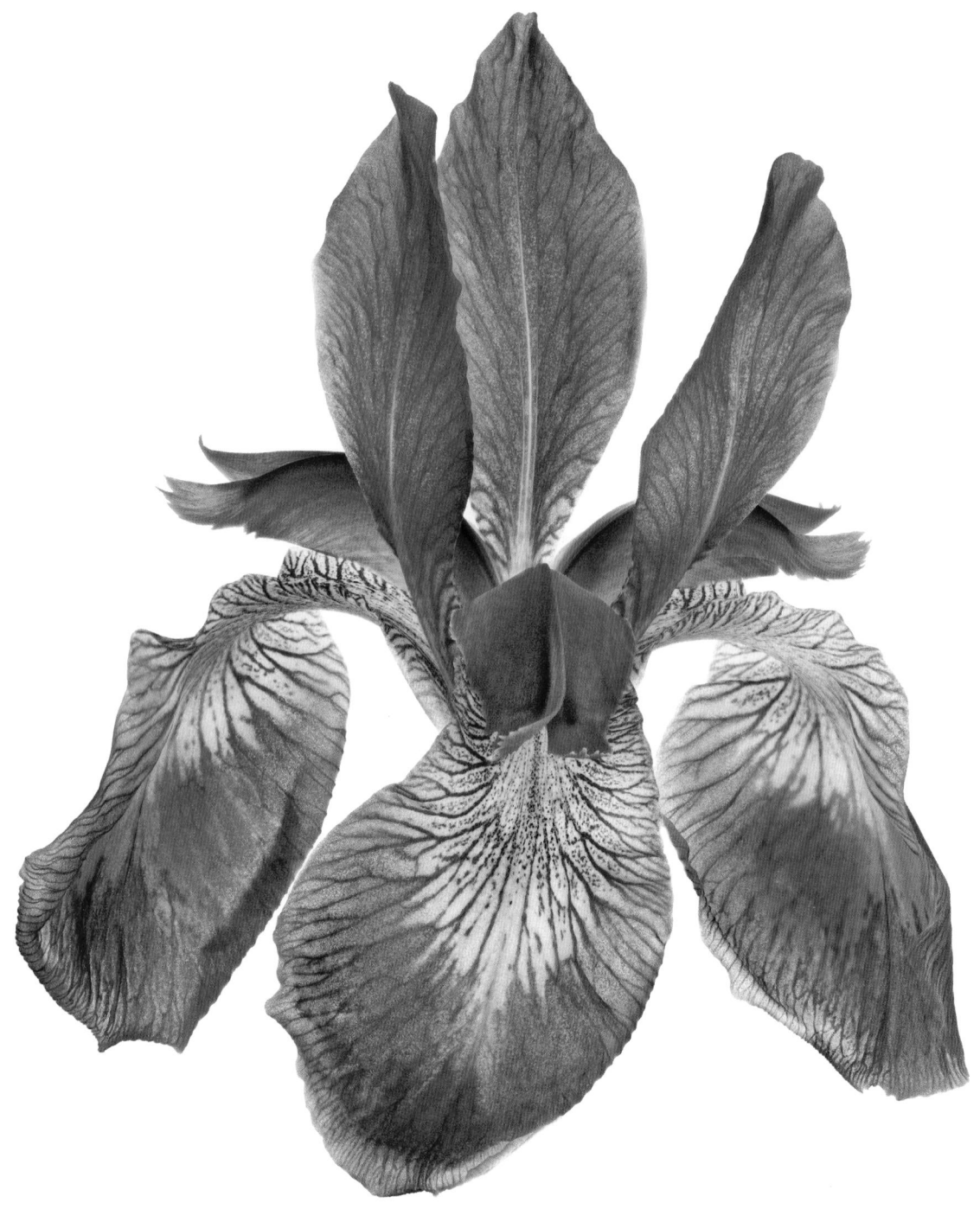

ABOVE Iris petals have veins that are sometimes masked by heavy coloration, and other times are easily visible thanks to contrasting hues. These markings help direct pollinators toward the flower's center. In *Iris sibirica* 'Caesar's Brother', this quality is beautifully highlighted.

RIGHT A species iris whose 2-inch flowers have been compared favorably to orchids, *Iris japonica* is indeed a treat to behold. Fringed, pale blue petals are marked purple, with yellow throats. Like all irises, all parts are in threes.

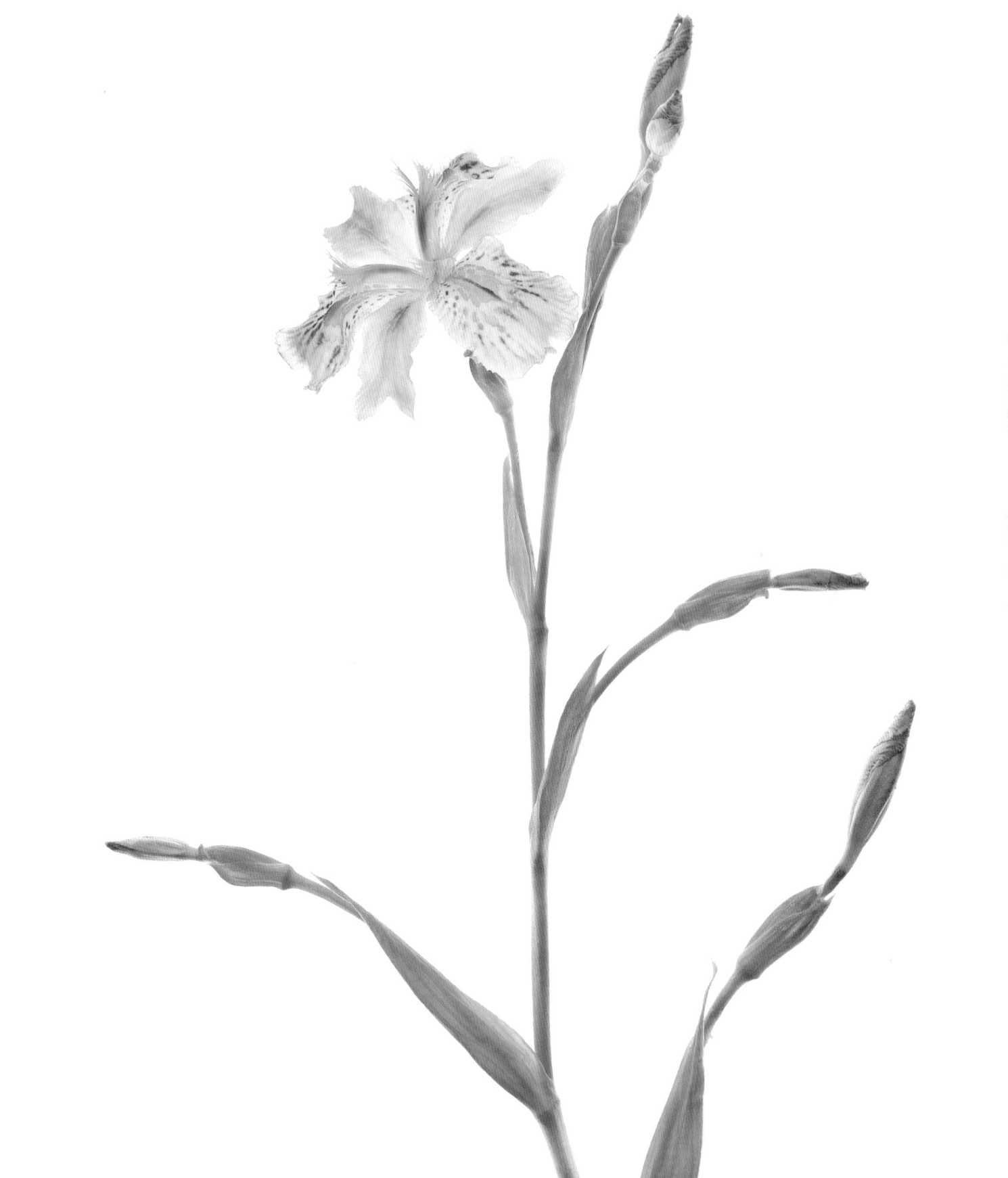

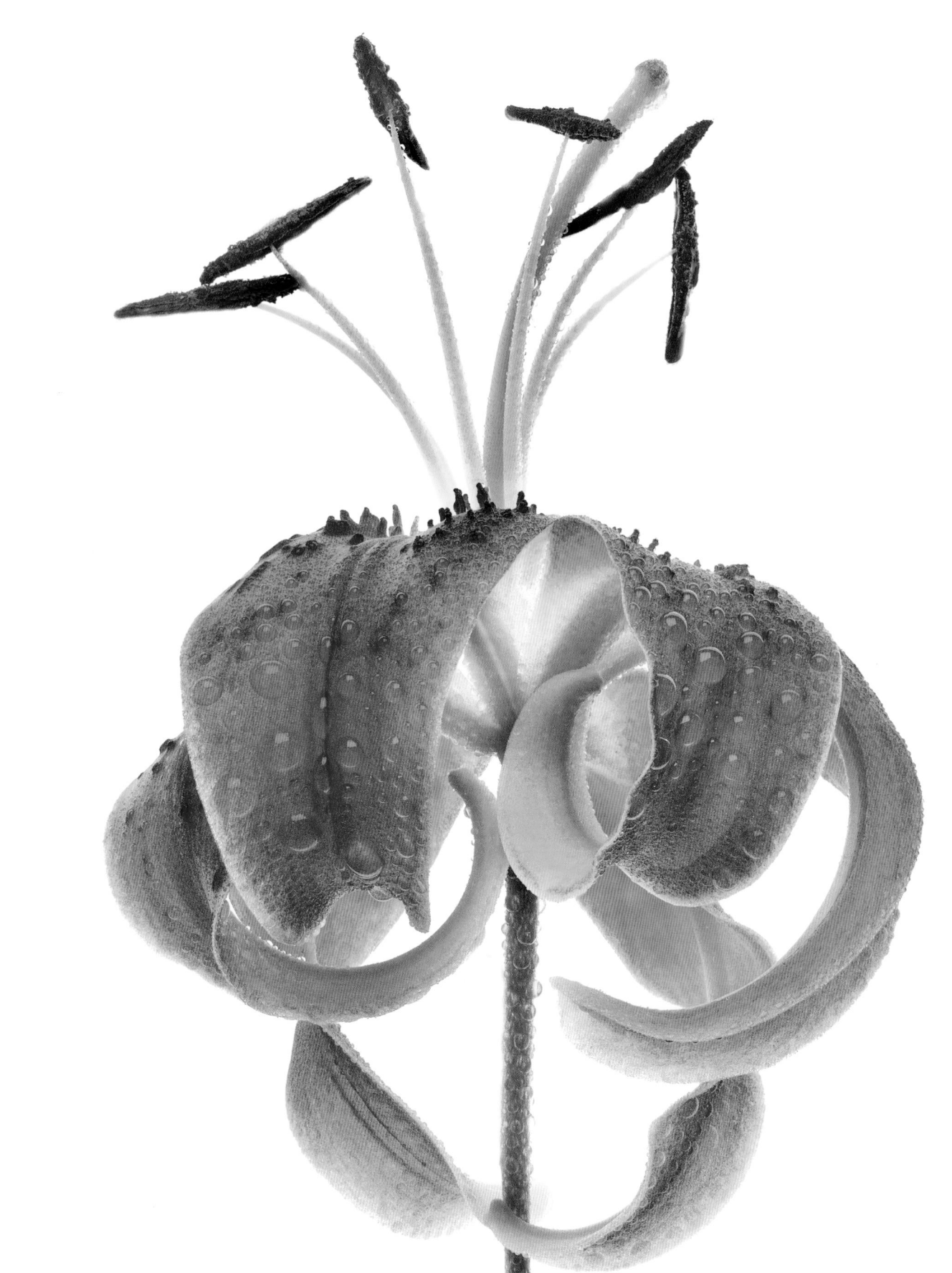

"Consider the lilies, how they grow:
they neither toil nor spin,
yet I tell you, even Solomon in all his glory
was not arrayed like one of these."

LUKE 12:27

What are those tiny protrusions or whiskers inside some lily blossoms? They are often prominent enough to be visible, as is seen here in some flowers from a Japanese lily, *Lilium speciosum*. The technical term is papillea, but their purpose is not known.

PAGE 164 Getting this close to an allium—in this case, the accurately named 'Purple Sensation'—is almost a voyeuristic experience. Small, individual flowers, each showing off classic lily family flower parts in sets of six, jumble together in exuberant disarray. And yet, if you stand back, the density turns out be symmetrical and orderly enough to create nice round, baseball-size flowerheads, 4 to 5 inches in diameter.

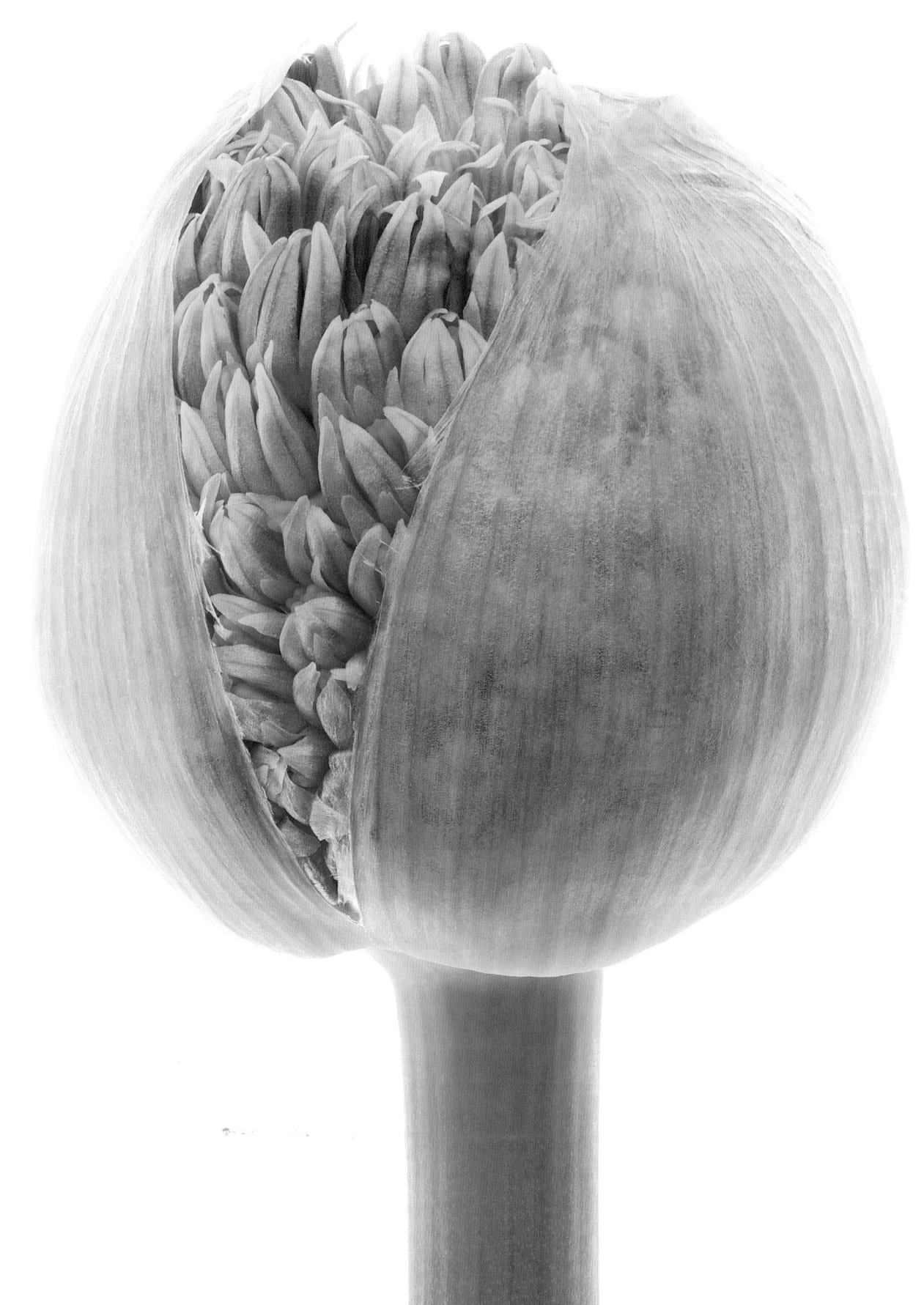

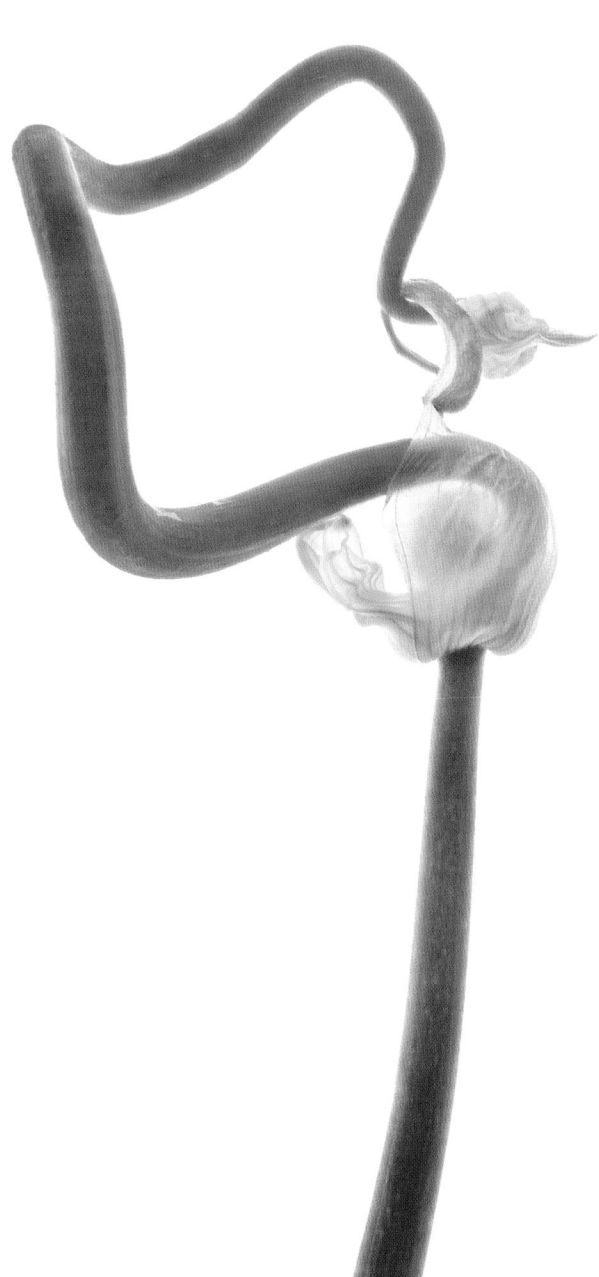

A rguably the world's most easily recognized and beloved flowers, lilies form a large, sprawling family of around 4000 species. It includes, first and foremost, true lilies, the glorious trumpet-shaped flowers of garden and florist and flower show. Experts cite a total of nine distinct divisions, notably the brightly hued Asiatics, and the richly fragrant Oriental hybrids, such as the glamorous 'Casa Blanca'. Related are intergeneric beauties such as trumpets, "Orienpets," and robust tetraploids, the results of extensive and adventurous breeding and selection projects by nurseries and enthusiasts. And let's not forget the headily scented Easter lilies, *Lilium longiflorum* and its hybrids, and waxy-bloomed, fragrant Madonna lilies, *L. candidum* and its hybrids.

There are also scads of lovely species to delight flower lovers, from the towering Chinese lily, *Lilium henryi*, spangled all the way to the top with gold-orange flowers, to the more modest, waist-high Canada lily, *L. canadense*, which sports a good show of black-speckled orange to red flowers, candelabra style. You may have seen the evocatively named Turk's cap ones, which have recurved petals; these originate from *L. martagon* and have been widely hybridized. The speckles, dots, or lines on some of these flowers function as air-traffic control for pollinators, guiding them toward the pollen in the center.

LEFT Behold a garlic "scape," which forms a few weeks after the first leaves and is actually a flower stalk. Gardeners and farmers cut these off so they won't divert plant energy away from the bulbs burgeoning belowground. If allowed to remain, they don't make flowers, but instead form little bulbils that can generate new plants.

OPPOSITE The developing flowerhead of *Allium* 'Purple Sensation' bursts forth from within a papery membrane, technically a two-part bract. Upon splitting apart, the thin bract's usefulness is done and it will hang below the spectacular inflorescence for a time, ultimately dropping away, forgotten—like a stage curtain after the players have appeared.

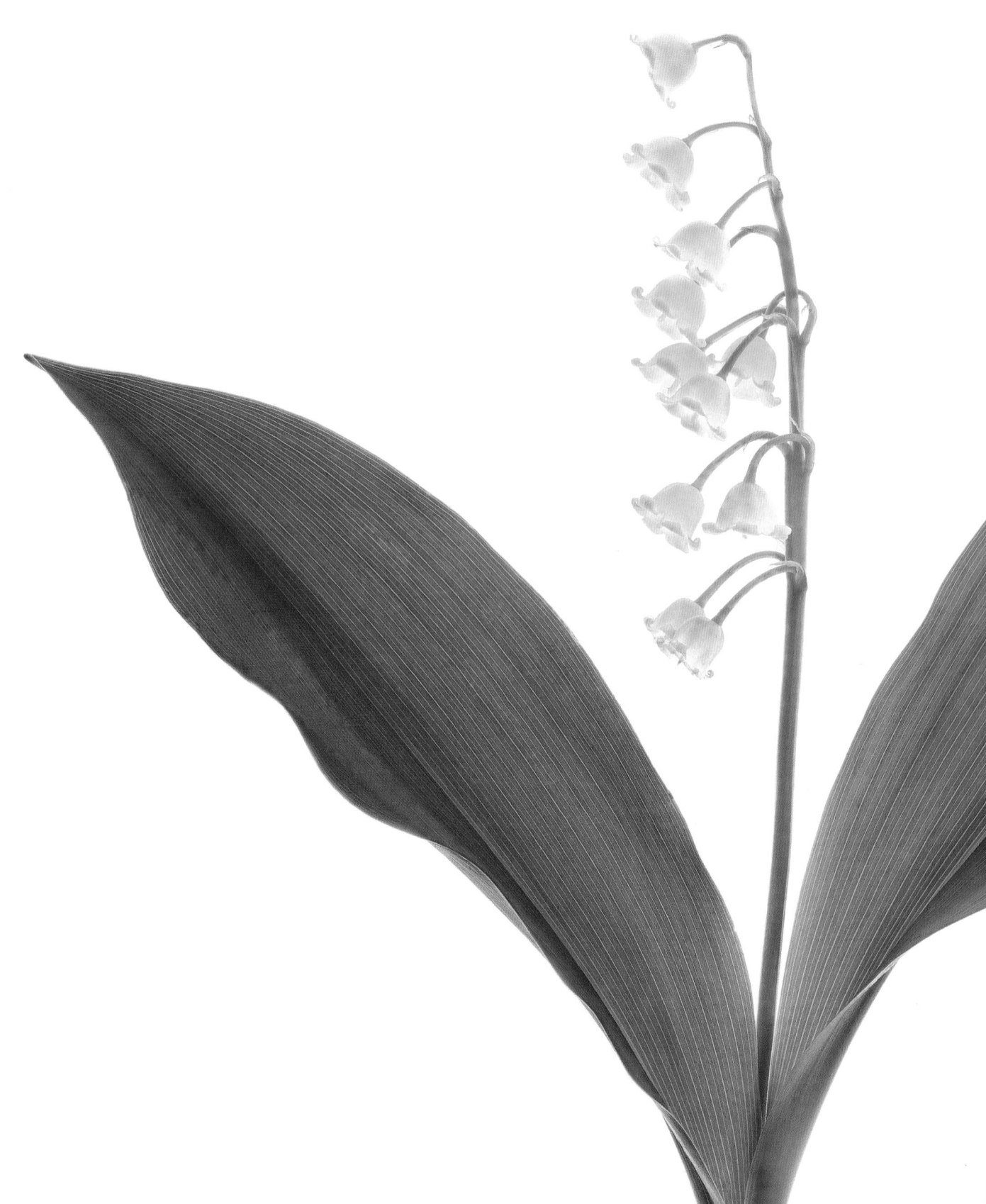

Additionally, lots of favorite wildflowers are in Lilaceae, including *Trillium*, *Clintonia*, trout lily (*Erythronium*), and the fall-blooming curiosity toad lily (*Tricyrtis*). Of these, trillium's three-petaled, three-sepaled flowers are the largest and prettiest; several species are also sweetly scented.

Some spring-flowering bulbs are housed here as well, particularly all the tulips (*Tulipa*) and crown imperials (*Fritillaria*). Traditionally, hyacinths (*Hyacinthus*), star-of-Bethlehem (*Ornithogalum*), bellworts (*Uvularia*), and grape hyacinths (*Muscari*) have also been included, although given the bloated size of this family, modern-day botanists argue for transferring some of these into their own or related families.

Furthermore, the lily family has traditionally encompassed all onion-type plants, from scallions and garlic to fantastic hybrids meant for ornamental use, such as the ever-popular 'Globemaster' and its offspring. Recent genetic research argues for putting these into Amaryllidaceae, although those who sell and grow the plants have not widely embraced this technical change. Daylilies, *Hemerocallis* species and cultivars, however, are now relegated to a separate family. Even without these groups, though, the lily tribe remains huge.

A lily is broadly defined as an herbaceous perennial with a six-part flower, composed of three petals and three sepals or six petal-like tepals. Typically six stamens emerge from the center. When a flower is fertilized and goes to seed, the resultant fruit is divided into three segments. There are a few exceptions, but overall anyone can easily peg a lily family plant with these criteria. Pollination is generally via bees, although other insects and hummingbirds may play a role, depending on the

Dear and familiar lily-of-the-valley, *Convallaria majalis*, keeps getting shuttled away from the lily family. It was in the asparagus family for a while; evidently the latest is that it belongs in the butcher's broom family, or Ruscaceae. A more practical factoid: Plants are self-sterile, which means that no matter how many bees stop by, a patch of a single clone will not set any seed.

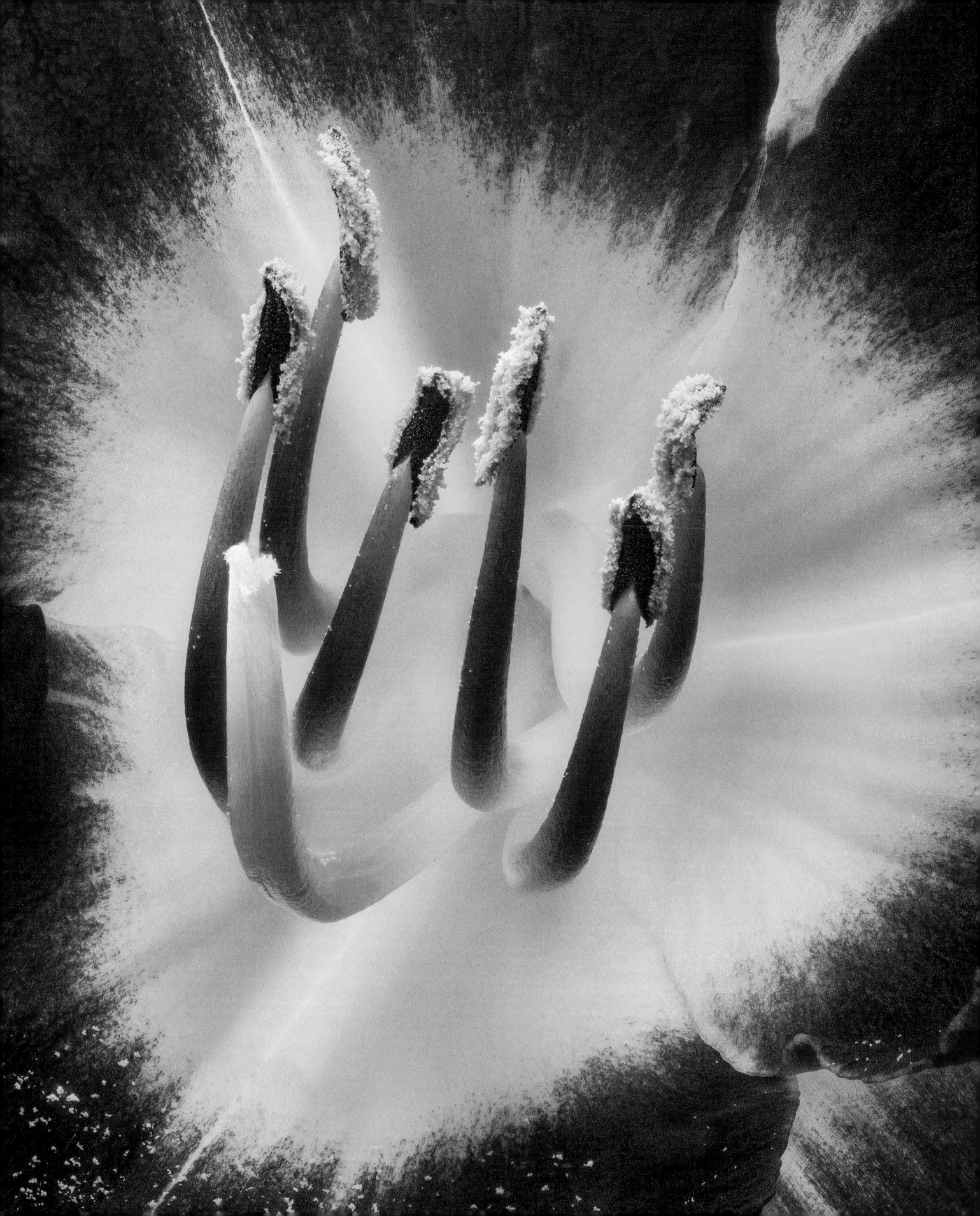

plant and the setting. Certainly the gorgeous, colorful, and often scented flowers evolved to woo pollinators.

For further confirmation, you can check the foliage. Lilies are monocots, so their growth habit is primarily upward (rather than outward) and their leaves are parallel veined and usually narrow. When a lily sprouts, there is one leaf initially. As the plant develops, leaves are either basal or alternate up the stem, or exhibit both characteristics.

Most lilies grow from bulbs, rhizomes, or at least small tubers. Essentially storage rootstocks, they allow the plants to go dormant during the hottest part of the growing season, regenerate year after year, and recover from dry spells or fires. These abilities account for the fact that many species originated in areas with stressful growing conditions. Needless to say, when pampered in a garden setting with good soil, regular water, and perhaps even some fertilizer, lilies tend to be generous

LEFT The daylilies, *Hemerocallis*, have been extensively hybridized, resulting in all manner of glorious blooms. Contrasting bicolors, like this one, must be especially irresistible to pollinators because they seem to blare out their presence while offering tantalizing pollen-thick stamens. Note the six prominent stamens and one pistil; fertilization can occur when the pistil is "receptive" with sticky stigmatic fluid at its tip. Needless to say, it's a fairly easy process, for humans as well as insects (though daylily experts remind us that crosses only lead to viable seed among genetically similar plants).

BELOW The glorious Easter lily, *Lilium longiflorum*, offers a clear, characteristic view of lily flower parts. Emerging from the middle of six stamens (male parts) is the plump, sexy, knoblike stigma (female). A sticky fluid bathes it when the flower is ready to receive, hold, and begin to germinate pollen.

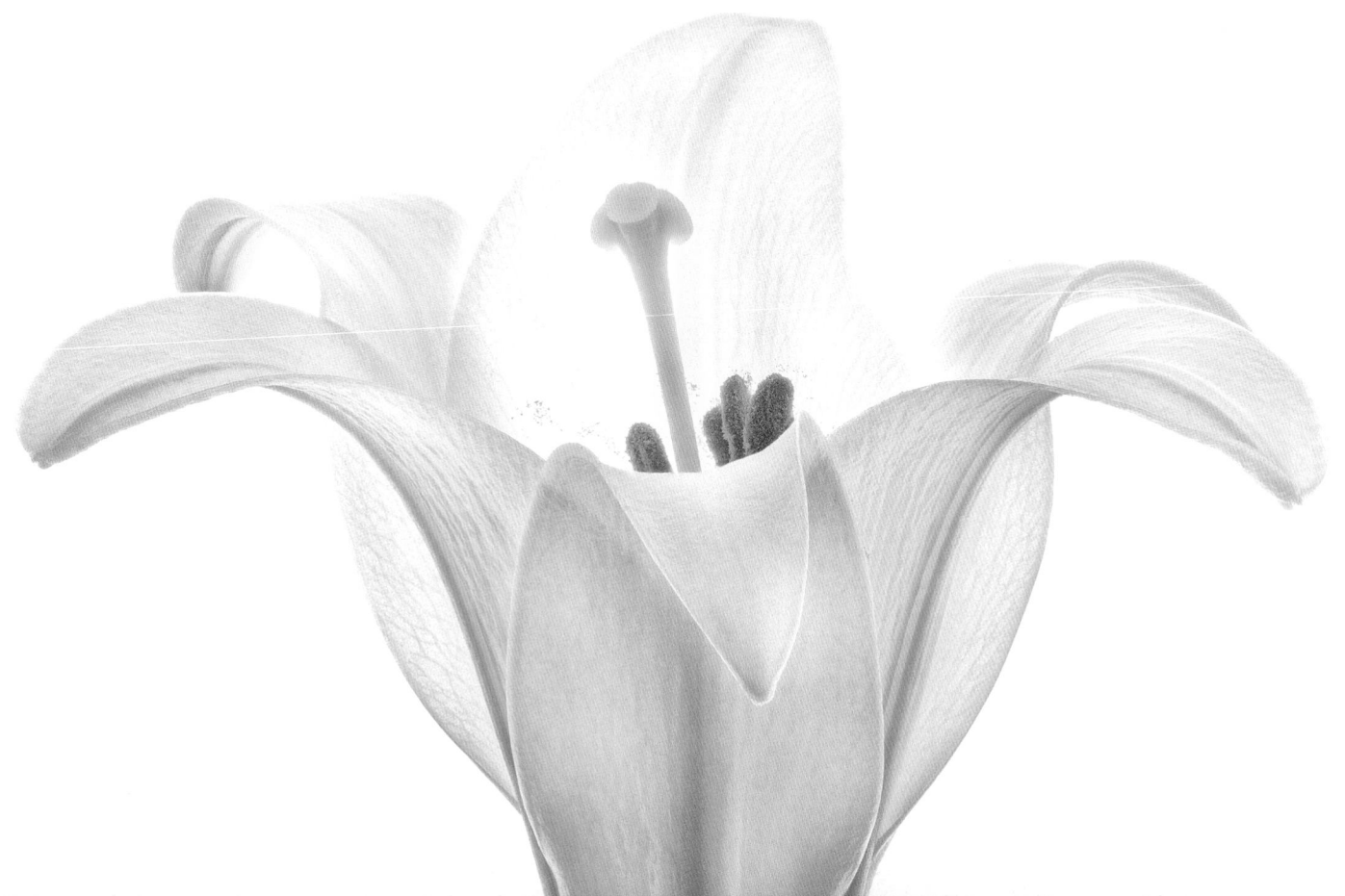

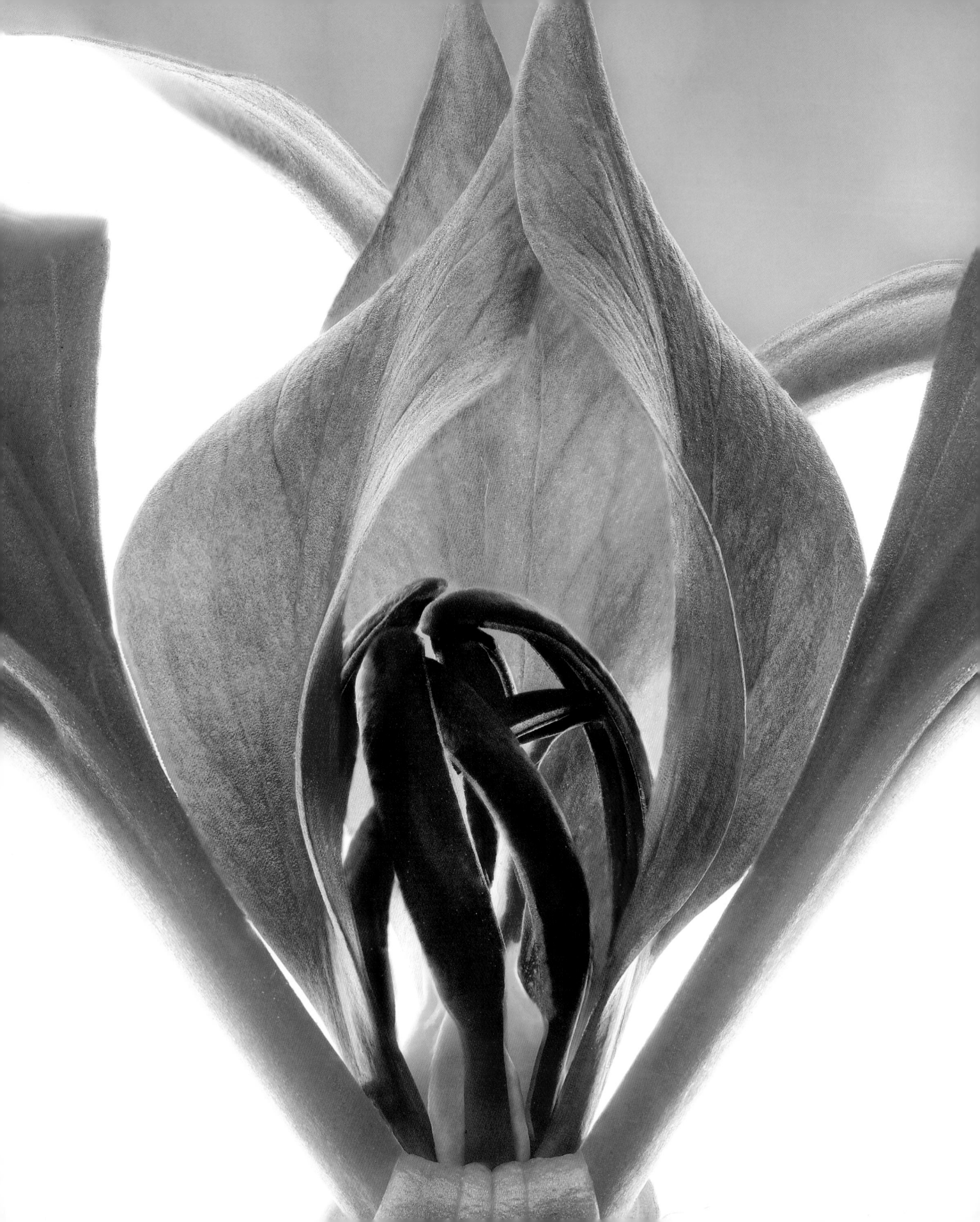

with growth and bloom. Most lilies like to have their heads in the sun and their feet in the shade, but many will adapt to less ideal conditions.

In Asia, where there is a long tradition of discovering uses for all sorts of plants, the starchy roots of some species are used in cooking. In taste and texture, they are a bit like a potato, with perhaps a bit of sweetness. But lilies, especially Easter lilies, are evidently toxic for cats, who can experience kidney failure if they chew or bite plants, lick pollen off their fur, or even drink water that has drained out of lily pots. If you have feline friends, find a way to remove the temptation (block access, crowd other plants in front of your bare-kneed lilies)—or avoid growing them altogether.

Beloved and graceful spring wildflowers, trilliums, are so named because they are amply tri-parted. They all have a whorl of three leaves atop an erect stem and a solitary, three-petaled, three-sepaled flower. In the interior, you will find six stamens with long anthers set between the six lobes of the ovary. Yellow *Trillium luteum* (PAGE 176) has a soft citrusy scent; maroon to deep brown *T. sessile* (LEFT) wafts a muskier fragrance; and the unmistakable white *T. grandiflorum* (PAGE 177) has bigger, noble flowers whose petals have wavy edges and a waxy texture.

Native Americans occasionally used various parts of native lilies in their foods and medicines. Fresh and dried bulbs have soothing as well as astringent properties. The main use that survives is for the oil extracted from the leaves, stems, and flowers, which has healing and softening qualities. Major cosmetic firms Procter and Gamble and Revlon use it in some of their skin-care products.

If you like lilies for bouquets, avail yourself of the tricks of the florist trade. Pinch or snip off the dangly stamens altogether—sooner rather than later. Gently brush off excess pollen with a toothbrush or pipe cleaner, lest it stain your clothing or fingers (this works like using a lint brush on your favorite sweater). If you're using white lilies, add a few drops of bleach to the water to prolong the show. Nudge partially opened blooms wider by putting their stems in lukewarm water. Conversely, preserve flowers or slow down the opening process by placing the stems in cold water and/or a refrigerator or at least in a darkened room. Fortunately, lilies of all kinds tend to last a wonderfully long time in the garden as well as in the vase. Savor them, for in all the wide world of flowers, few blooms are as accommodating and spectacular as those of lilies.

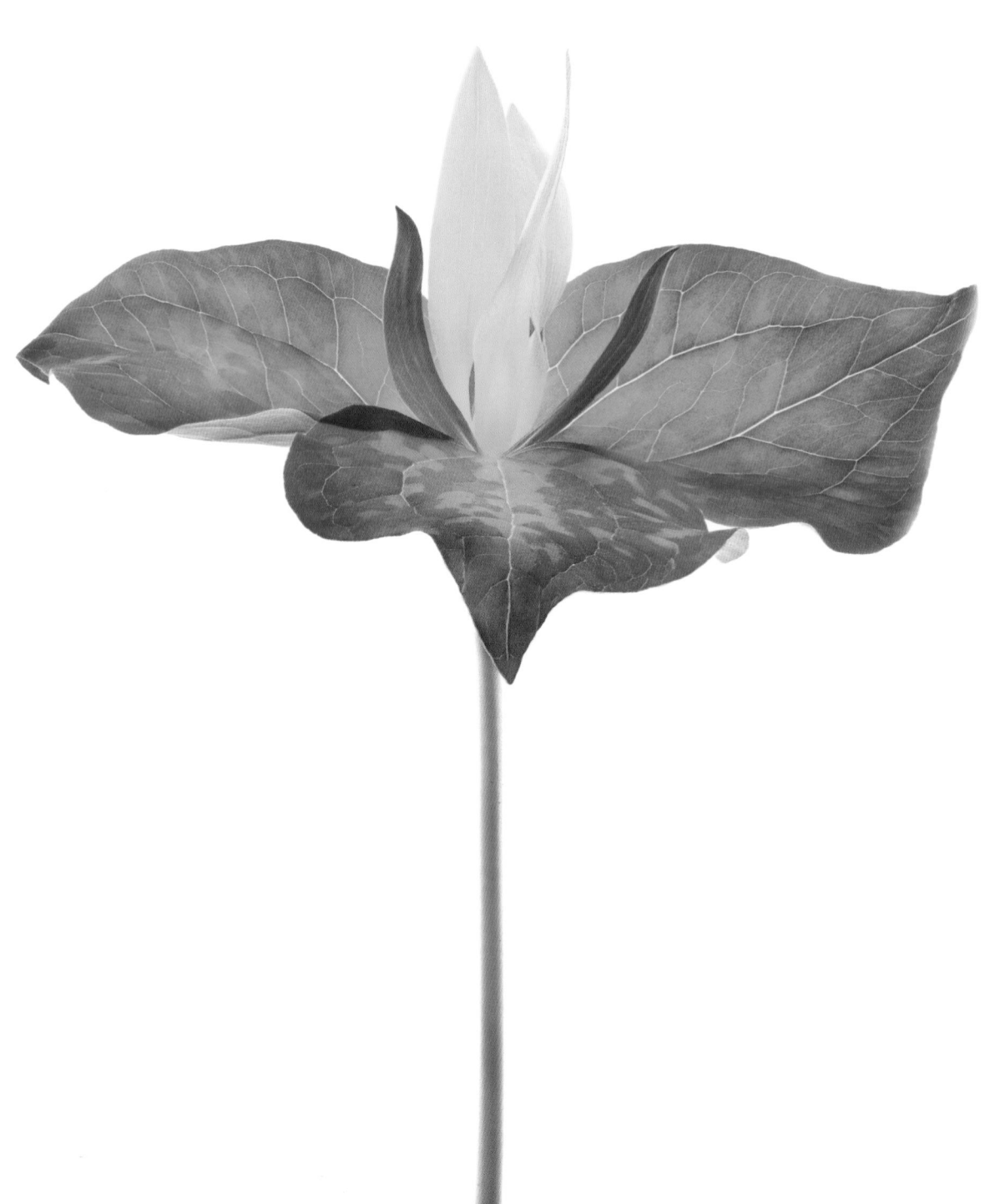

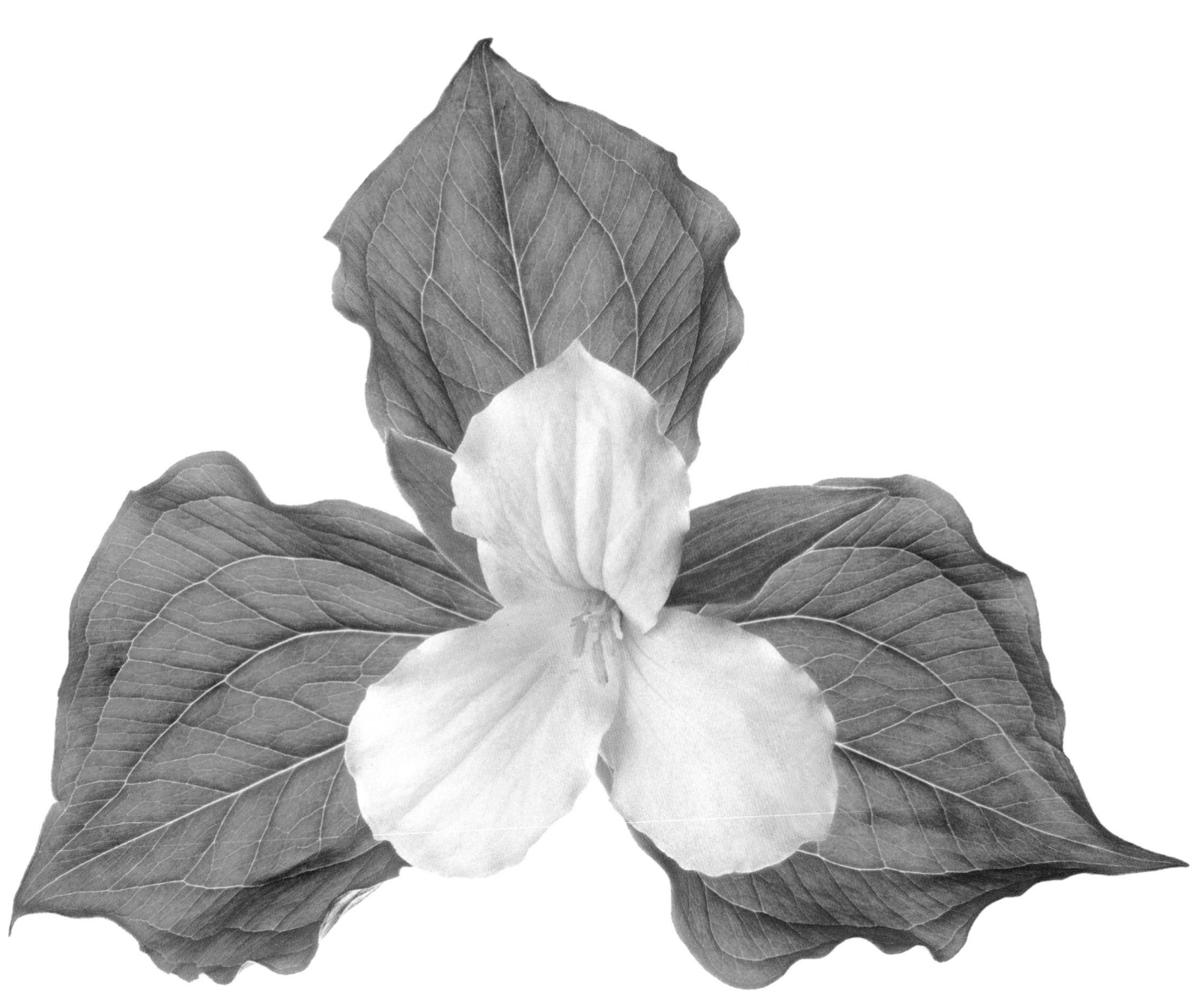

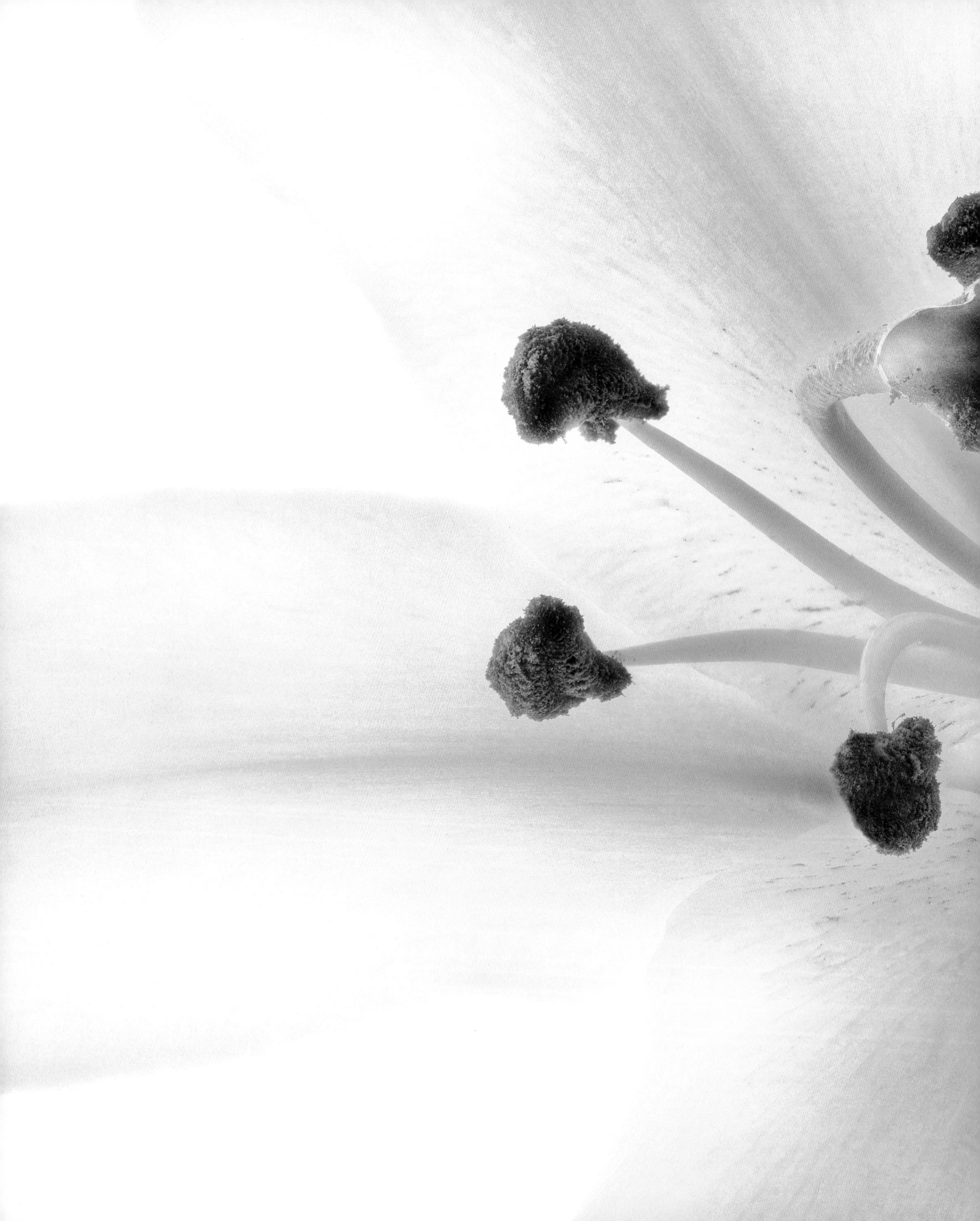

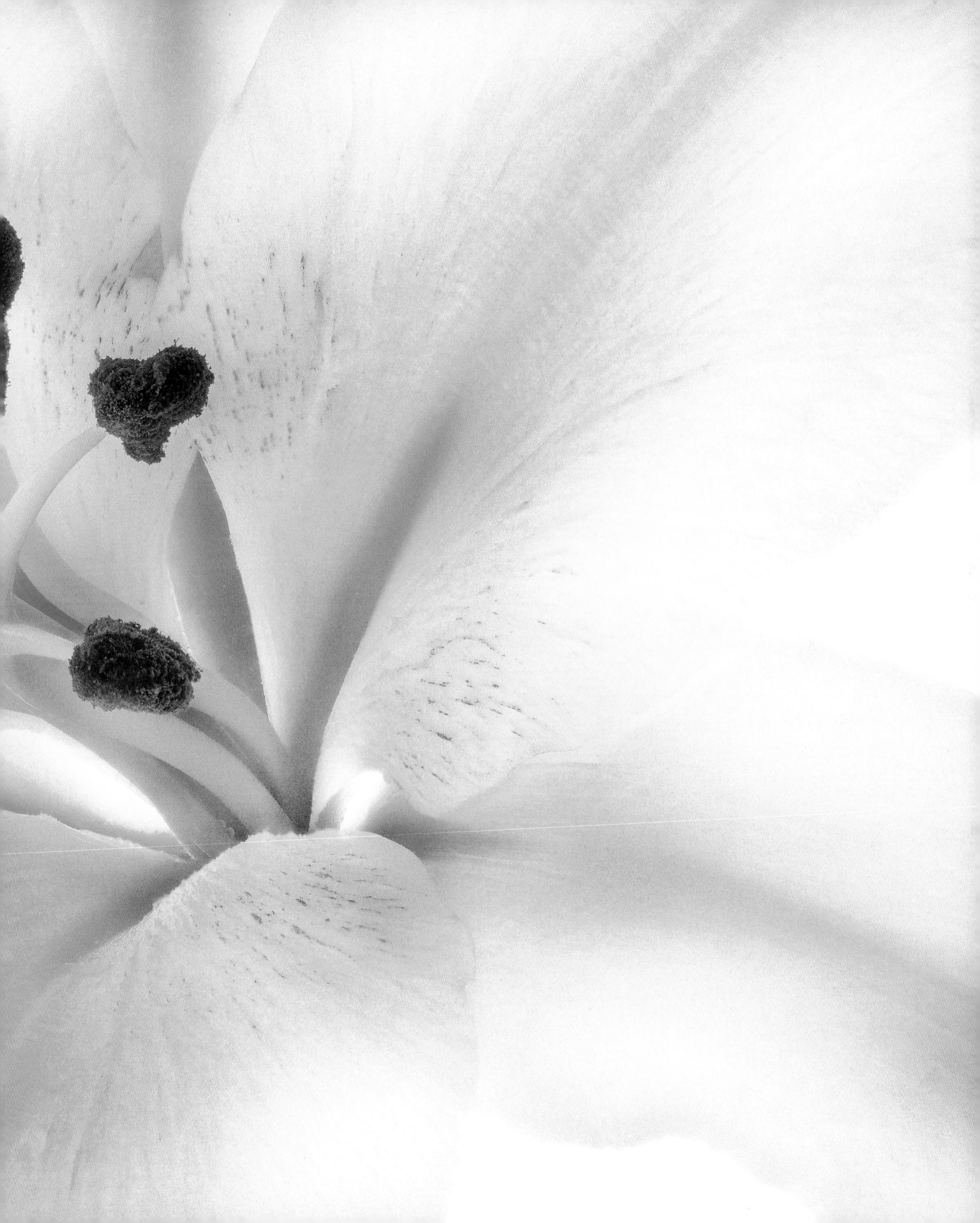

BELOW The wild trout lilies, *Erythronium*, sport (usually) yellow flowers reminiscent of a miniature Asiatic lily. Both the sunny blooms and mottled leaves of the robust hybrid 'Pagoda' are larger than any of the species.

PREVIOUS SPREAD Peer into the throat of a bold and brassy trumpet lily and you'll see what a pollinator does: colorful lines or "nectary furrows" directing your attention downward, and anthers thick with pollen. Lily pollen is heavy and sticky, and it contains protein—all good reasons why insects, not wind, pollinate these flowers.

RIGHT Dainty Spanish bluebells, *Hyacinthoides hispanica*, earn their membership in the lily family because their bells are six lobed (their petals, or tepals, are fused near their bases) and their clappers, or stamens, also number six.

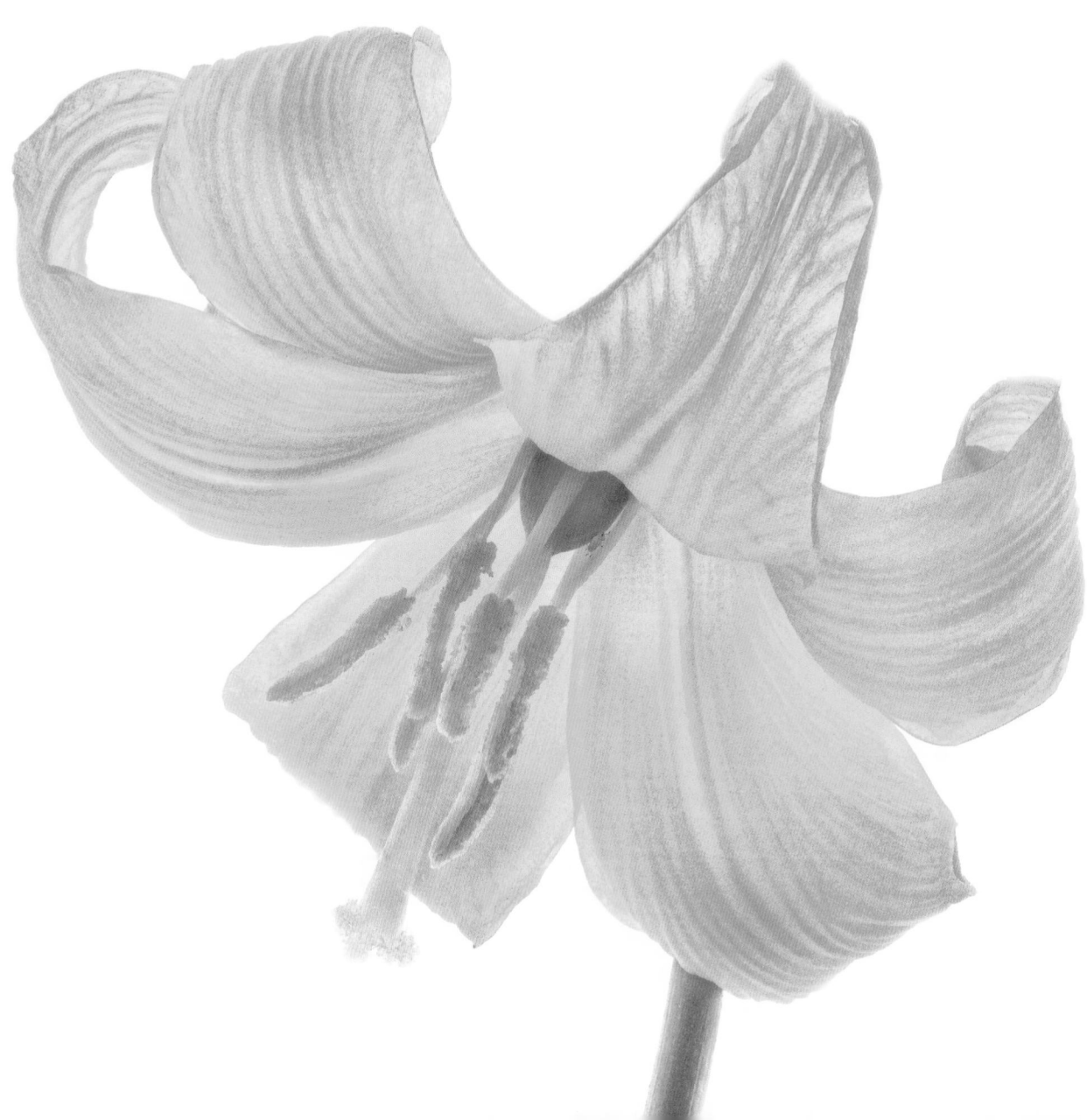

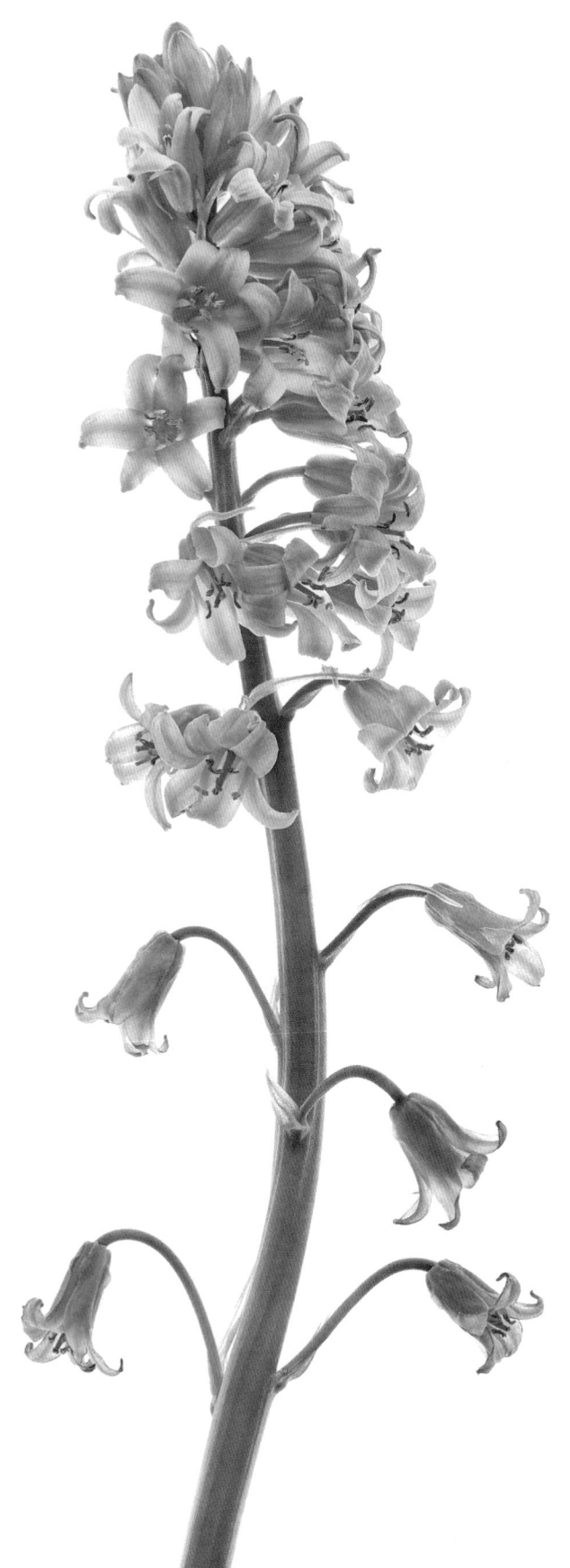

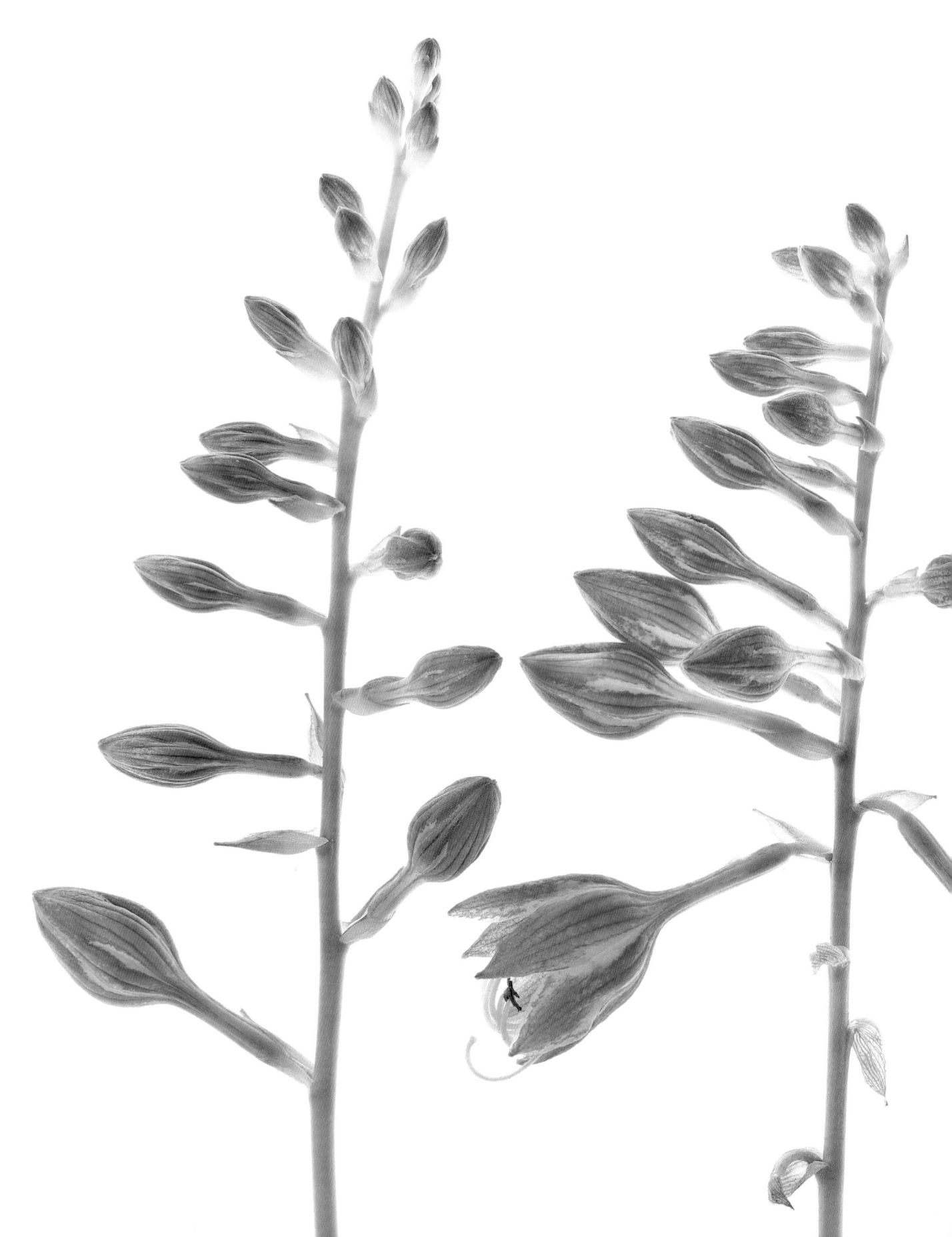

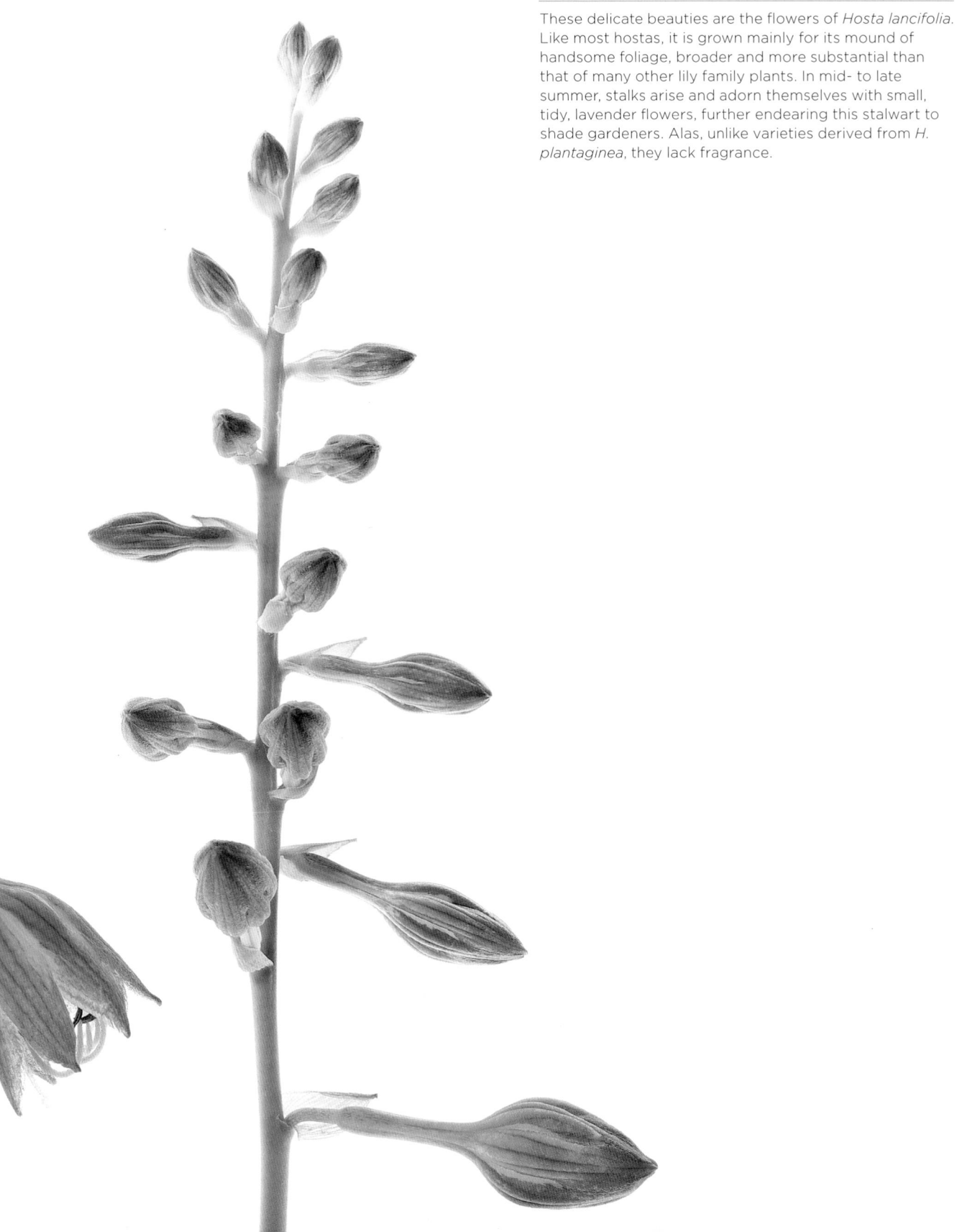

These delicate beauties are the flowers of *Hosta lancifolia*. Like most hostas, it is grown mainly for its mound of handsome foliage, broader and more substantial than that of many other lily family plants. In mid- to late summer, stalks arise and adorn themselves with small, tidy, lavender flowers, further endearing this stalwart to shade gardeners. Alas, unlike varieties derived from *H. plantaginea*, they lack fragrance.

BELOW Aptly named spring snowflake, *Leucojum vernum*, serves up a generous early season helping of tiny, chubby white bells with green accents. Sometimes placed in the Amaryllis family, sometimes placed with the lilies, it is especially valued for its ability to prosper in soggy ground.

RIGHT Exuberant yet classy, *Nectaroscordum siculum* deserves a better common name than honey garlic. With its umbels of nodding creamy flowers infused with pink or deep red seguing to light green toward the base, its elegance is undeniable.

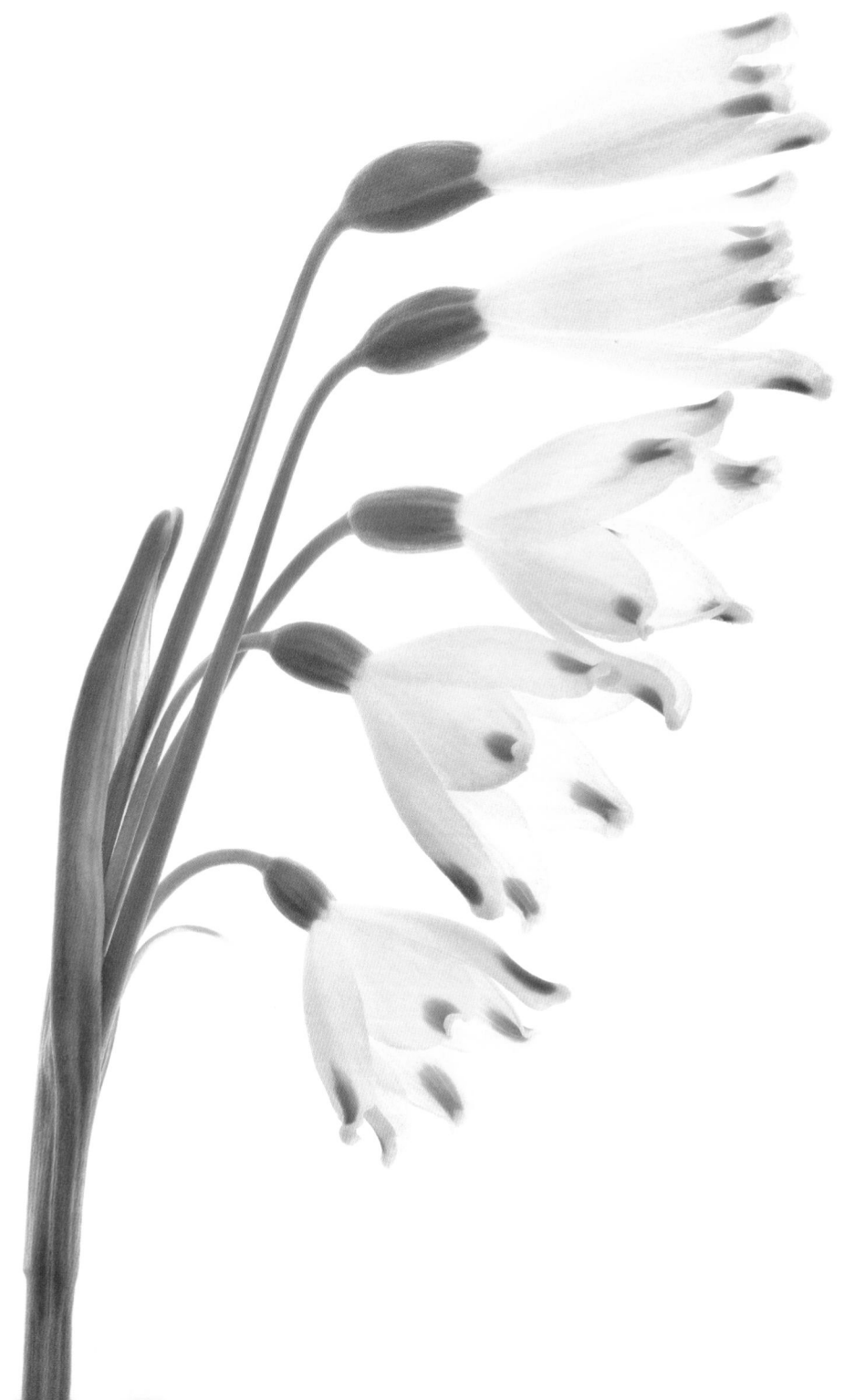

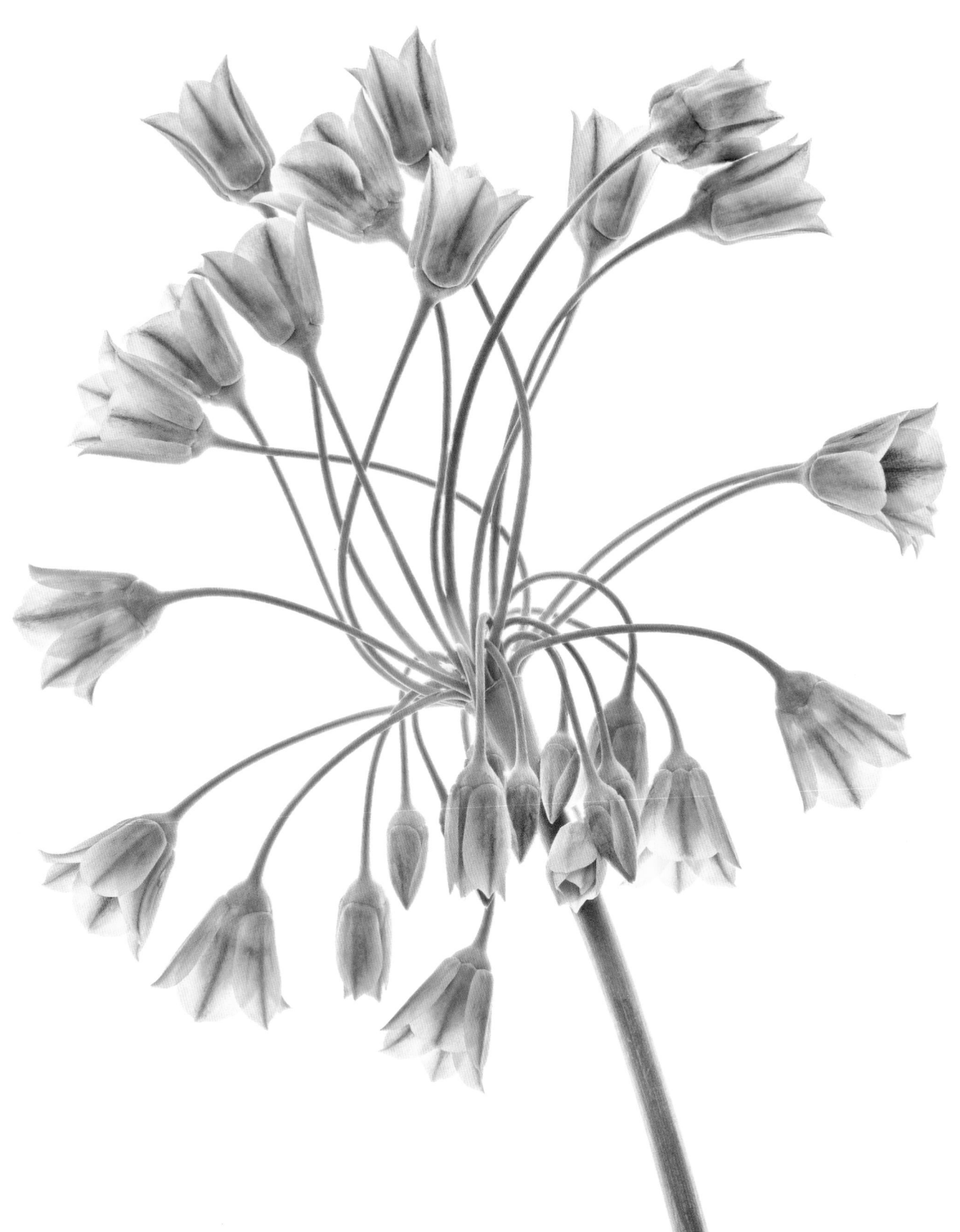

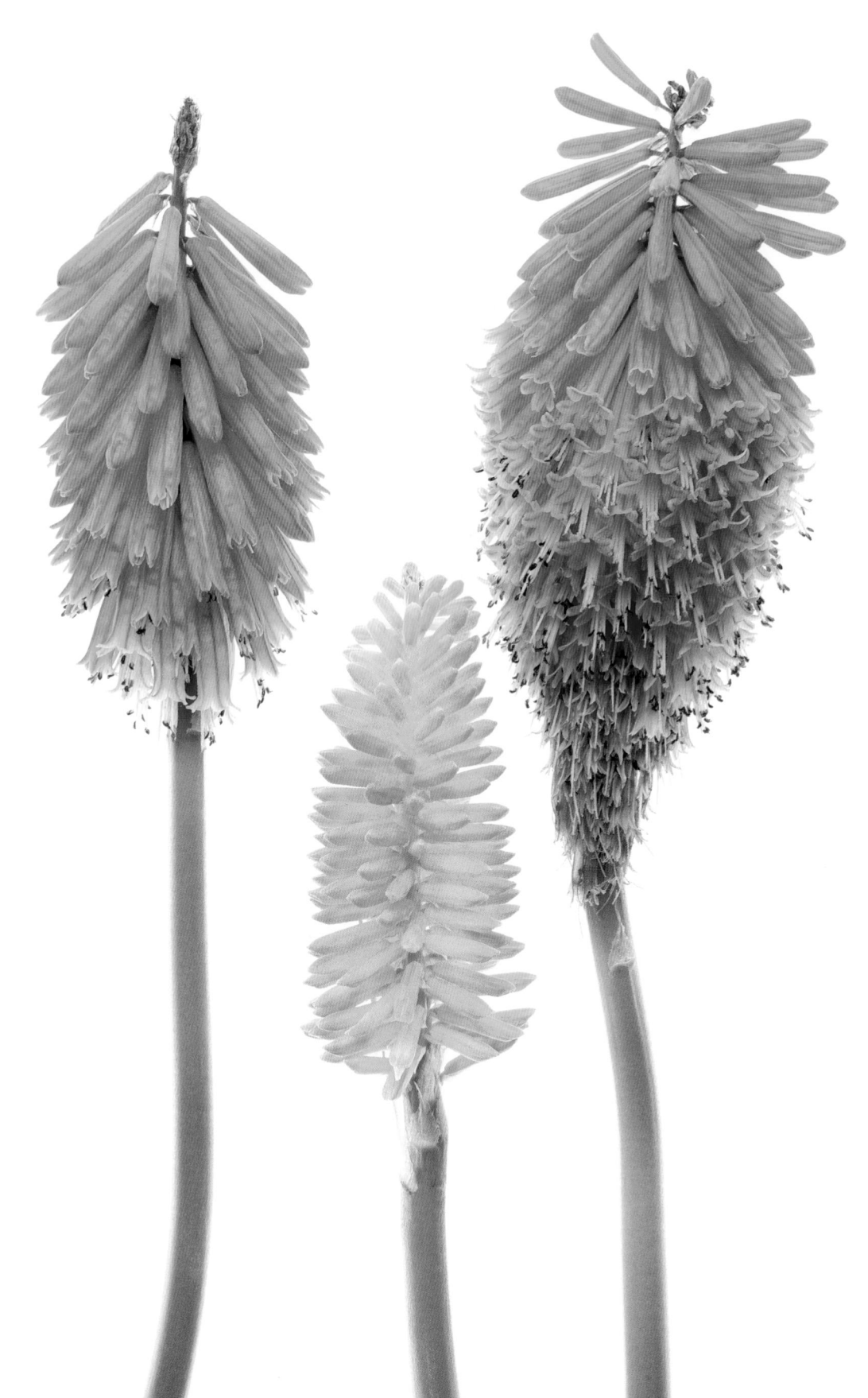

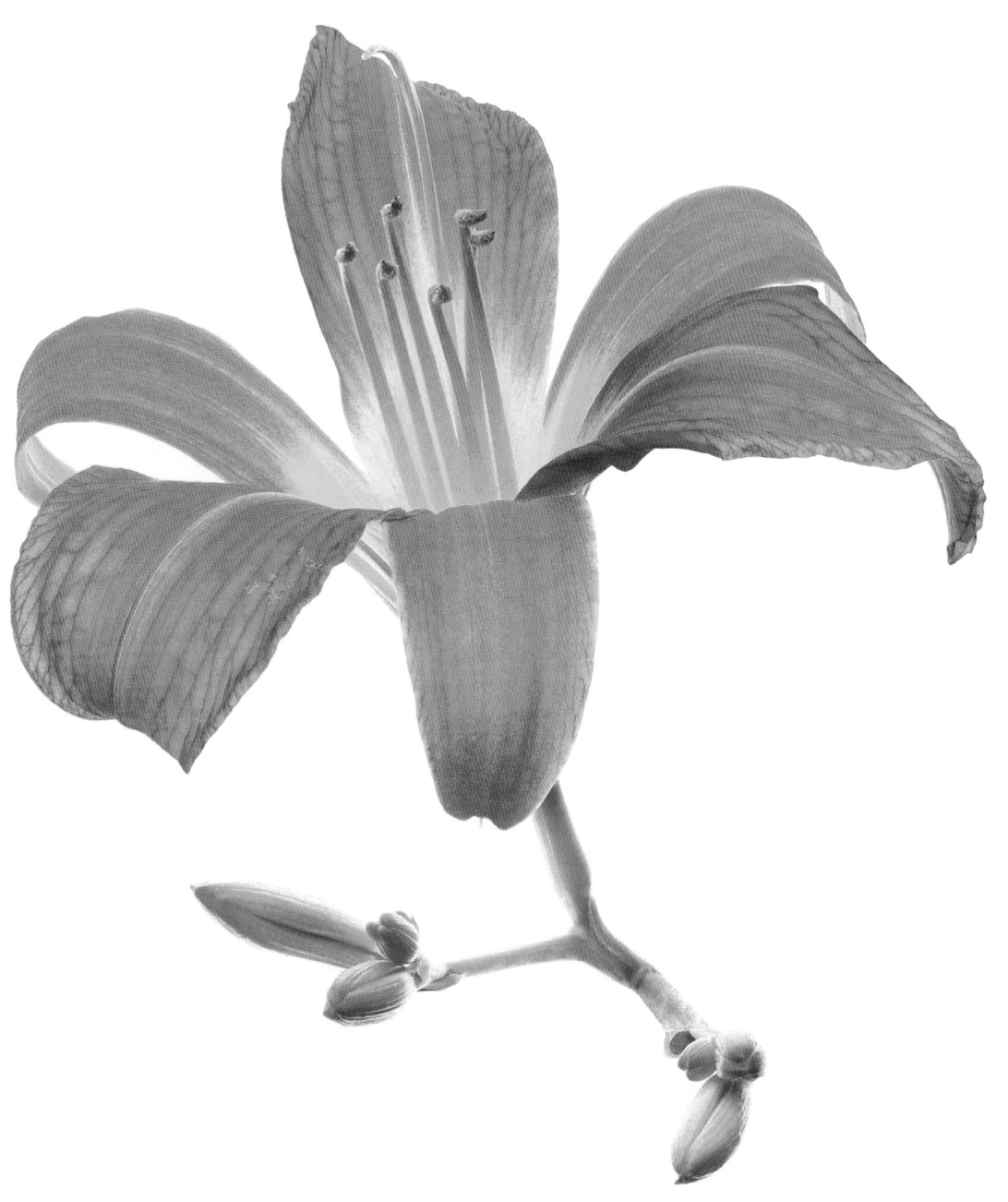

LEFT Answering to several evocative common names—red hot poker, torch lily, poker plant—the *Kniphofia* hybrids are big, tough, and dramatic. They produce so much nectar that hummingbirds and even bigger birds, notably orioles, visit. Spikes bloom from the bottom up.

ABOVE The orange or tawny daylily, or "tiger lily," *Hemerocallis fulva*, is an Asian import that has spread widely in gardens, roadsides, fields, and other sunny spots. Curiously, seed production is rare. Once the flowers have bloomed, they simply fall off. So no wonder the plants have developed—some might exclaim, overdeveloped—the ability to reproduce vegetatively.

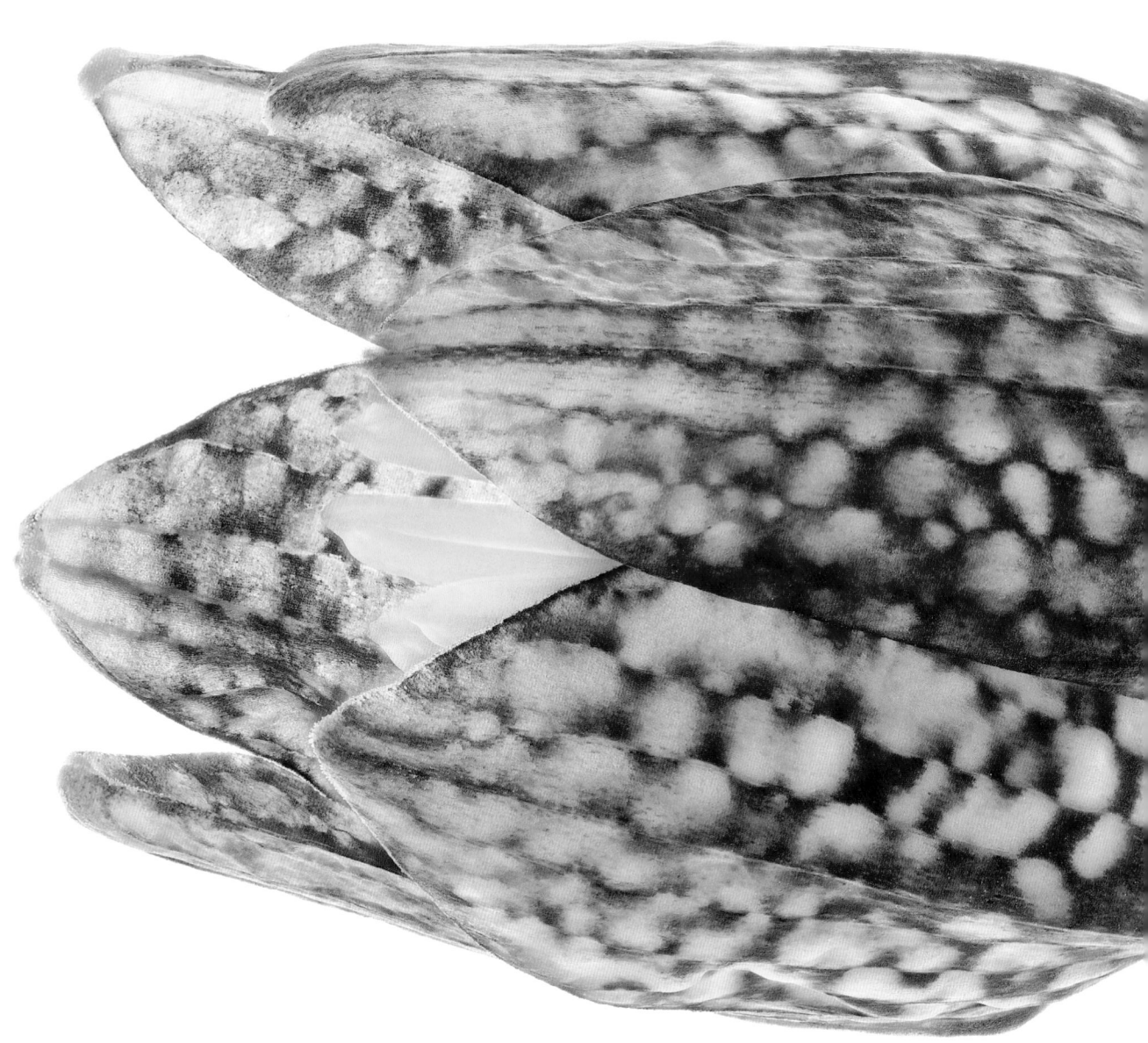

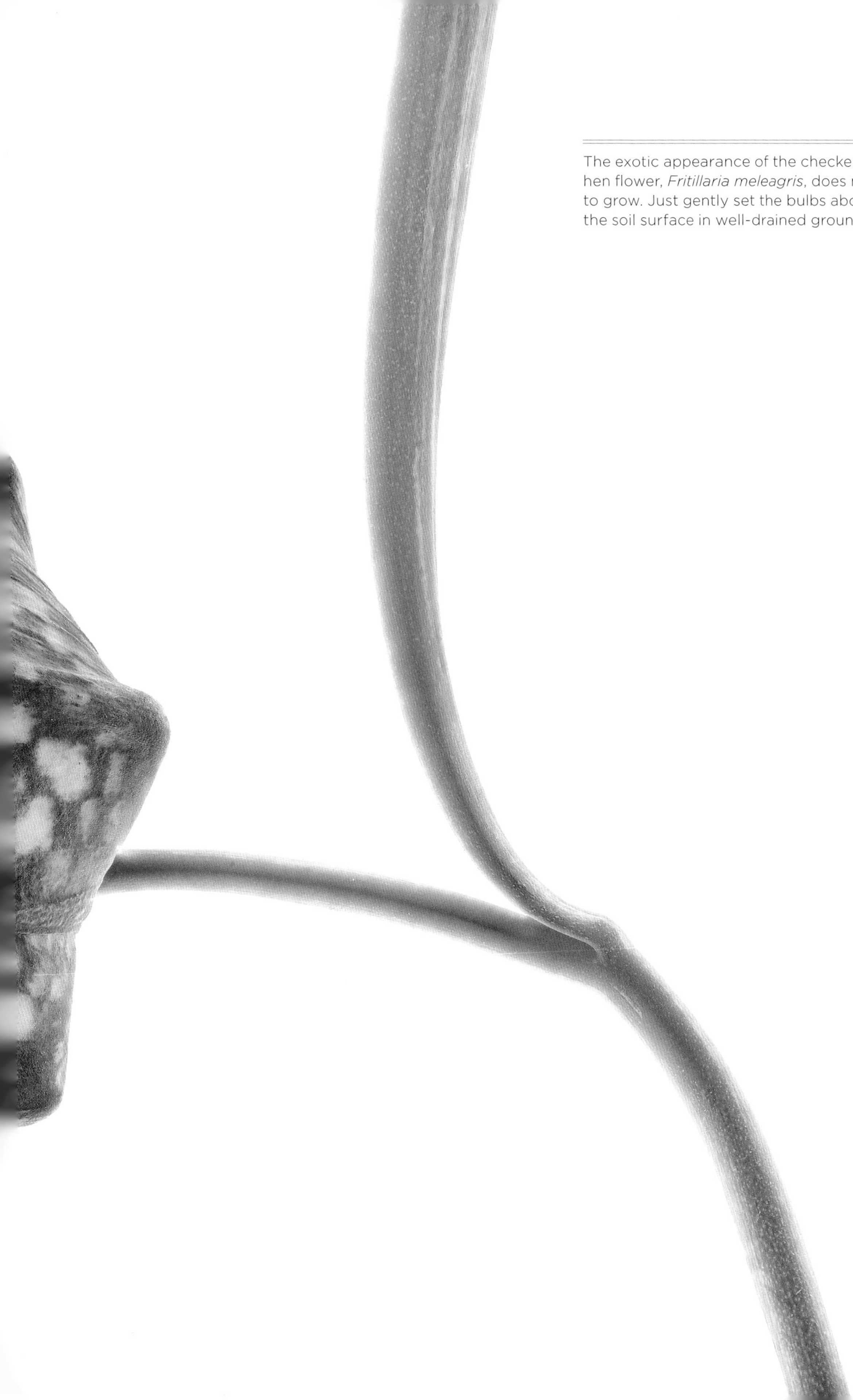

The exotic appearance of the checkered lily or Guinea-hen flower, *Fritillaria meleagris*, does not mean it is tricky to grow. Just gently set the bulbs about 6 inches below the soil surface in well-drained ground in the fall.

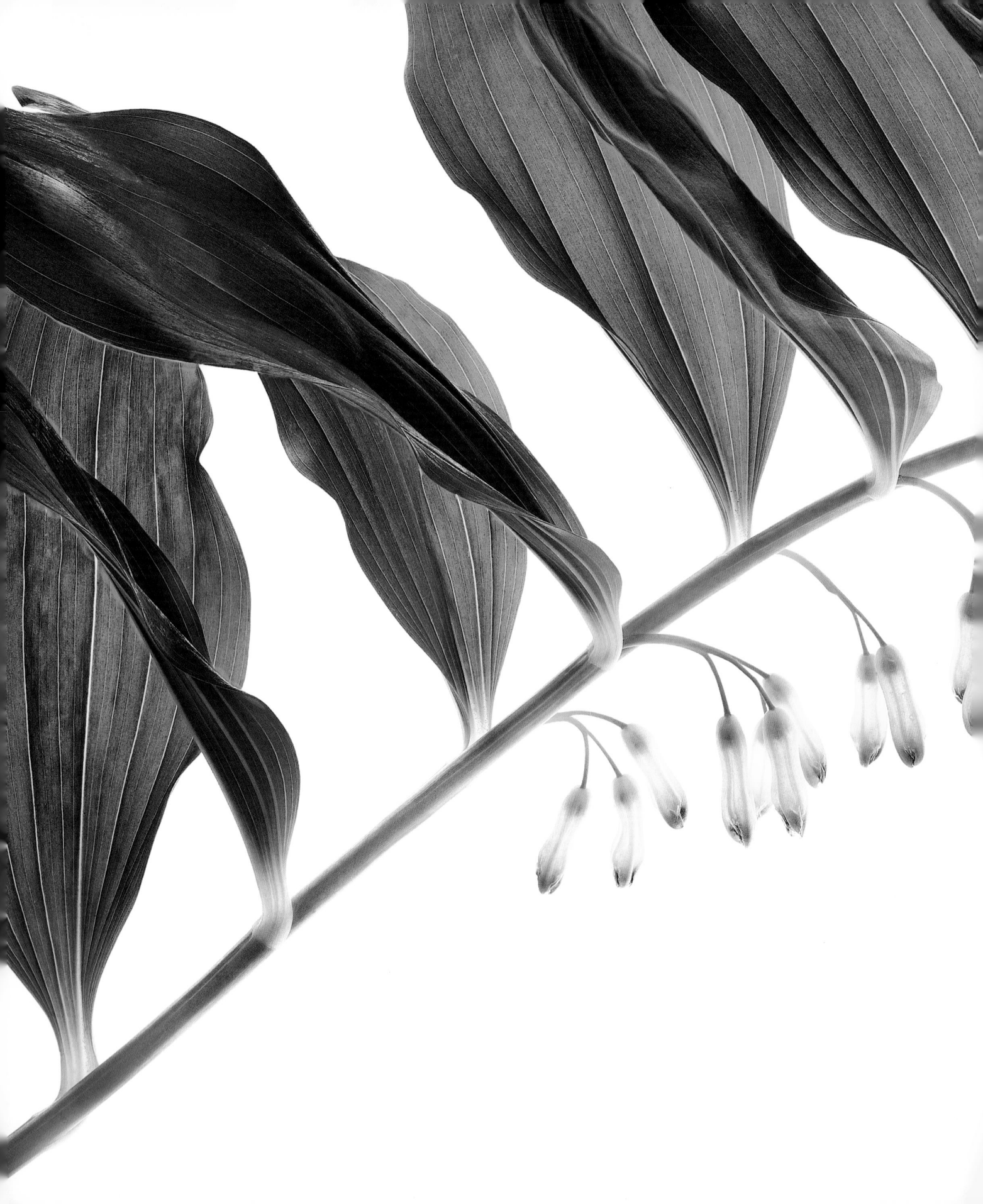

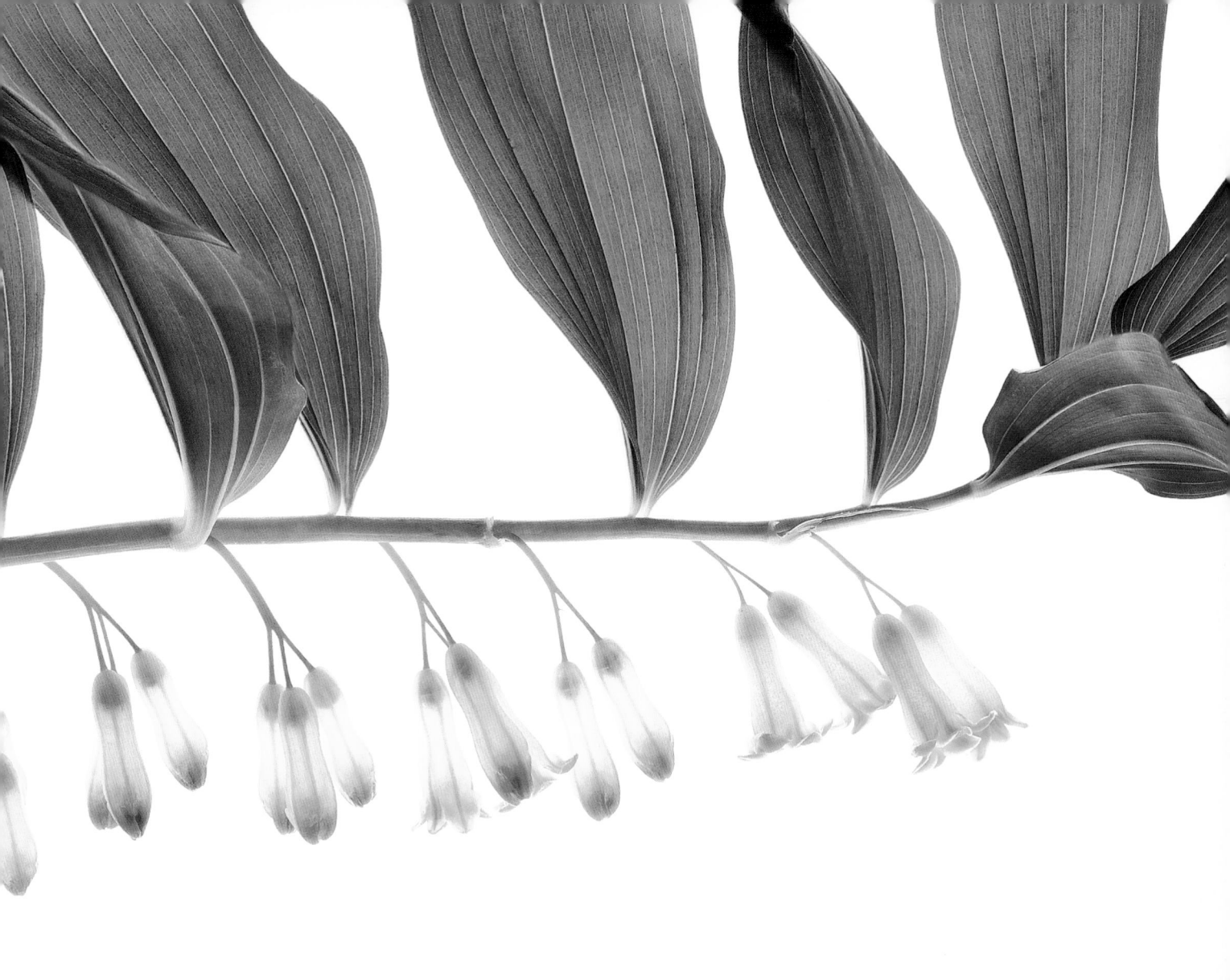

Solomon's seal, *Polygonatum odoratum*, is a springtime favorite. Theories about the origin of common name abound, including the round seal-like scars left by stalks from previous years, and the plant's long use and efficacy in treating, or sealing, wounds. Another possible explanation is that the six-petaled flowers, when they open fully, are pointed at the tips to form a Star-of-David-like configuration, which in early days was known as Solomon's seal.

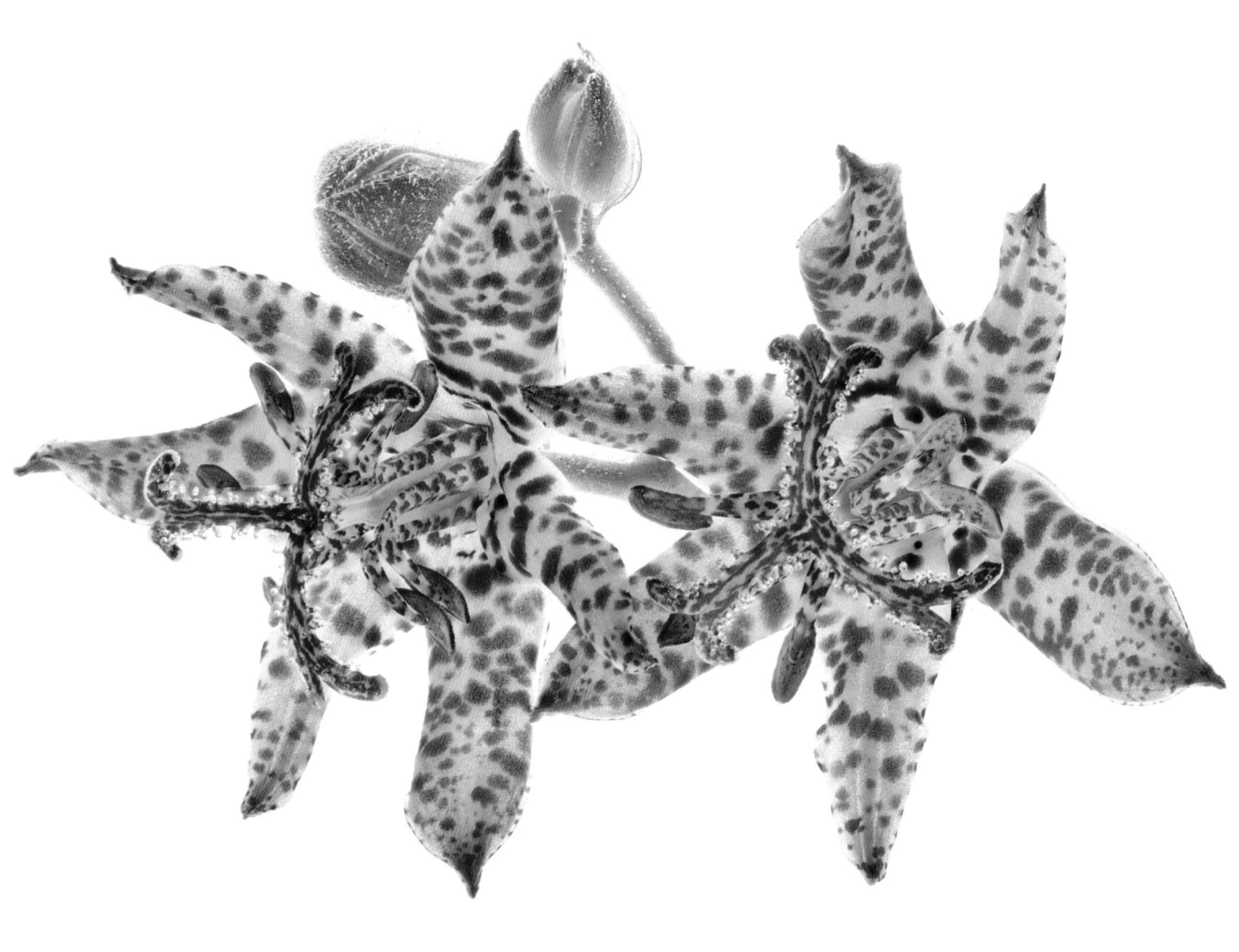

ABOVE The garish flowers of the toad lily (this species is *Tricyrtis formosana*) bloom in late summer to fall. Their wild colors and spots manage to attract pollinating insects active at that time.

RIGHT Fruit-scented and dependable, common grape hyacinth, *Muscari botryoides*, deserves a closer look. Here you can glimpse, just inside the urn-shaped flowers, the jazzy contrast of dark-hued stamens.

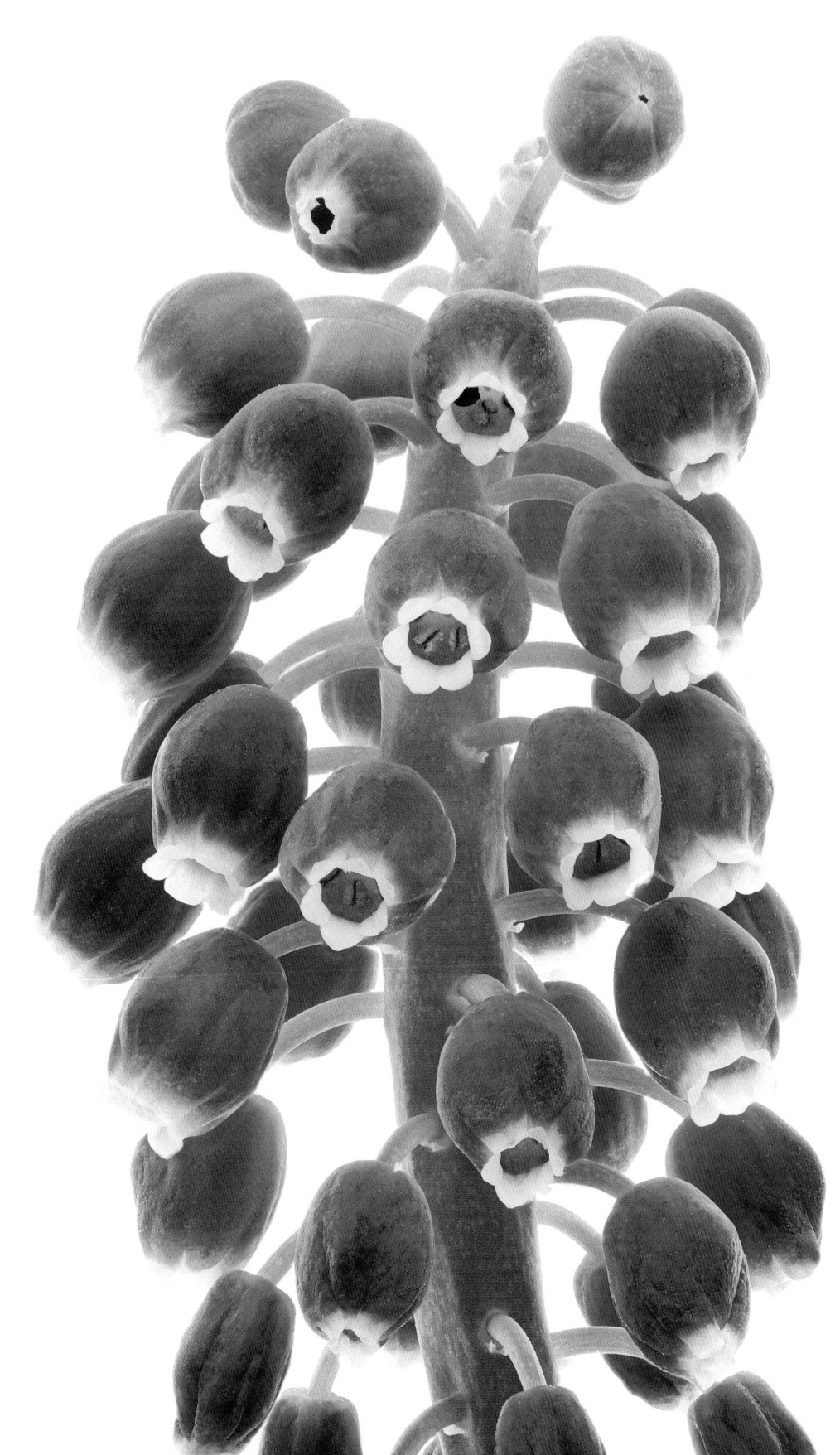

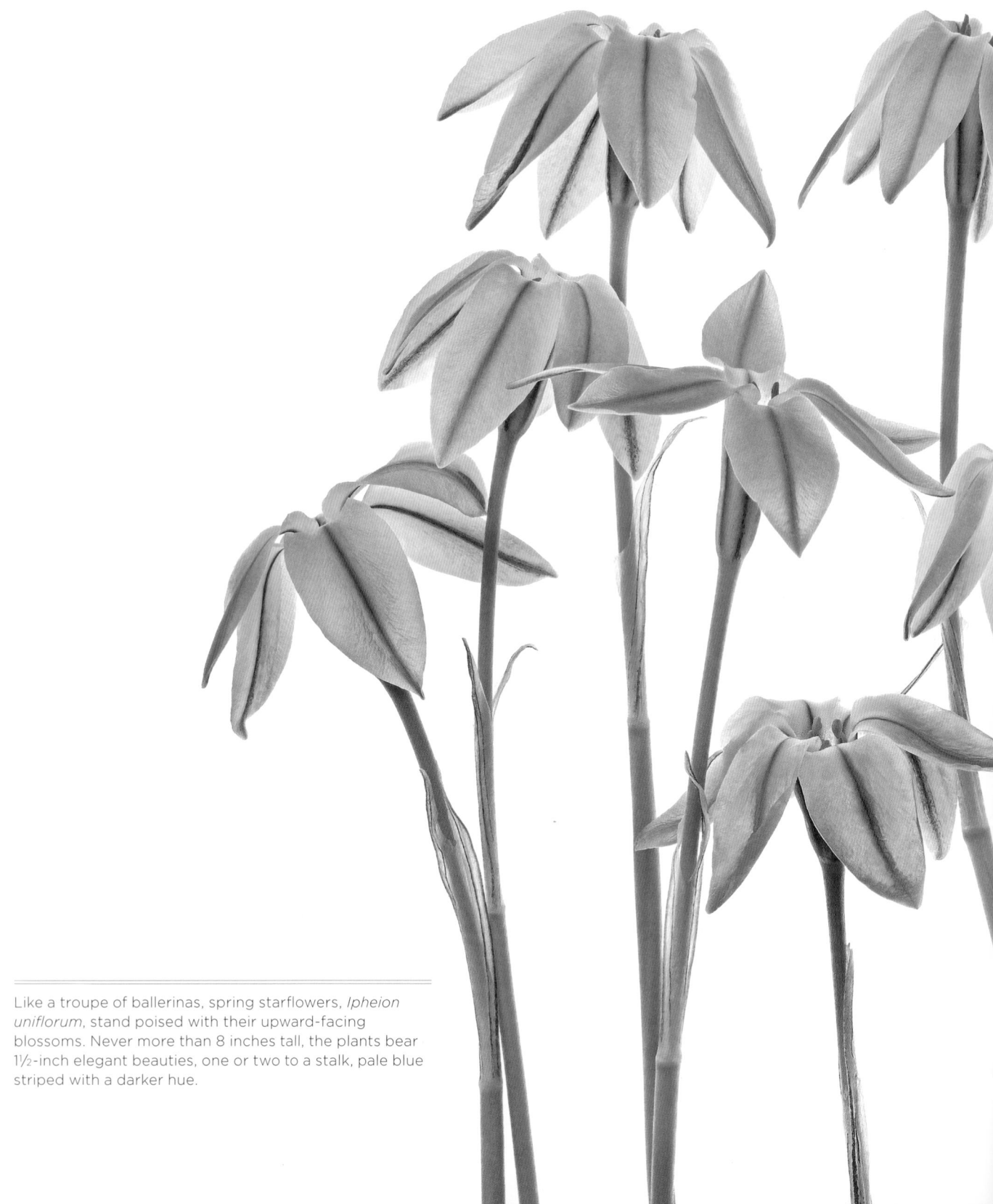

Like a troupe of ballerinas, spring starflowers, *Ipheion uniflorum*, stand poised with their upward-facing blossoms. Never more than 8 inches tall, the plants bear 1½-inch elegant beauties, one or two to a stalk, pale blue striped with a darker hue.

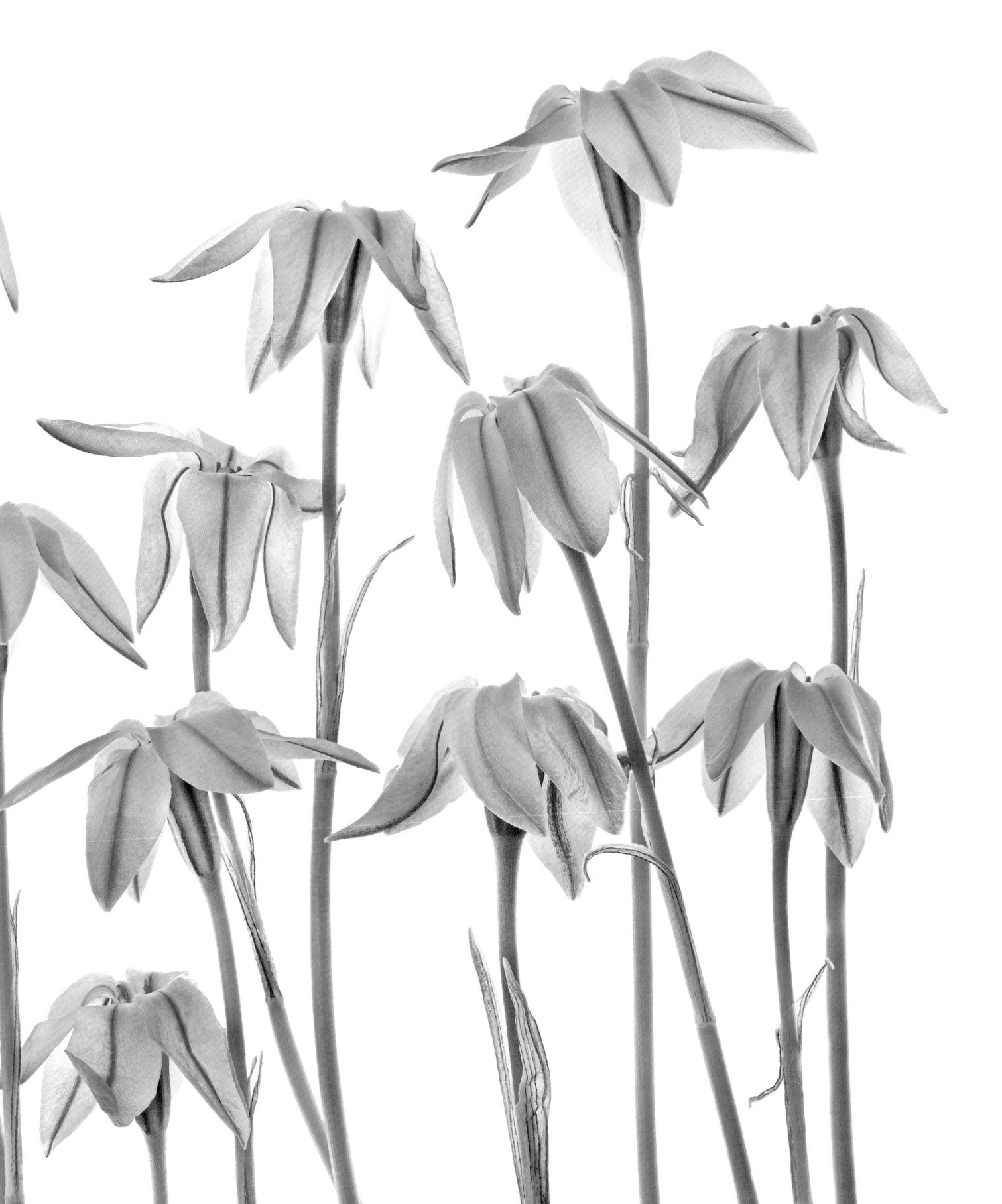

Drooping star-of-Bethlehem or silver bells, *Ornithogalum nutans*, has the six-parted flowers characteristic of the lily family and its immediate kin. Notice the green stripes on the petals, unique to this species.

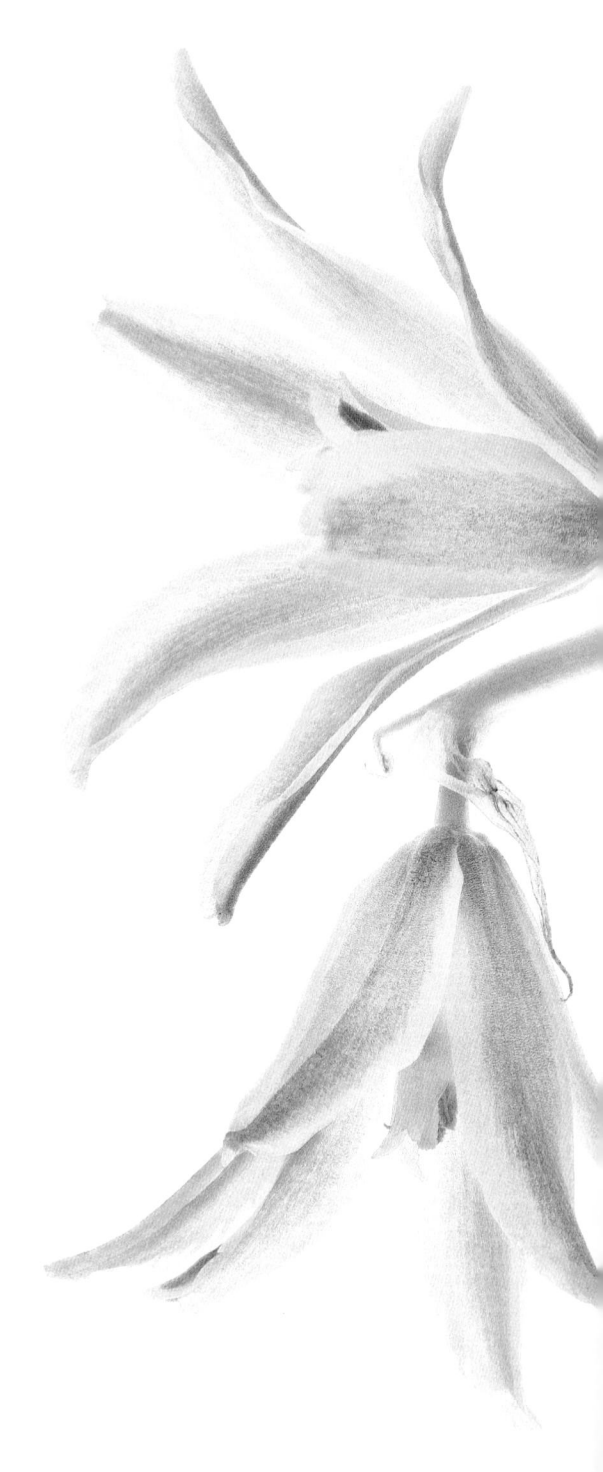

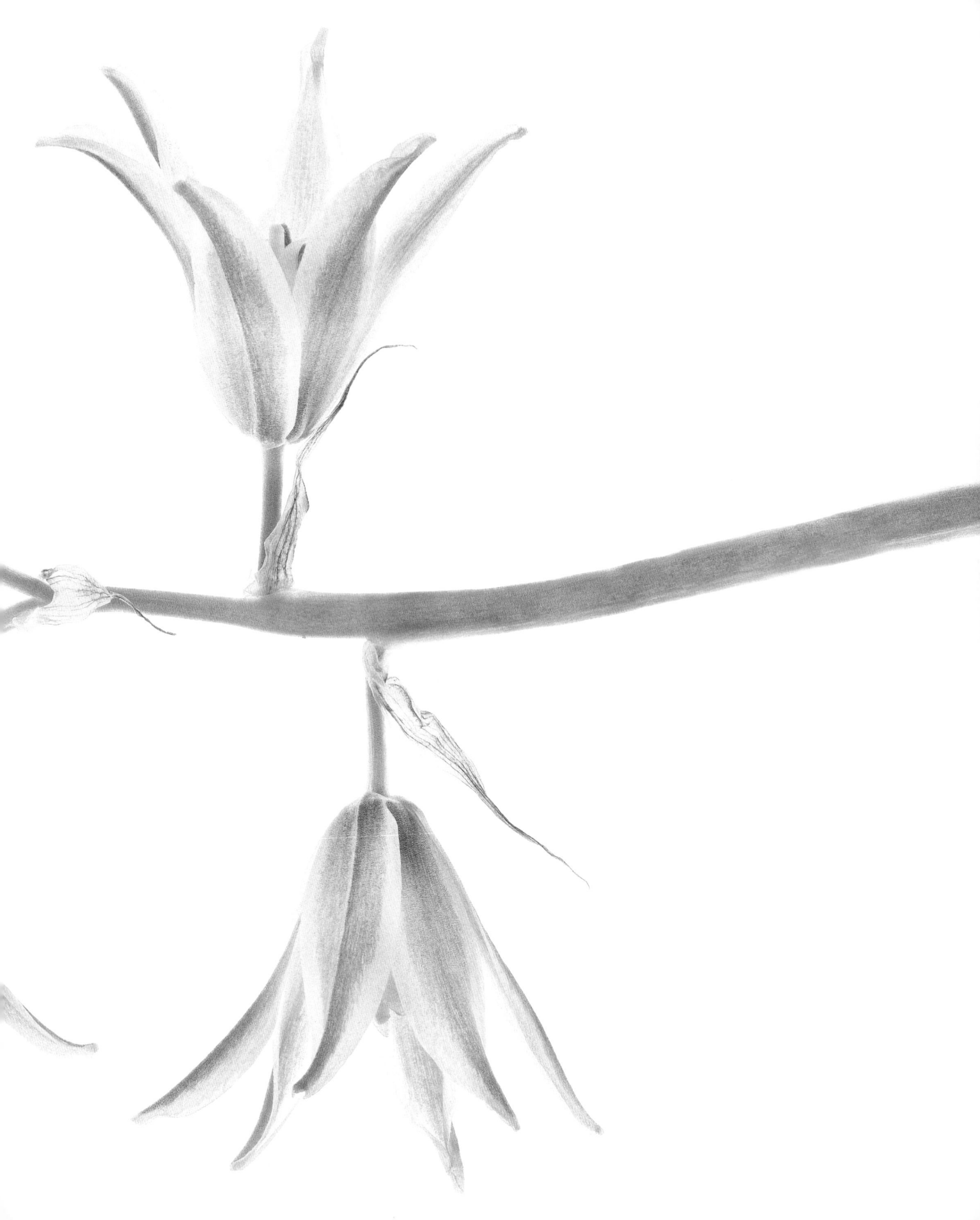

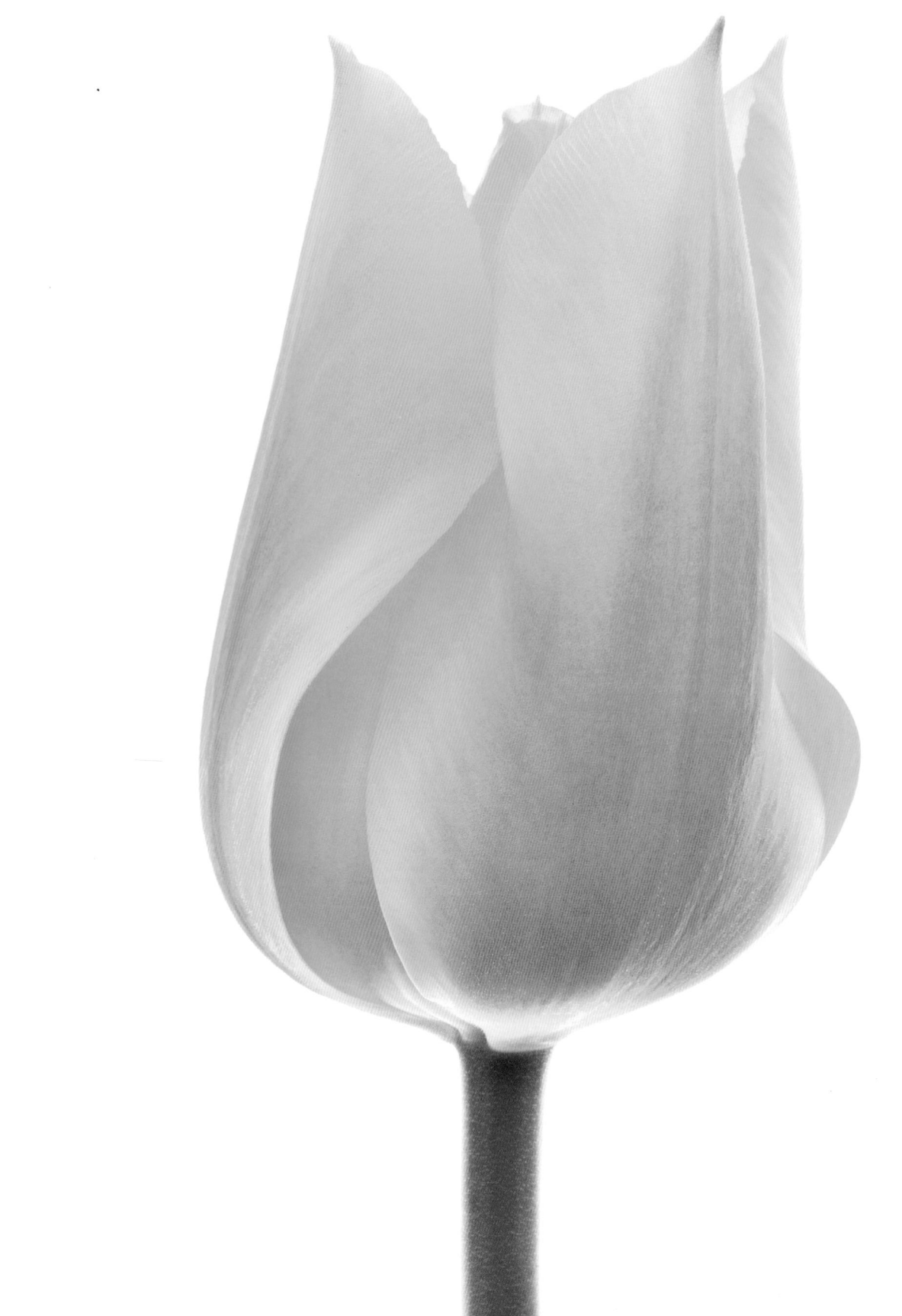

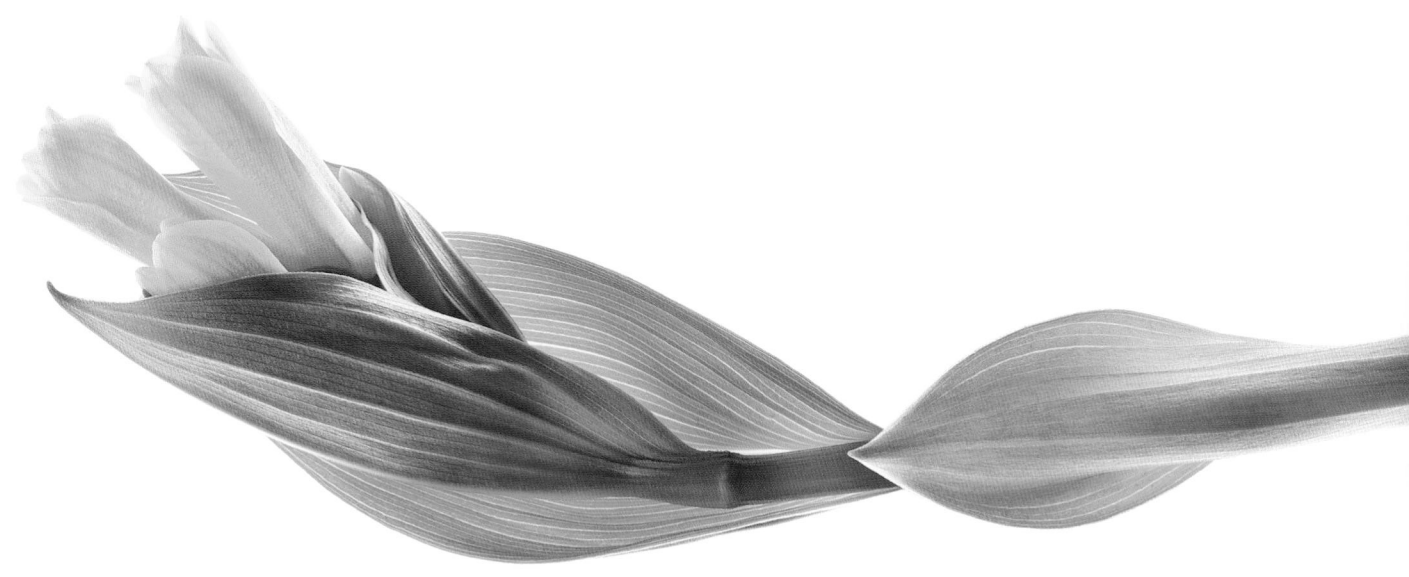

OPPOSITE All tulips are in the lily family, thanks to their six sleek petals, six sepals, and six prominent stamens inside (note also that tulip stigmas usually have three lobes). Beyond that, there is tremendous variation in colors, markings, petal as well as plant form, and bloom times, so much so that the experts have divided them into at least a dozen discrete groups. Thus there are Single Early and Double Early ones, plus Parrots, Kaufmannianas (Waterlily), Rembrandts, and more. Gardening tip: Do not let your tulips go to seed, which drains energy from the plant that would be better sent down into the bulb to fuel next spring's show.

Six slender petals and sepals, slightly twirled or twisted, form the sweet, delicate yellow blooms of the large-flowered bellwort, *Uvularia grandiflora*. The light green leaves, meanwhile, are smooth above and downy underneath.

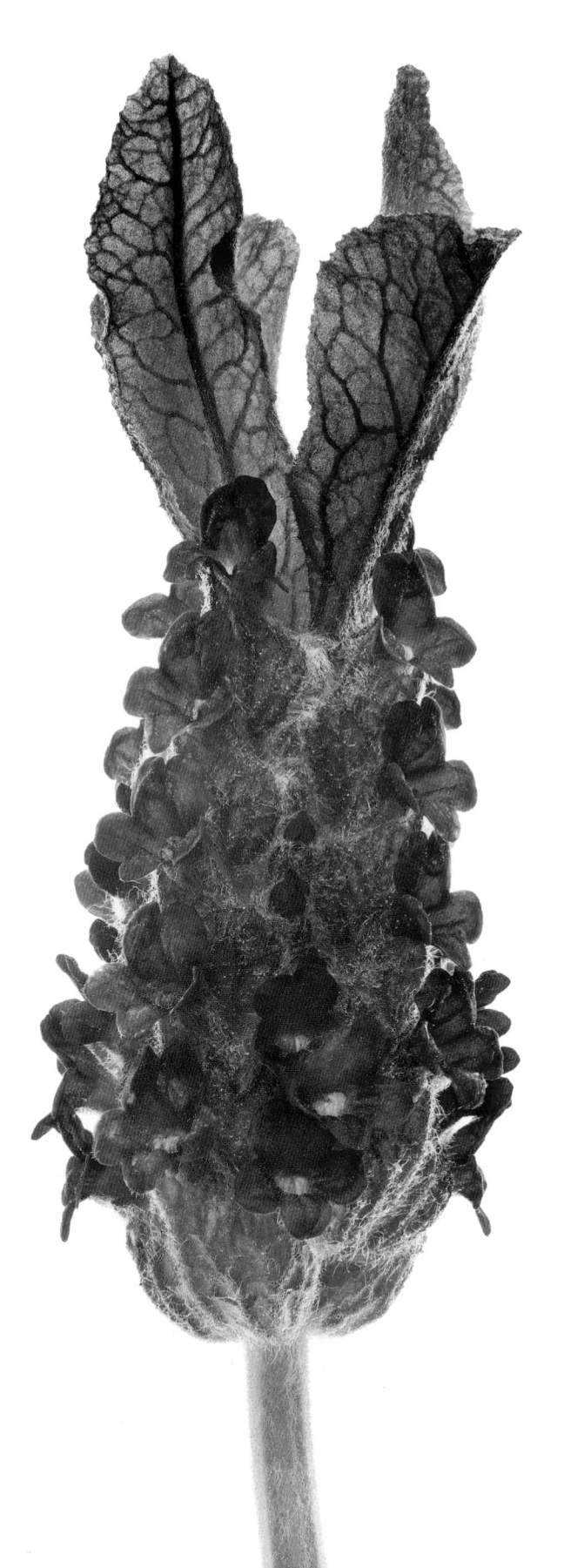

THE MINT FAMILY

Lamiaceae (or Labiatae)

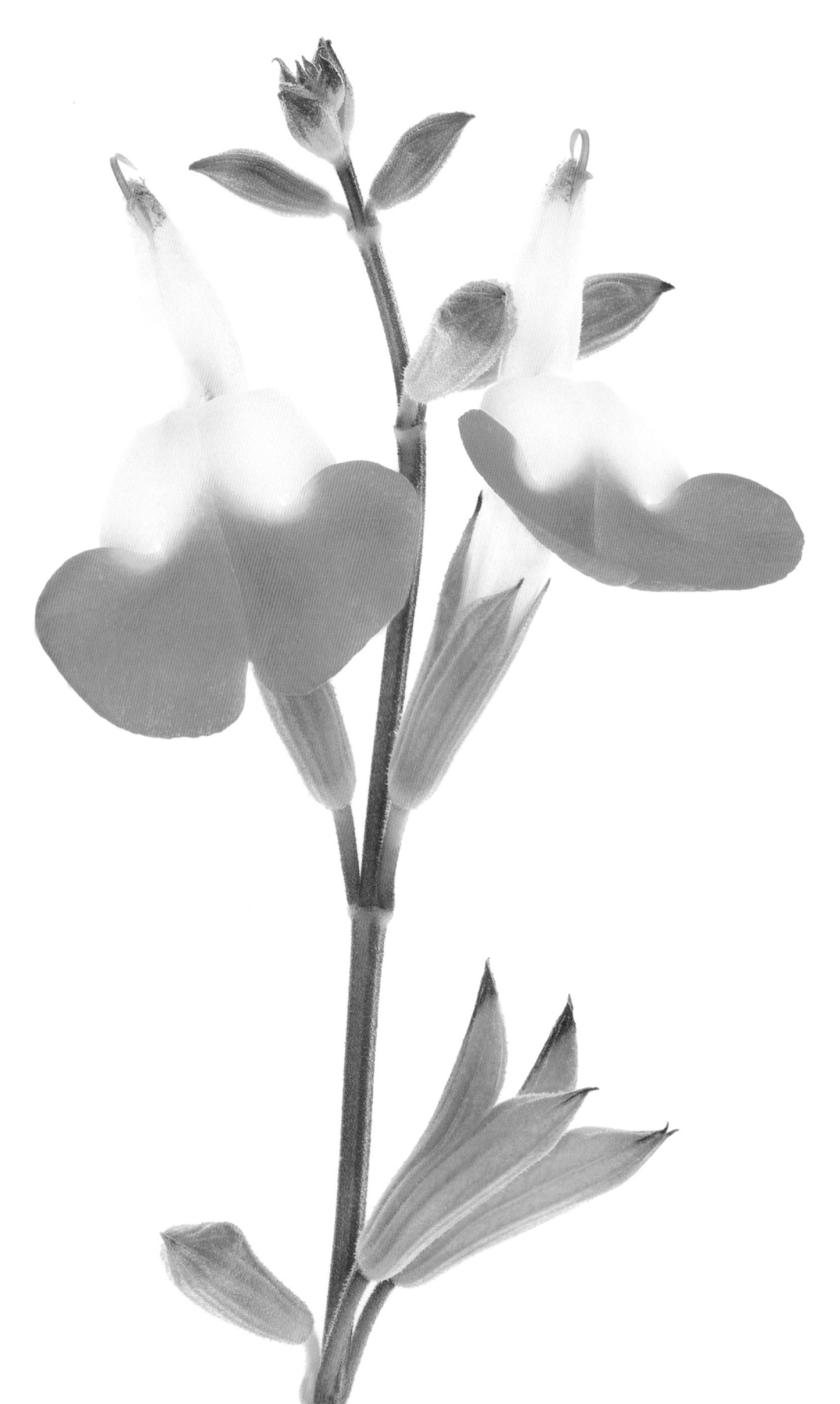

"Tennesseeans, famous for their mint
juleps, have a tradition of presenting
a new bride with a gift of mint plants
dug from the family garden."

SHARON LOVEJOY, *Trowel and Error*, 2002

'Hot Lips' is a snazzy bicolor selection of *Salvia microphylla*. The contrast allows us to appreciate the hooded, three-lobed upper lip and the spreading, two-lobed lower lip so characteristic of these flowers. They have two delicate stamens each.

PAGE 200 French lavender, *Lavandula stoechas*, looks different from other lavenders, but those perky extra petals emerging from the top are actually (sterile) bracts. If you watch, the bees browse lower, among the true flowers.

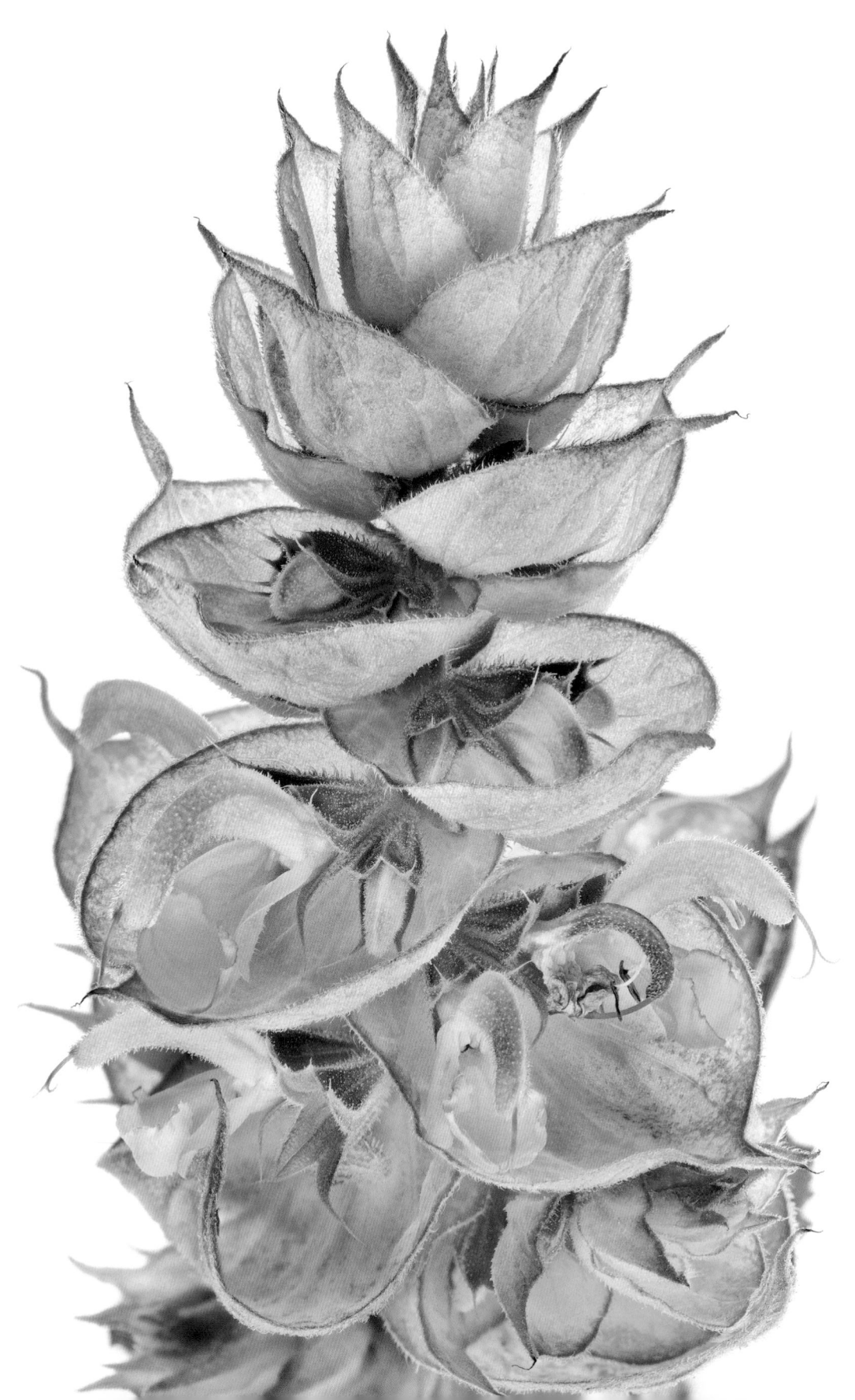

Chef Mario Batali's recipes typically call for the freshest of ingredients, including his famous Mint Love Letters, which are ravioli stuffed with chopped mint, shelled sweet peas, and a generous amount of Parmigiano-Reggiano cheese. The mint makes all the difference, as its refreshing, sweet tang gives the dish a cool elegance.

Because *Mentha* grows almost everywhere in the world, it is found in the foods and drinks of many cultures, from India to America. Mint tea, mint punch, mint jelly, mint sauce, and mint shredded into green salads, fruit salads, and soups hot or cold—the possibilities are legion. Closely related peppermint, *Mentha ×piperita*, and spearmint, *M. spicata*, are the two most commonly used species. Others include apple mint, Corsican mint, curly mint, woolly mint, orange mint, water mint, ginger mint, pineapple mint, mountain mint, yerba buena, and many other closely related and allied species, varieties, and odd hybrids.

It is interesting that mints have been in cultivation, passed around, and tampered with for so long that their seeds simply do not "come true"; that is, if you harvest

LEFT Most unusual looking among the many sages and mints is clary, *Salvia sclarea*. Heart-shaped, balsam-scented leaves may capture your attention first, but the blooms are intriguing. Loose spikes feature true salvia flowers interspersed with large, colorful bracts.

BELOW These are the flowers of Japanese catmint, *Nepeta subsessilis*. Typical of the mint family, they are in the blue to purple color range, tubular, and small, gaining impact only because they are borne in spires. All plant parts contain the compound nepetalactone, the scent of which interacts with the nasal passages of most cats to inspire temporary euphoria.

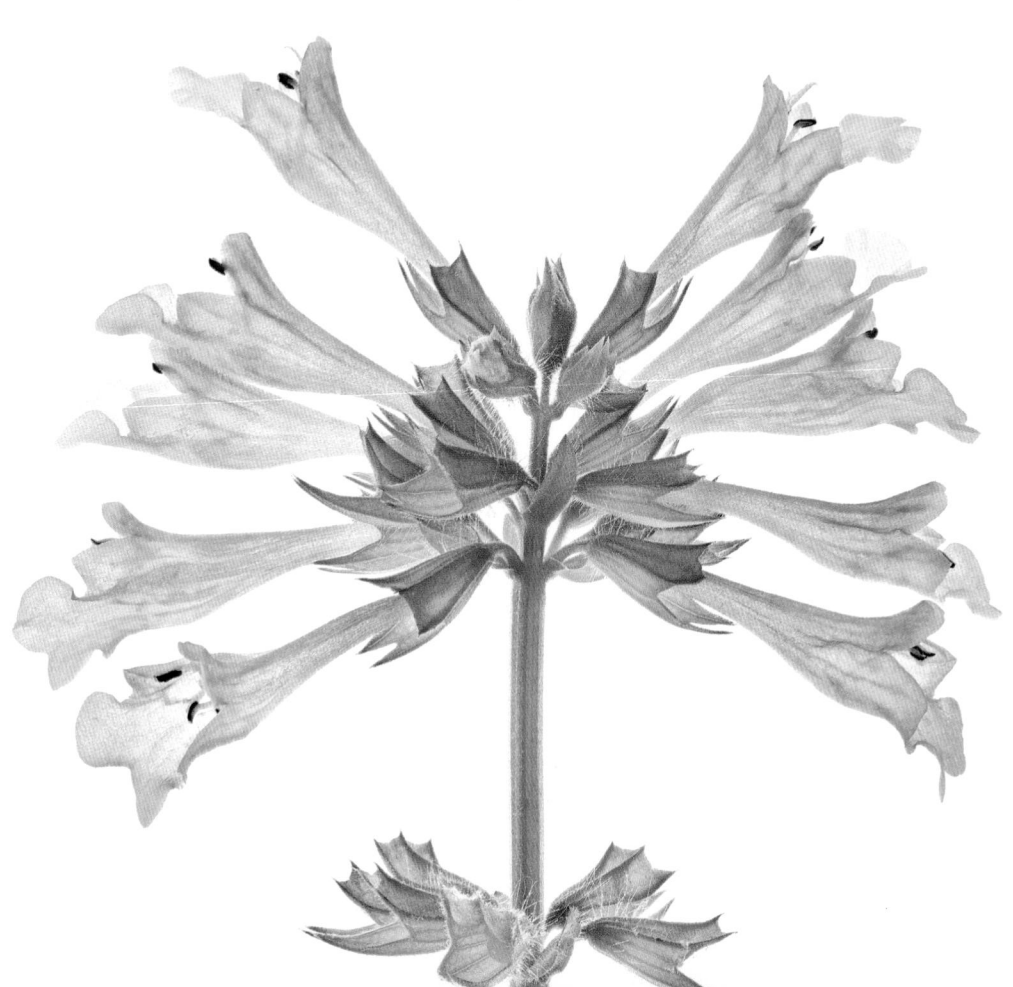

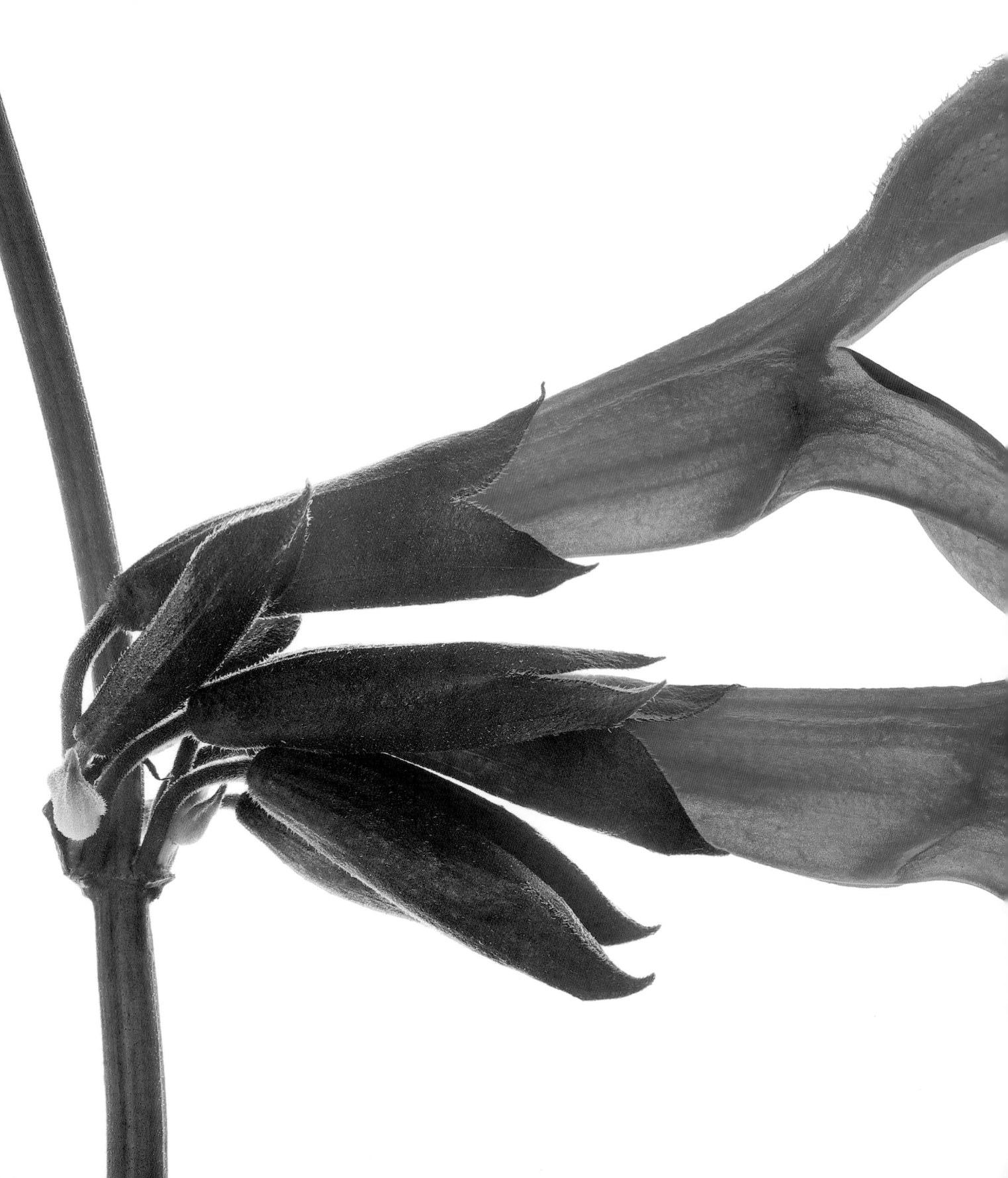

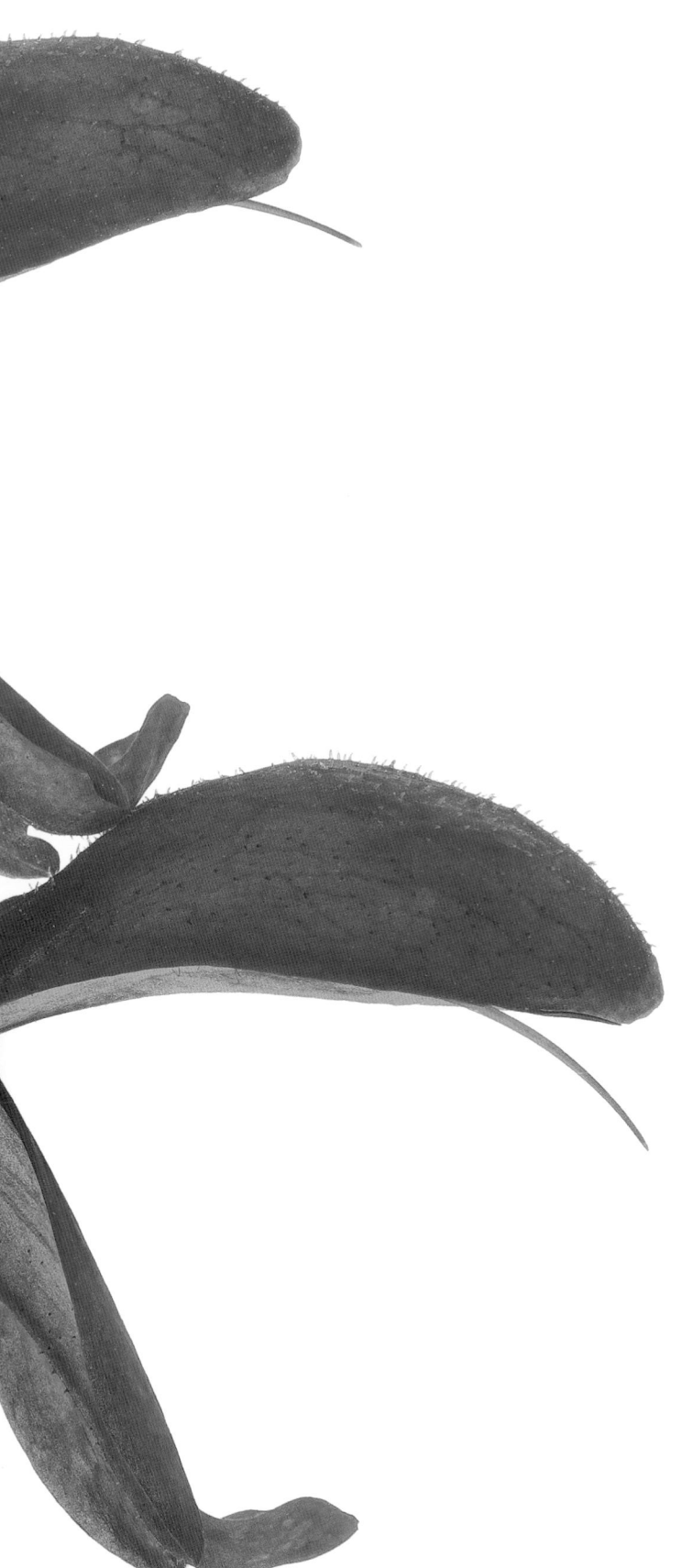

and plant mint seeds, you won't get the same plant. For this reason, nurseries and avid gardeners simply raise favorites from root or stem cuttings. Luckily, this is easy.

Rugged, handsome, aromatic mint leaves also have medicinal benefits, which adds to their appeal. Herb books typically list mint as a soothing digestive aid, gas reliever, and stimulant. The volatile oil, thymol, acts as a mild anesthetic to the stomach wall and helps reduce nausea and vomiting. Mint also freshens breath, which is why it is a popular ingredient in toothpaste and mouthwash. These plants are both good and good for you.

Mints have landscape value, provided you are able to control their often rampant ways. Most spread by stolons (leafless horizontal stems) above or below the soil surface. The plants tend to thrive in full sun and moist ground, often to the point of outgrowing their bounds and wearing out their welcome. Where abundant growth becomes a problem, you can harvest mints for the above-mentioned uses, give plants away, or simply tear them out and discard them. Rein in overeager growth by raising your favorite mints in pots, either aboveground or sunk into the garden, or by sinking down deep, substantial barriers to confine the spread of their root systems. Alternatively, try growing mints in somewhat drier settings. But overall, their lush, green growth is an attractive sight. They look fine in the company of almost any colorful flower, annual or perennial—as well as with other herbs, of course, in the garden, in pots, or in window boxes.

Mints are the flagship members of the large family Lamiaceae or Labiatae. Quite a few aromatic, delicious, and/or medicinal herbs are grouped with them. Some of these are quite familiar, including basil, lavender, marjoram, rosemary, sage, and thyme. Others may be less

So-called hummingbird sage, *Salvia guaranitica*—hummingbirds seem to like all salvias—has the most vivid, true-blue flowers imaginable. Like all mint family members, it has square stems and aromatic leaves.

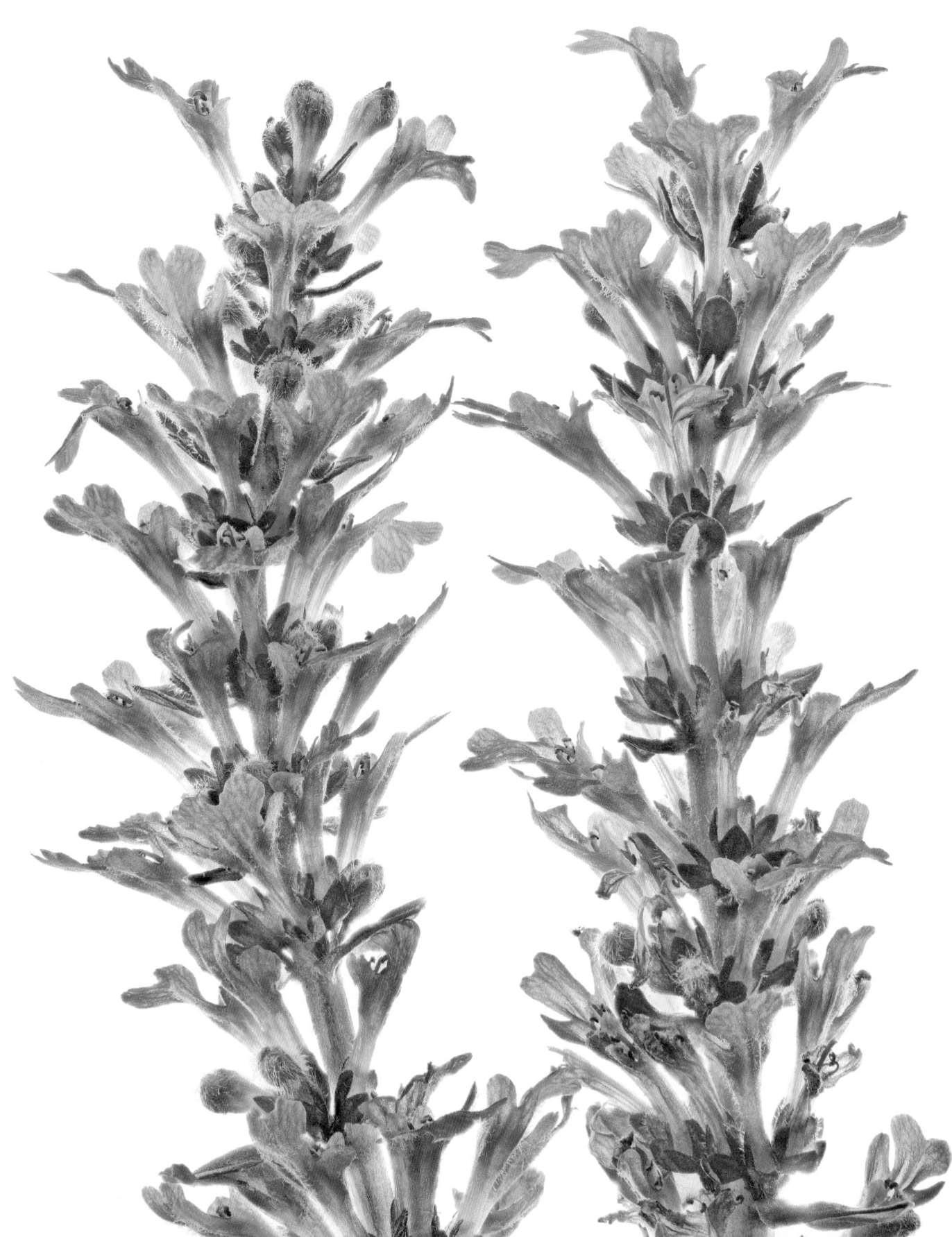

so, such as lemon balm, horehound, hyssop, catmint, and Oswego tea.

With few exceptions, certain distinguishing physical characteristics are common to members of this family. Practically all have square stems, a feature easily checked. If someone remarks that your lamium (*Lamium maculatum*), lamb's ears (*Stachys*), or coleus is a member of the mint family, bend down and run your fingers along a stem to verify the shape. Does lavender (*Lavandula angustifolia*) fit the bill? Yes. How about beebalm (*Monarda didyma*)? Again, yes. As further confirmation, you can check that the leaves are paired in opposites along the stems, or whorled.

You probably won't grow mint, lamium, or lamb's ears (or coleus, for that matter) for the blooms, but it is worth noting that the flowers of this family also share common characteristics. They are often rather small, so you have to come in close to observe, but your curiosity will be rewarded. Mint family flowers are symmetrical, with parts in fives—five united petals, five united sepals. Typically petals are fused into an upper and a lower lip. Various types of bees thus find it easy to land, gain easy access to the nectar, and pollinate many of these plants. And if you grow the red-flowered monardas or salvias, ruby-throated hummingbirds will likely visit as well, and maybe even some flitting butterflies.

LEFT Another, less commonly seen catmint is the Siberian one, *Nepeta sibirica*, whose blossoms are displayed in whorls. Individual flowers are large, up to 1½ inches long, and dramatically flared. The plant smells strongly of cinnamon, which seems to repel deer, rabbits, and, yes, cats, but is enchanting to many humans.

BELOW "Screeching red salvia" is how one horticulturist sneeringly dismissed the popular bedding plant *Salvia splendens*. And yet up close, its bright, long-tubed flowers show off strong, keen color and admirable symmetry.

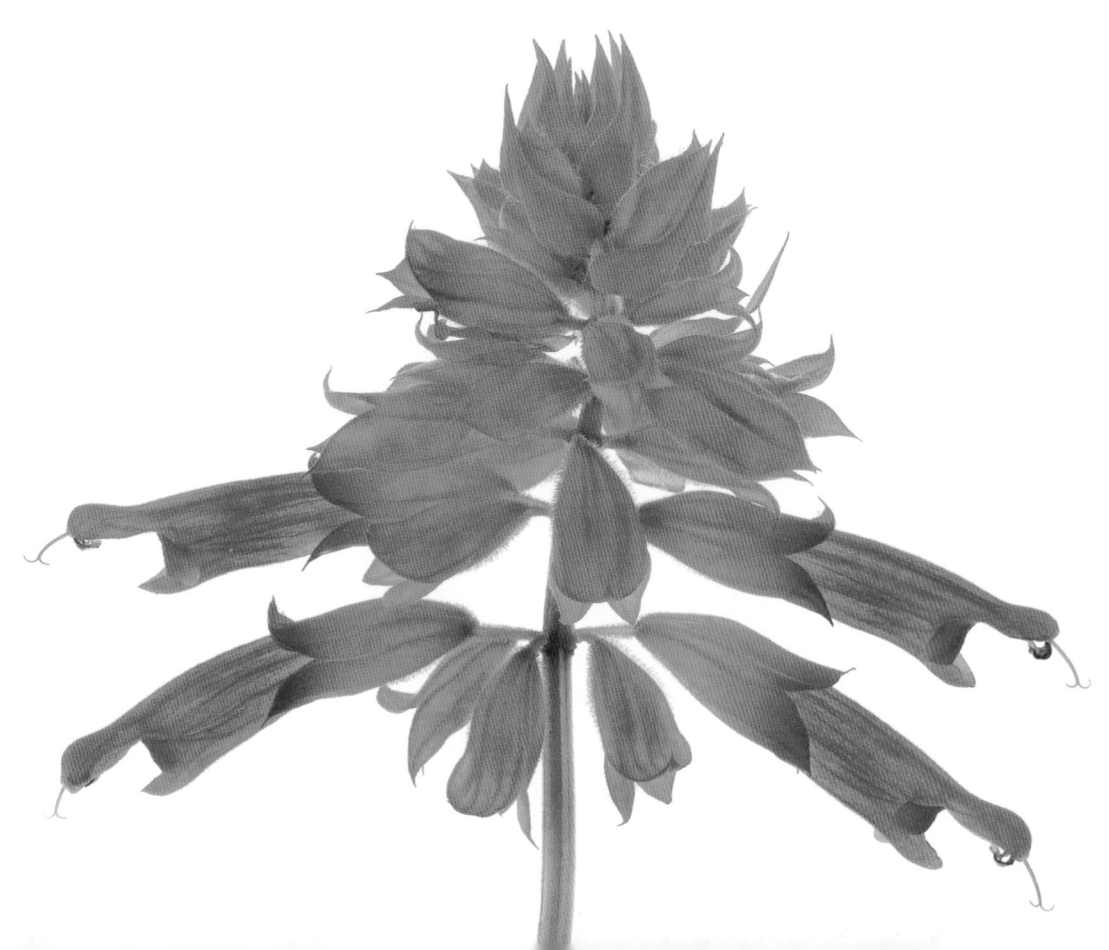

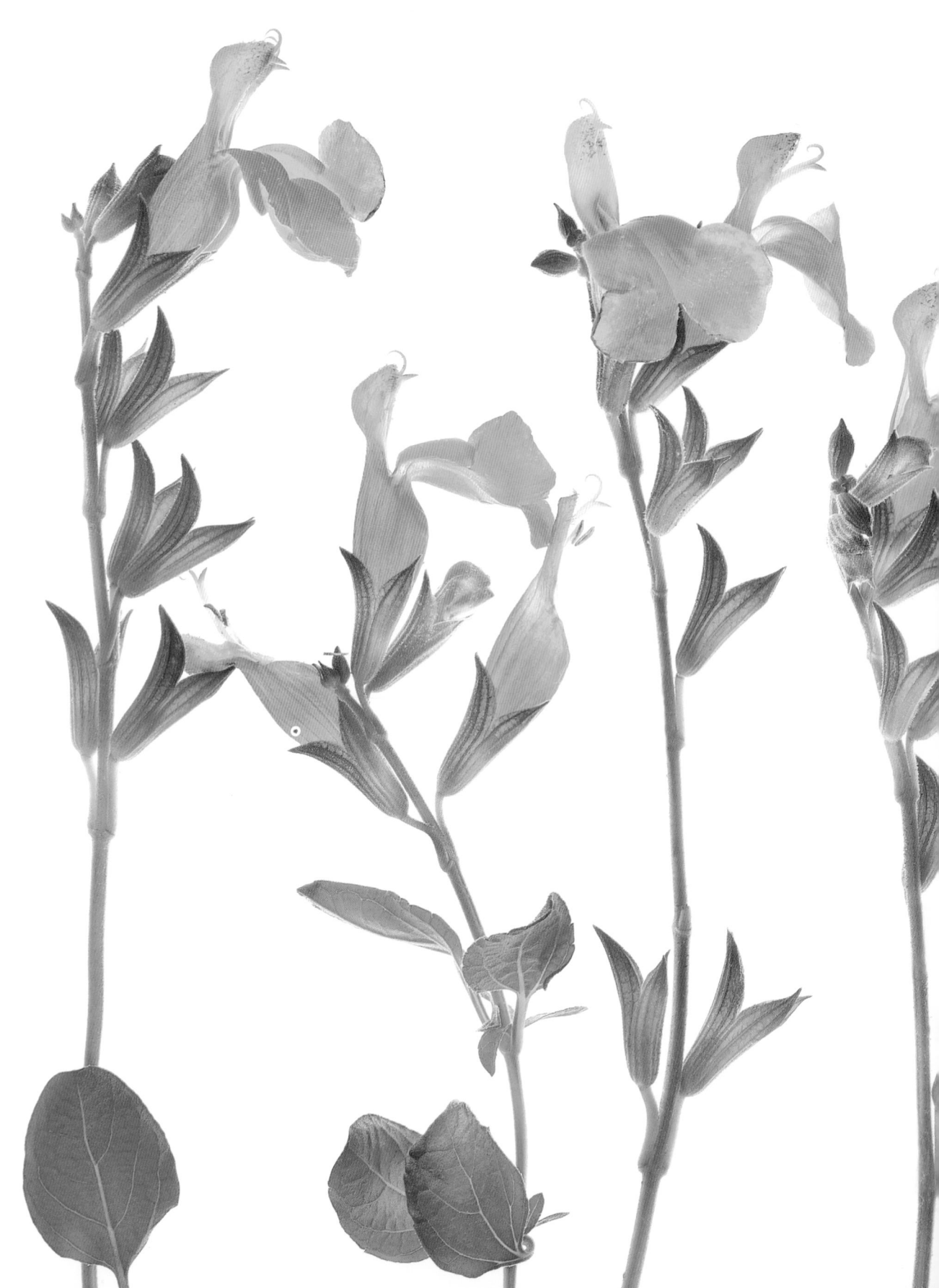

A close relative is *Salvia greggii*, autumn sage. In the wild, it is really variable in terms of color, running the gamut from the pink shown, to red, lavender, purple, rose-orange, and white.

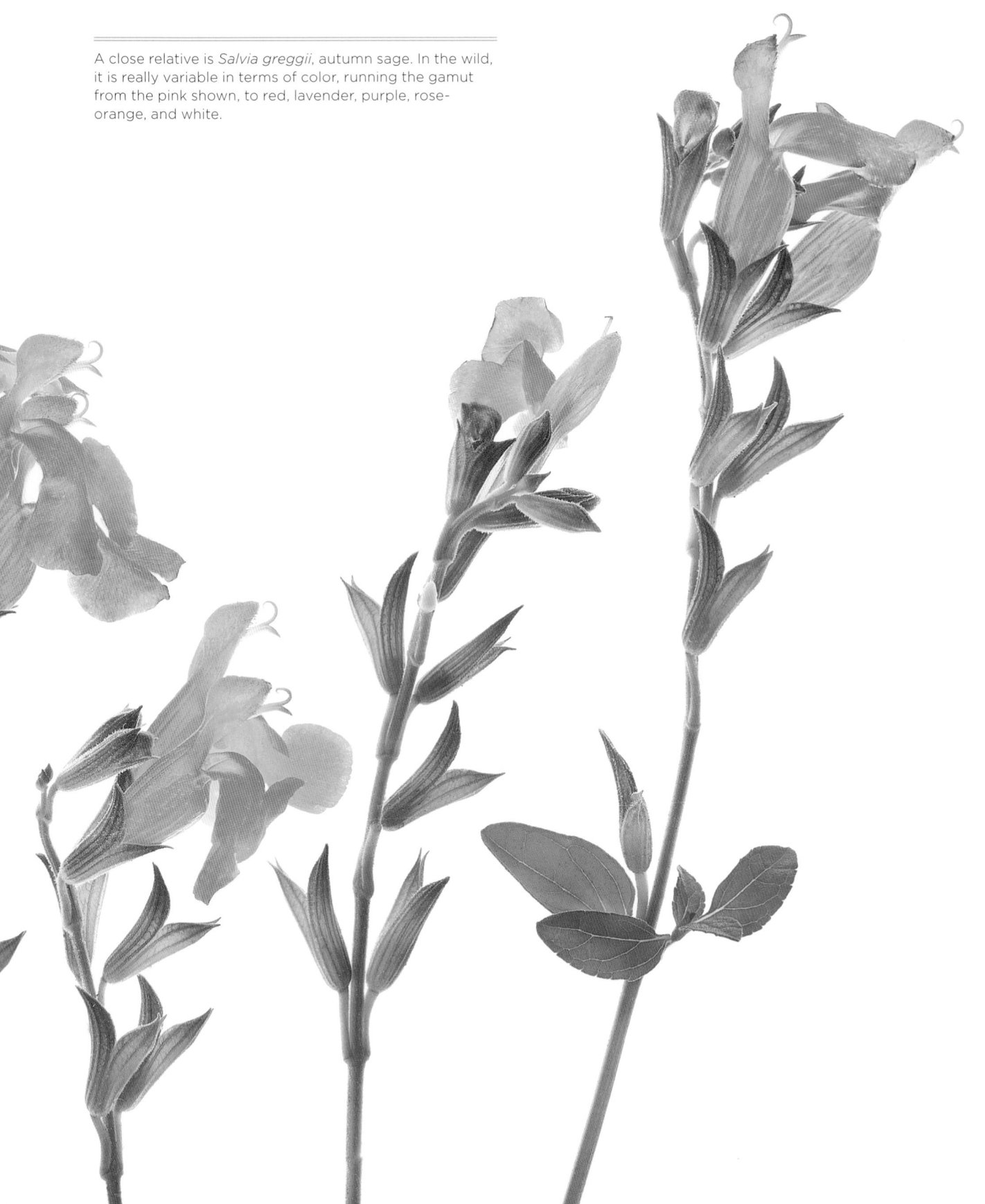

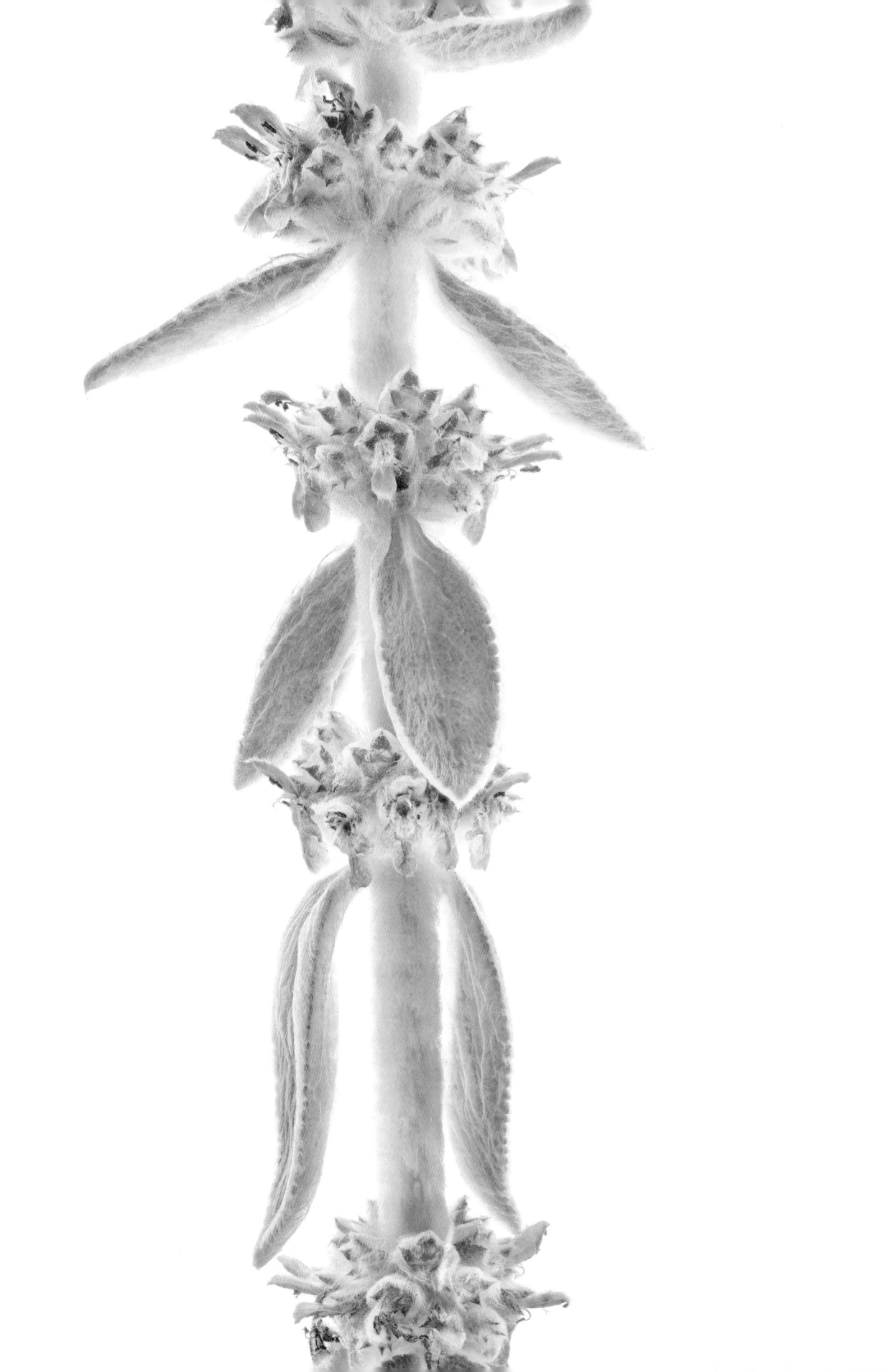

What about lavender's flower spikes? These are actually made up of numerous tiny flowers, which are easily separated off for closer inspection. Technically, each little lavender flower consists of the base, or calyx, that comes first and then opens to the whorl of petals collectively called the corolla. Peer in or use a magnifying glass: they are indeed two-lipped. Hobbyists who make lavender arrangements, pillows, and the like have discerned by trial and error that the flowers are at their most aromatic when the corollas are fully open. If you prefer to enjoy spires of fresh lavender in bouquets, however, pick a bit earlier so they can continue to open indoors.

With fragrance and easy growth as assets, it is not surprising that plant breeders have waded into the mint family to discover or encourage improvements. The splashy groundcover *Lamium galeobdolon* 'Hermann's Pride' comes to mind immediately. Its jagged strong green leaves are handsomely accented with silver, and joined by masses of cheery yellow flowers for weeks in late spring and early summer (most lamium blossoms tend to be in the pink-purple color range). The lamb's ears cultivar *Stachys byzantina* 'Big Ears' really earns its name with 8-inch velvety soft leaves (it also goes by the name 'Countess Helene von Stein', but her ladyship may not have been flattered). On the other hand,

LEFT Many gardeners snip off the flower spikes of lamb's ears, *Stachys byzantina*. But if you let them stay, those wee, purple-pink, mintlike flowers draw bees and the occasional hummer. Pollination is obviously a cinch, for an untrimmed plant sows offspring all around it.

BELOW The slender, tubular yet two-lipped flowers of beebalm, *Monarda didyma* (bright pink in this cultivar, 'Pink Lace') are crowded together in headlike clusters, with leafy bracts below. Each individual flower offers pollen on two projecting stamens, while the sweet nectar inside attracts bees, hummingbirds, butterflies, and others.

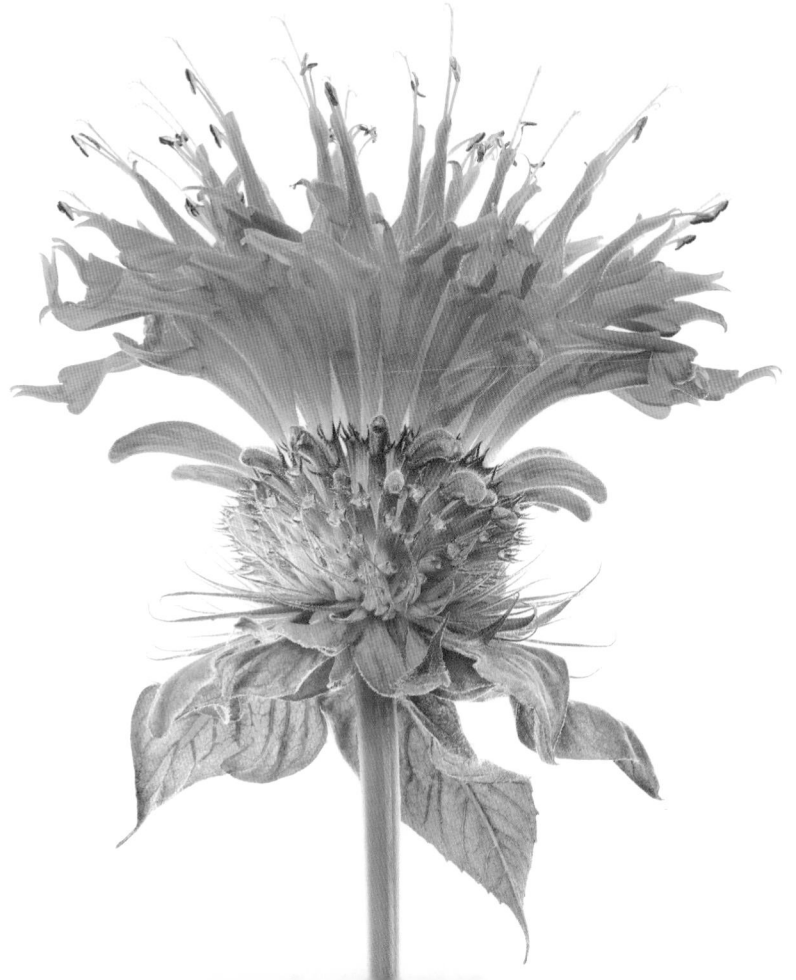

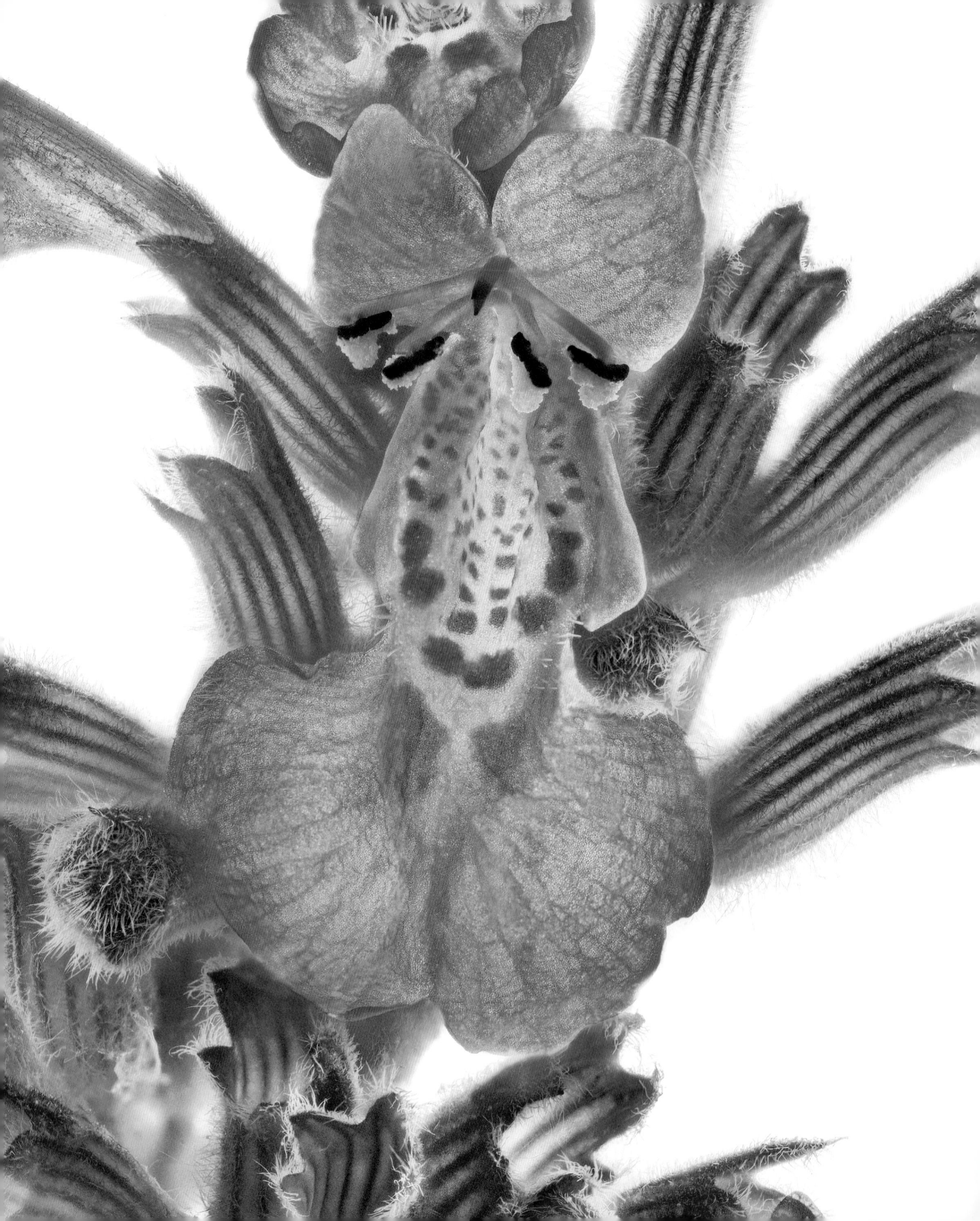

if you're always cutting the flowers off your lamb's ears, you might want to track down the flower-free cultivar 'Silver Carpet'. And for those who relish brilliant colors in their mint family plants, there are some sensational *Monarda didyma* cultivars, bigger and brighter and better than the original species in every way, such as

vivid pink 'Coral Reef' and deep red 'Jacob Cline'. Just remember, as with their vigorous lower-growing mint relatives, any of these plants can outgrow their bounds when supplied with ample sun and moisture. So try to place them where their lusty growth will be welcome, and a joy to behold.

LEFT The spikes of English lavender, *Lavandula angustifolia*, are actually whorls of numerous tiny flowers. These still conform to the mint family template of a two-lipped but five-lobed corolla backed by a five-part calyx. There are four miniscule stamens.

BELOW Another nepeta with the characteristic two-lipped flowers, *Nepeta racemosa*. Further tip-offs that it is a mint family member include square stems and aromatic foliage. The way the flowers are carried in this species are considered not whorls or racemes but rather "verticillasters," which simply refers to the fact that they are borne in intervals in the axils of opposite bracts.

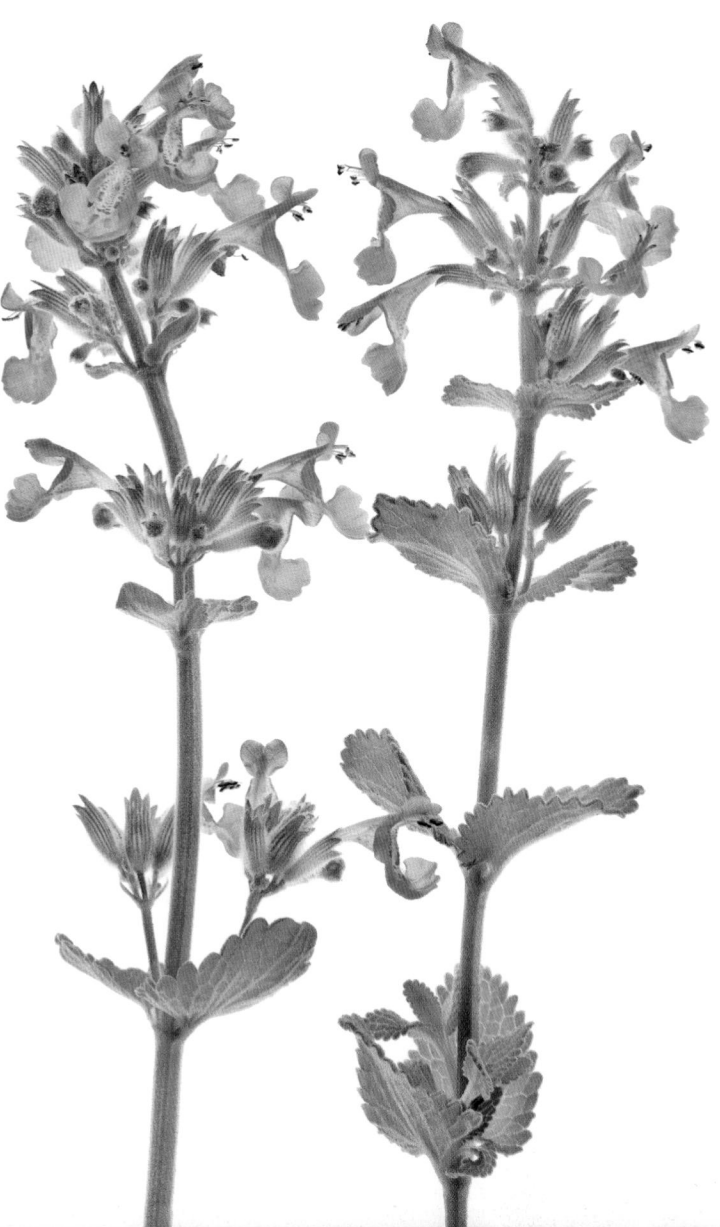

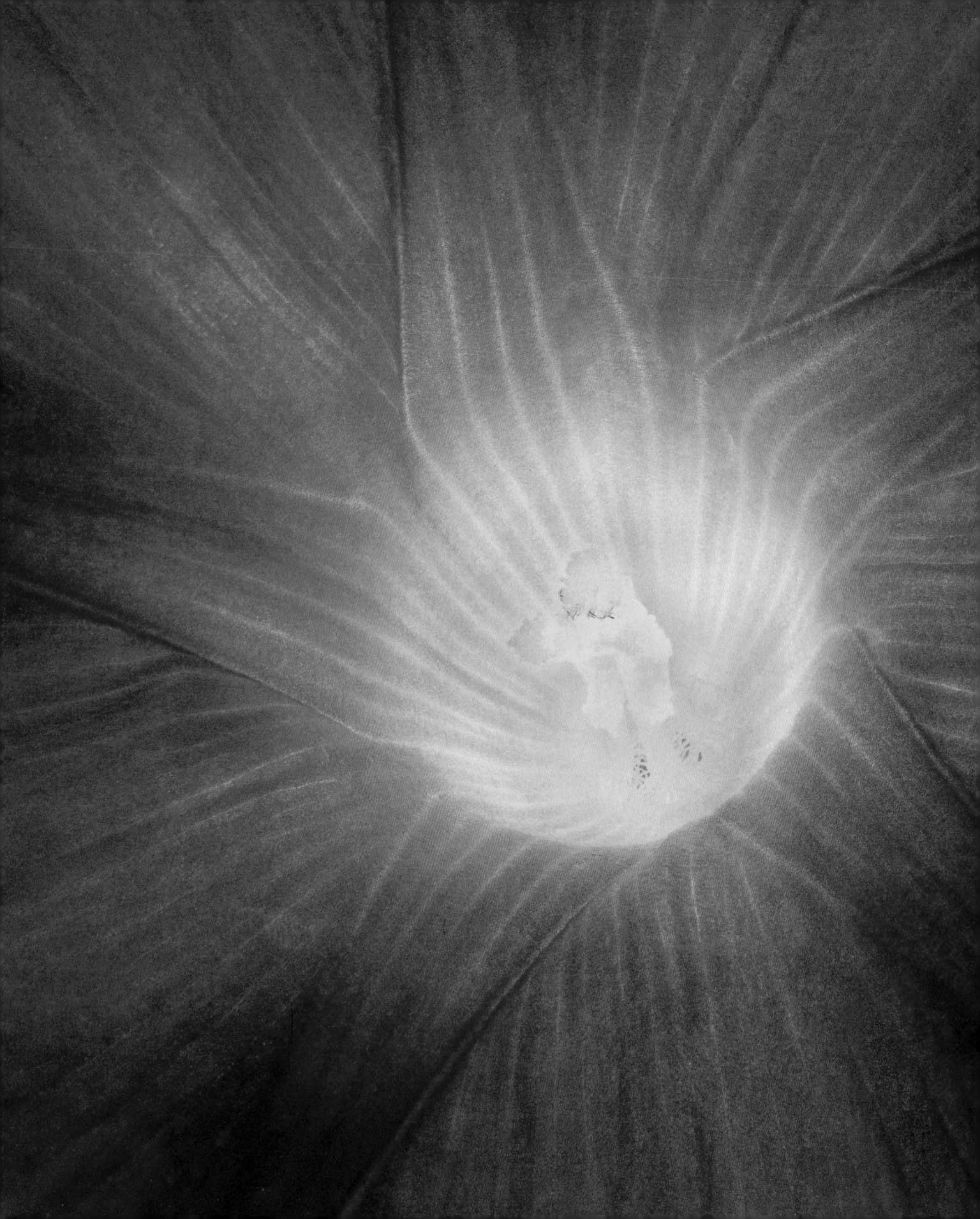

THE MORNING GLORY FAMILY

Convolvulaceae

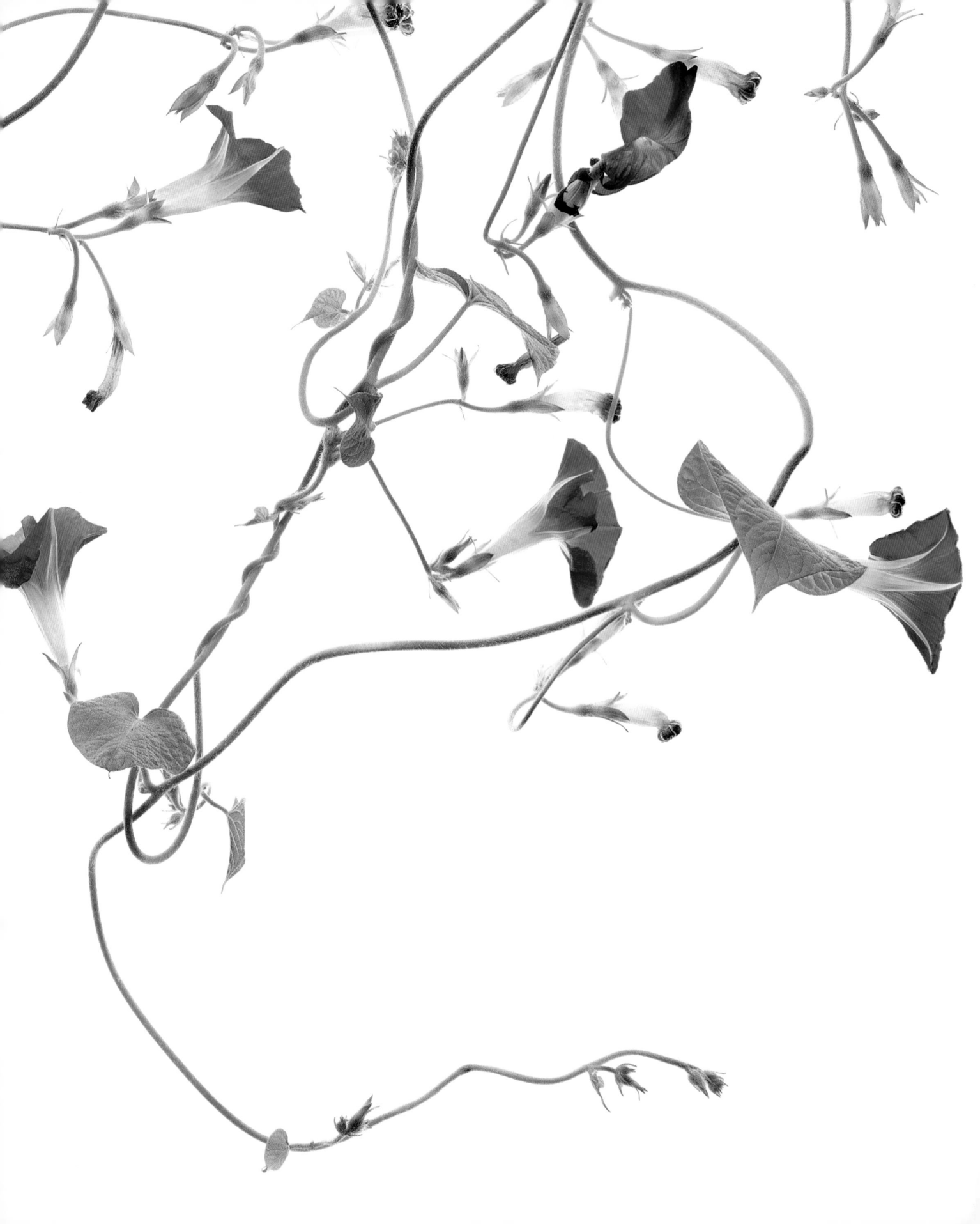

"A morning-glory at my window satisfies
me more than the metaphysics of books."

WALT WHITMAN, "Song of Myself," 1855

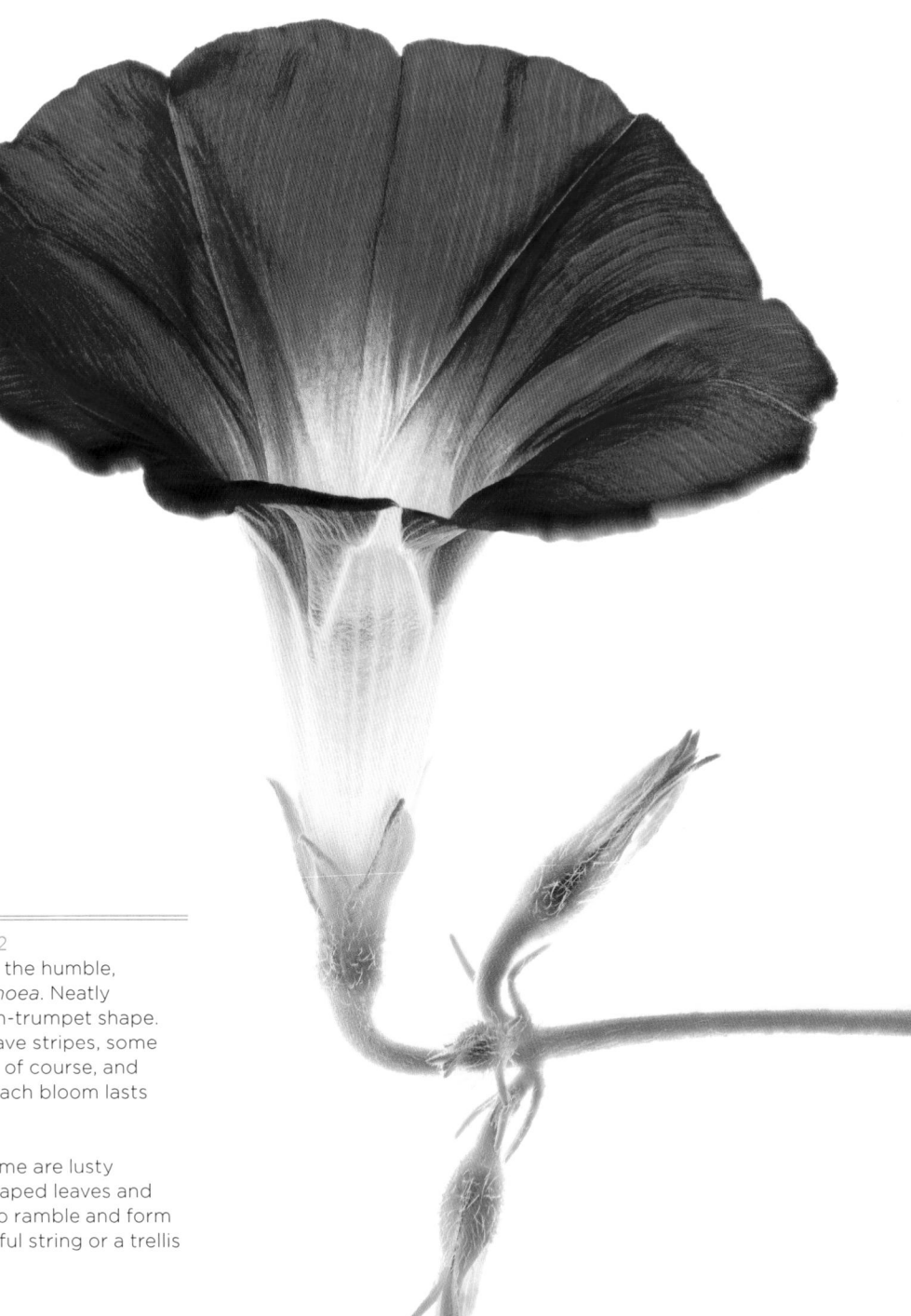

RIGHT AND PAGES 216, 220, AND 222
Magnificent symmetry is on display in the humble,
ubiquitous morning glory flower, *Ipomoea*. Neatly
spiraled buds unfurl to a flawless open-trumpet shape.
Some have two-tone flowers, some have stripes, some
have white throats—pollinator guides, of course, and
urgent calls for attention, given that each bloom lasts
only a day.

LEFT Morning glory plants in their prime are lusty
climbers, producing plentiful heart-shaped leaves and
twining tendrils. However, they tend to ramble and form
tangles unless you provide some helpful string or a trellis
to coax them on their way upward.

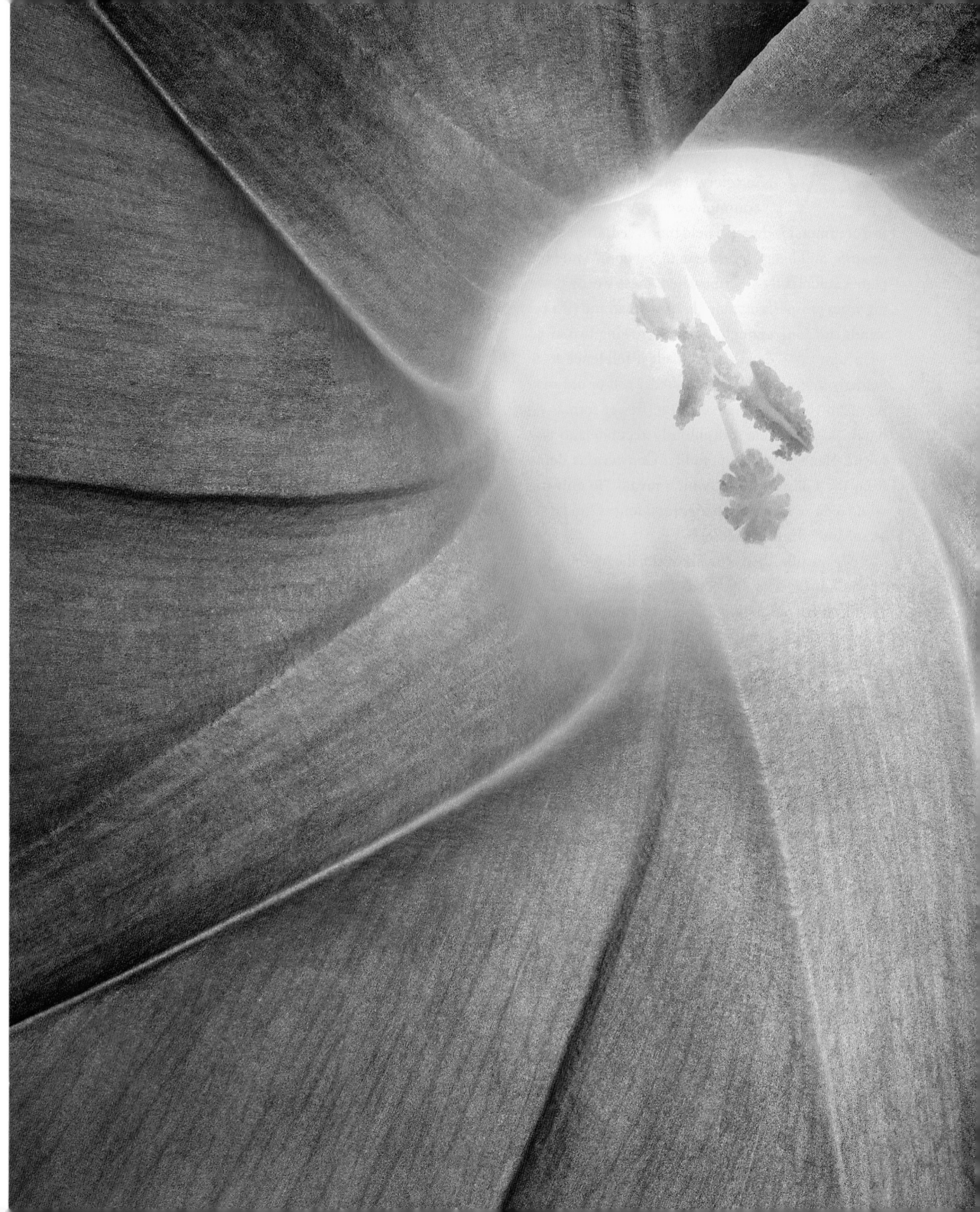

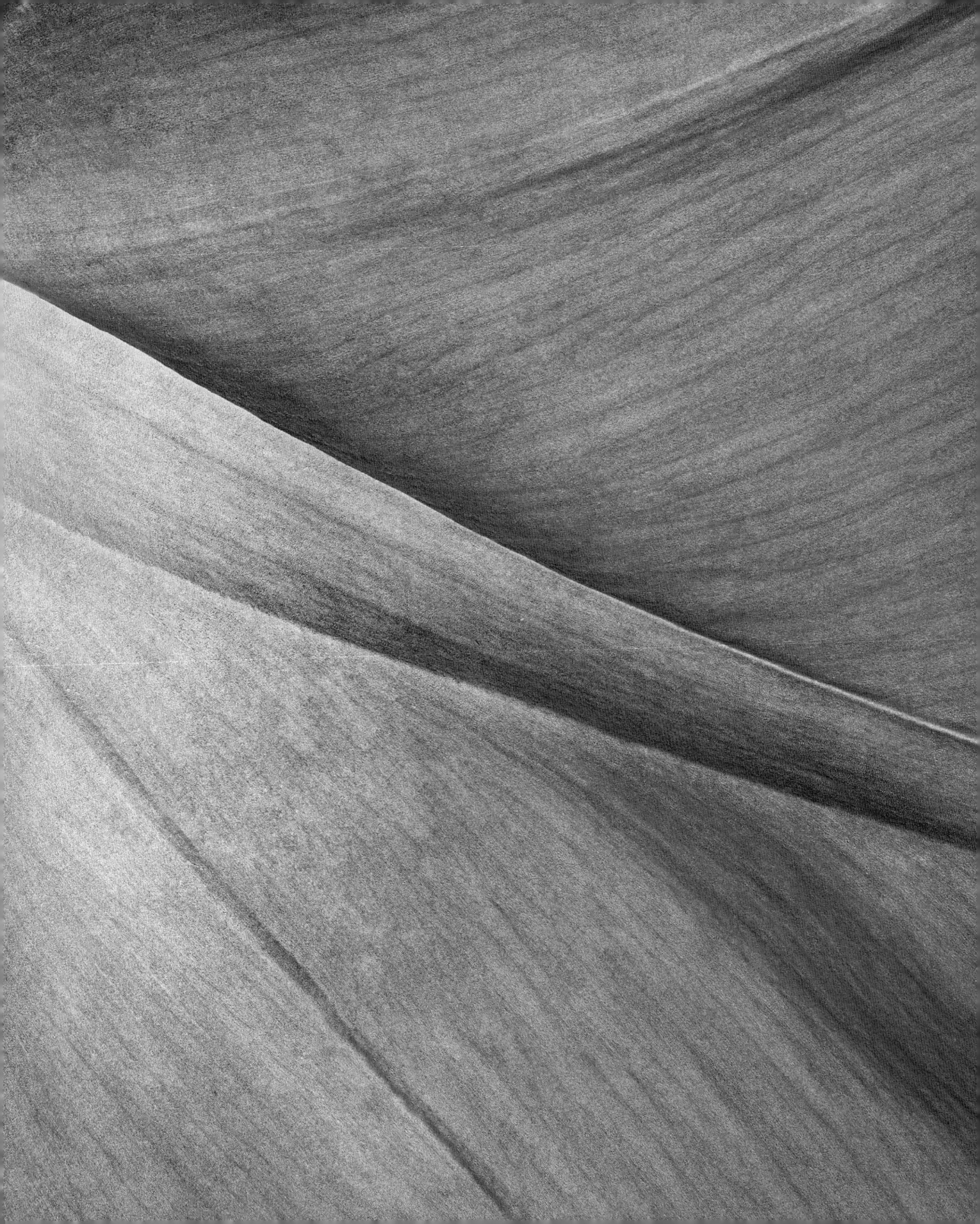

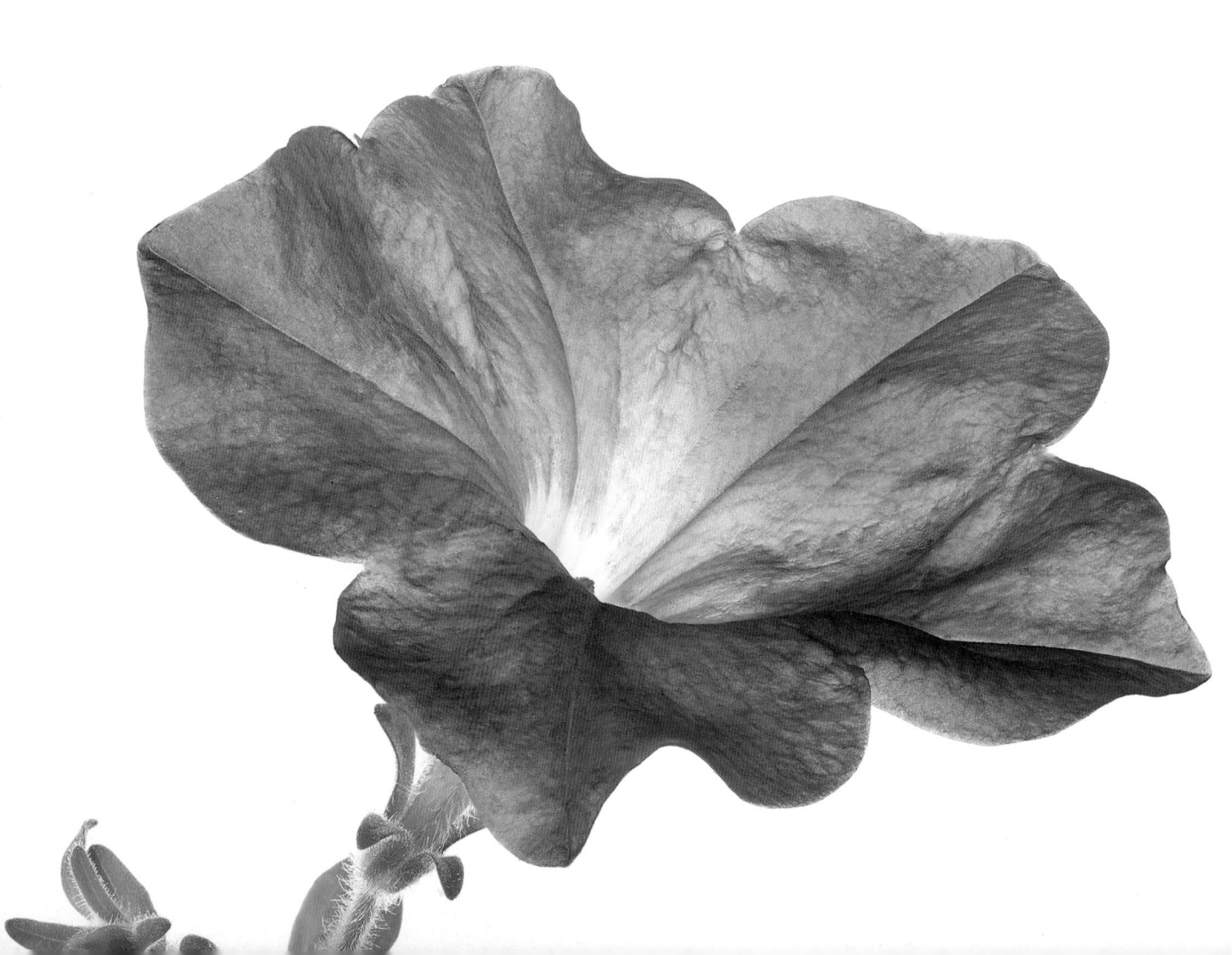

THE NIGHTSHADE FAMILY

Solanaceae

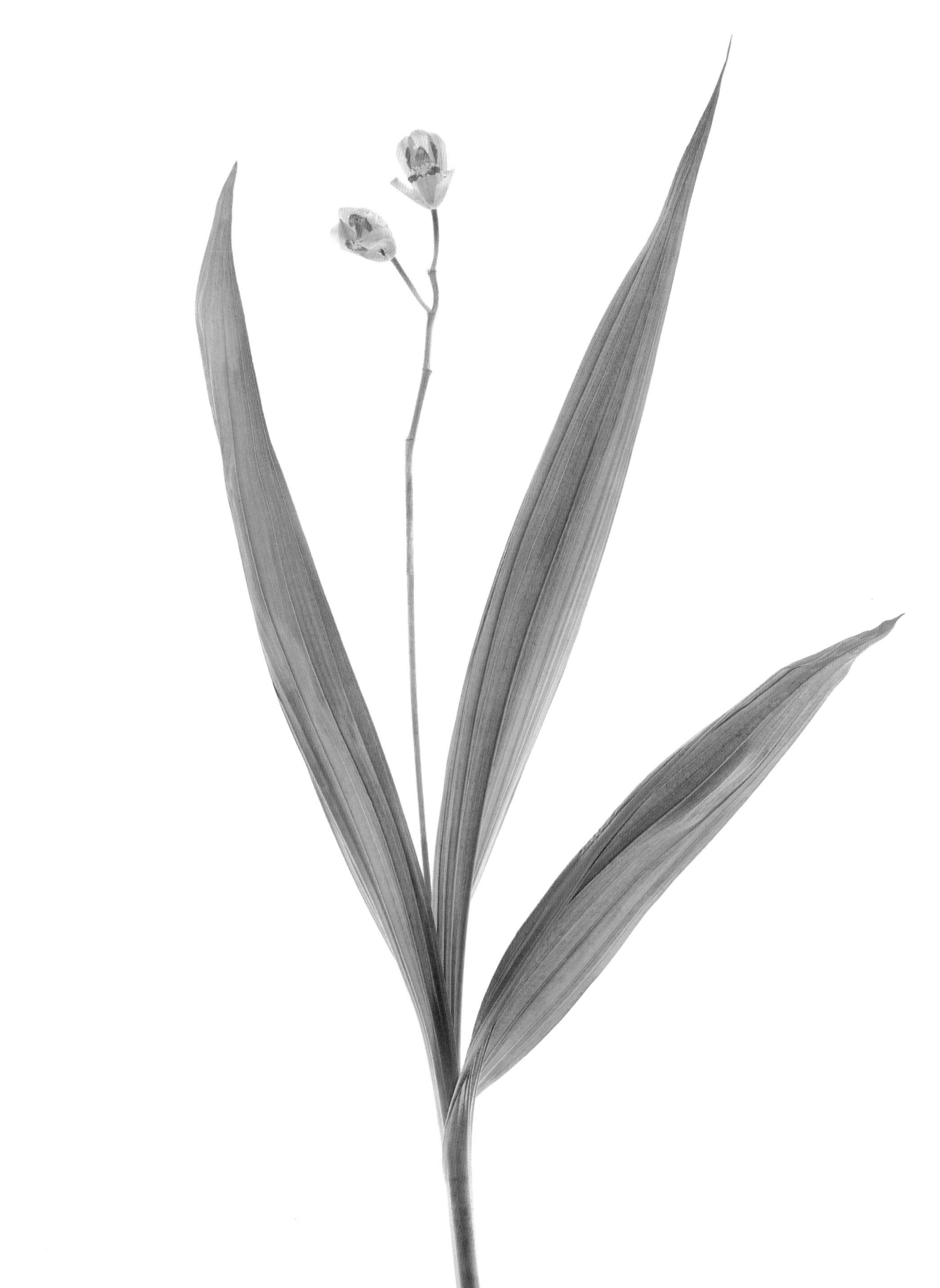

THE ORCHID FAMILY

Orchidaceae

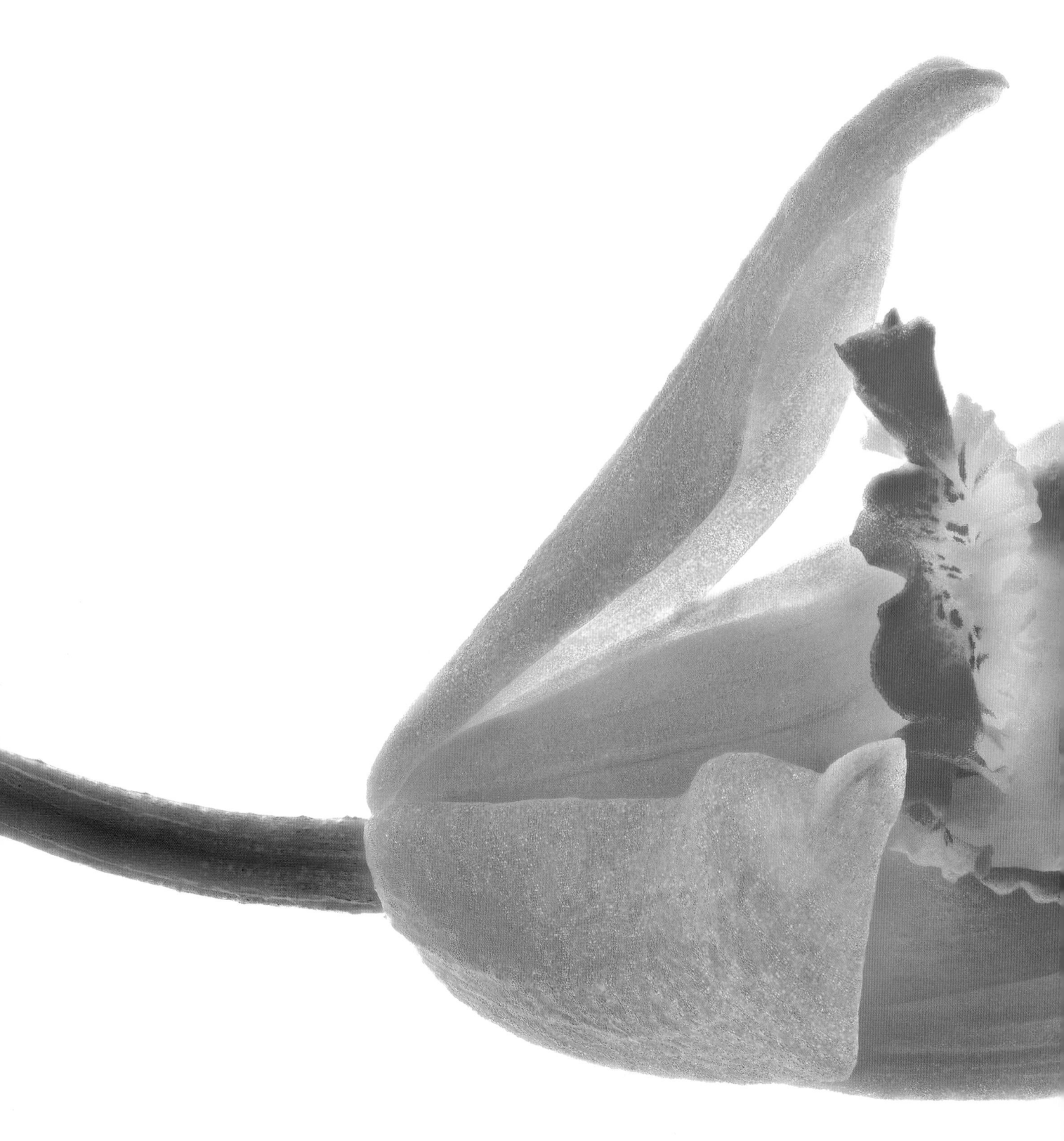

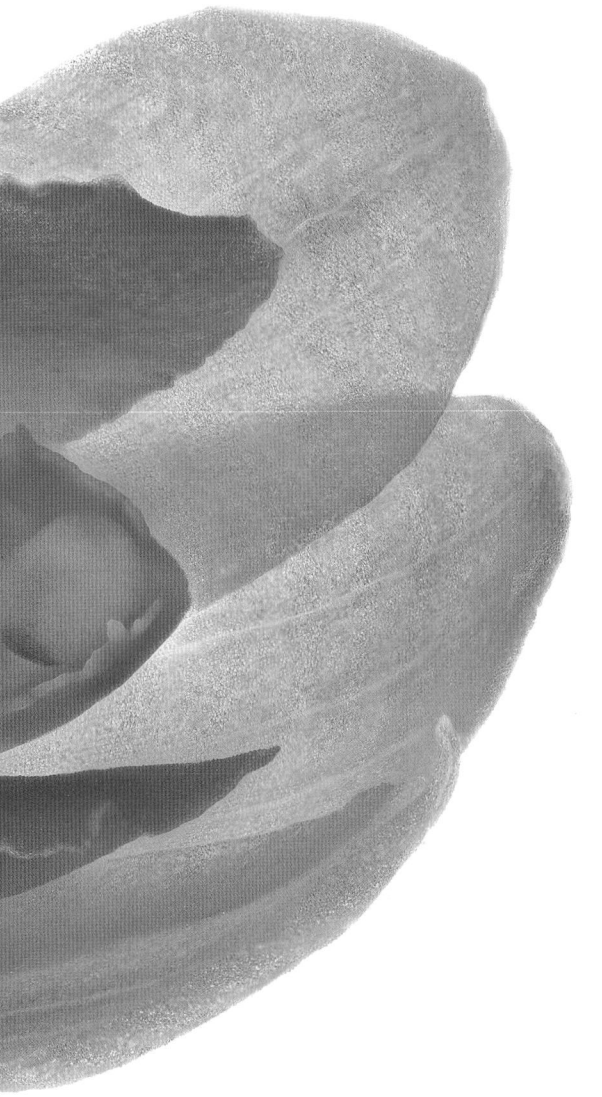

"The world is so huge that people are always getting lost in it. There are too many ideas and things and people, too many directions to go. I was starting to believe that the reason it matters to care passionately about something is that it whittles the world down to a more manageable size. It makes the world seem not huge and empty but full of possibility. . . . If I had been an orchid-hunter . . . I would have seen it as acres of opportunity where the things I loved were waiting to be found."

SUSAN ORLEANS, *The Orchid Thief*, 1998

Ever notice how long-lasting orchid flowers are in cultivation? Many have durable texture, but another explanation is that they wait in vain for a pollinator, looking fresh for as long as they are able (their natural pollinators are far away in their home habitat).

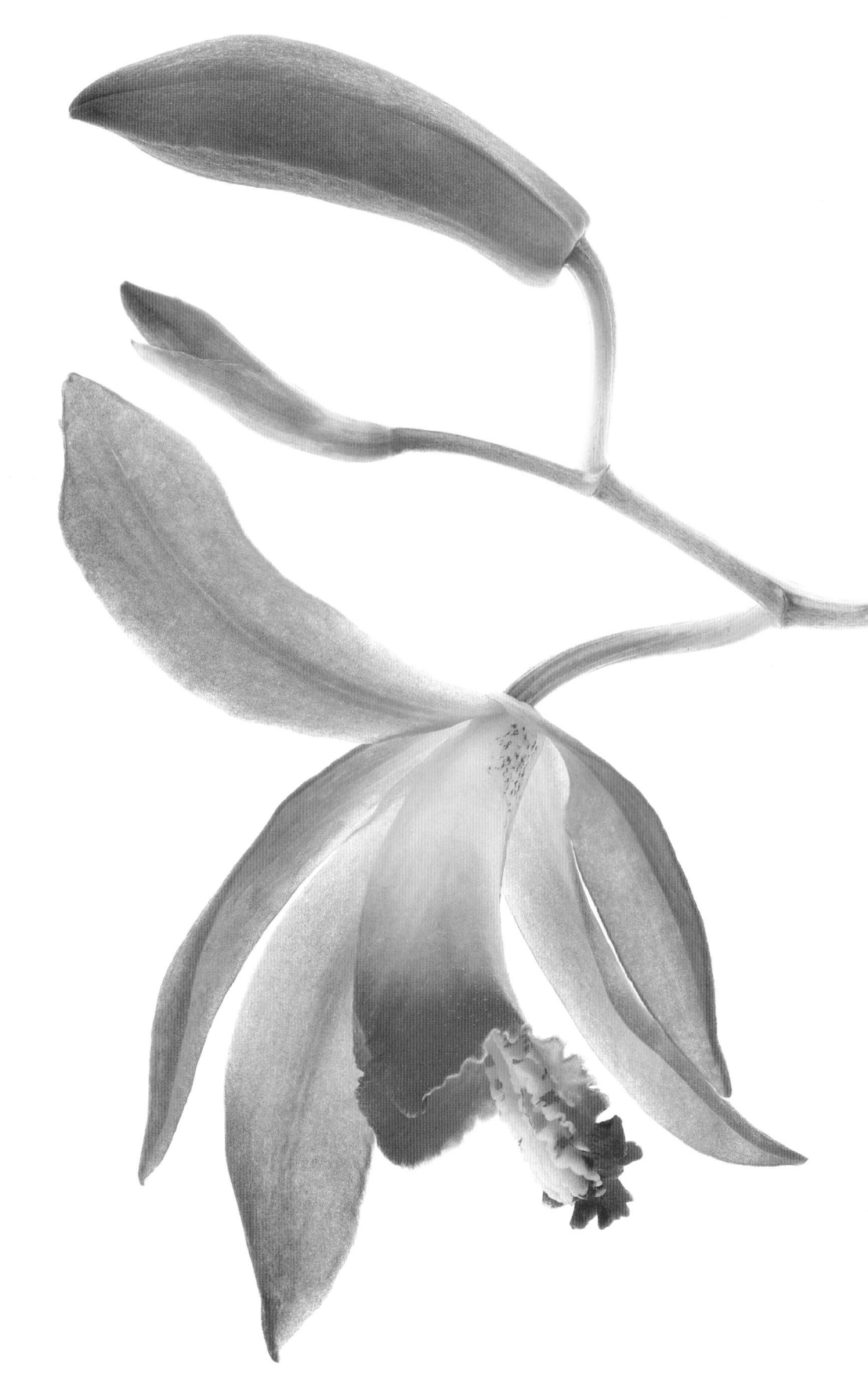

Entering the world of orchids is like entering an exotic foreign land. To the uninitiated, the diversity of colorful flowers and plant types seems both alien and overwhelming. To the hobbyist, there are always new delights to discover and caretaking skills to learn and perfect. To the orchid fanatic, no other pastime compares. You learn to understand and speak the complex language of orchid forms and nomenclature. The spare bedroom gets converted to a mini-greenhouse, or the back deck becomes a glorious jungle. Bookcases fill with reference works, conventions are marked on the calendar, and like-minded friends nearby and around the world share information and plants. But no matter your level of interest in Orchidaceae, there is no question that the opening of a flawless, strange, beautiful flower is an event. Get the camera and call over admirers!

This is indeed a colossal and complex plant family. Orchids are found on every continent except Antarctica. Many, but not all, are tropical. Some grow in soil like other, more familiar plants, but some are air plants, or epiphytes. Some even manage to grow on rocks or survive on a substrate of rotting organic matter. Some have big individual flowers, and some have tiny ones in sprays. Current taxonomic estimates place the number of genera at around 900, with upwards of 26,000 known species. As for named, registered hybrids, 150,000 is a fair guess. In all categories, there are more entries all

LEFT AND PAGE 228 *Bletilla striata* will prosper in the ground in mild-climate gardens, or in pots wherever frosts threaten. The exquisite pink-purple flowers measure no more than 1½ inches across. With their conspicuous ruffled lip, they're reminiscent of a miniature *Cattleya*.

the time. The diversity is incredible. So what on earth unites all these plants?

The answer is flower form. Yes, some are corsage beauties and some look like dangerous bugs. But they all share a handful of common features. The most important and obvious is bilateral symmetry. Cut any orchid flower vertically down the middle and you will have two halves, two mirror images.

Details within the flower further qualify it—and yes, sorry, they do have unfamiliar technical terms. The male and female reproductive parts are not separate, but are literally fused together into a (usually) white, waxy textured structure known as a column. Does this mean that orchids self-fertilize, or inbreed? No, although there are a handful of exceptions. The rostellum prevents that from happening. It is a small obstructive bit of tissue coming off the stigma that literally separates them. It also exudes and deposits a tiny amount of sticky fluid on a visiting insect's back or head as it passes by, which pollinia adhere to. Pollinia are not actually pollen grains, but rather an aggregate or little packet tucked under an anther cap. Actual orchid pollen grains are truly minute, so this is a good strategy to get them on their way. Pollinia also aid orchid identification, as their size, shape, and number vary from species to species. Finally, orchid flowers have a lip, or labellum, at the bottom, which you've surely noticed. It's actually a modified petal, and its purpose is simple and obvious: a landing pad for visiting pollinators.

You've learned some basic orchid vocabulary, just in case you were thinking of entering this wondrous land. Find more information and a mentor or an orchid club, try a few plants, and enjoy your journey.

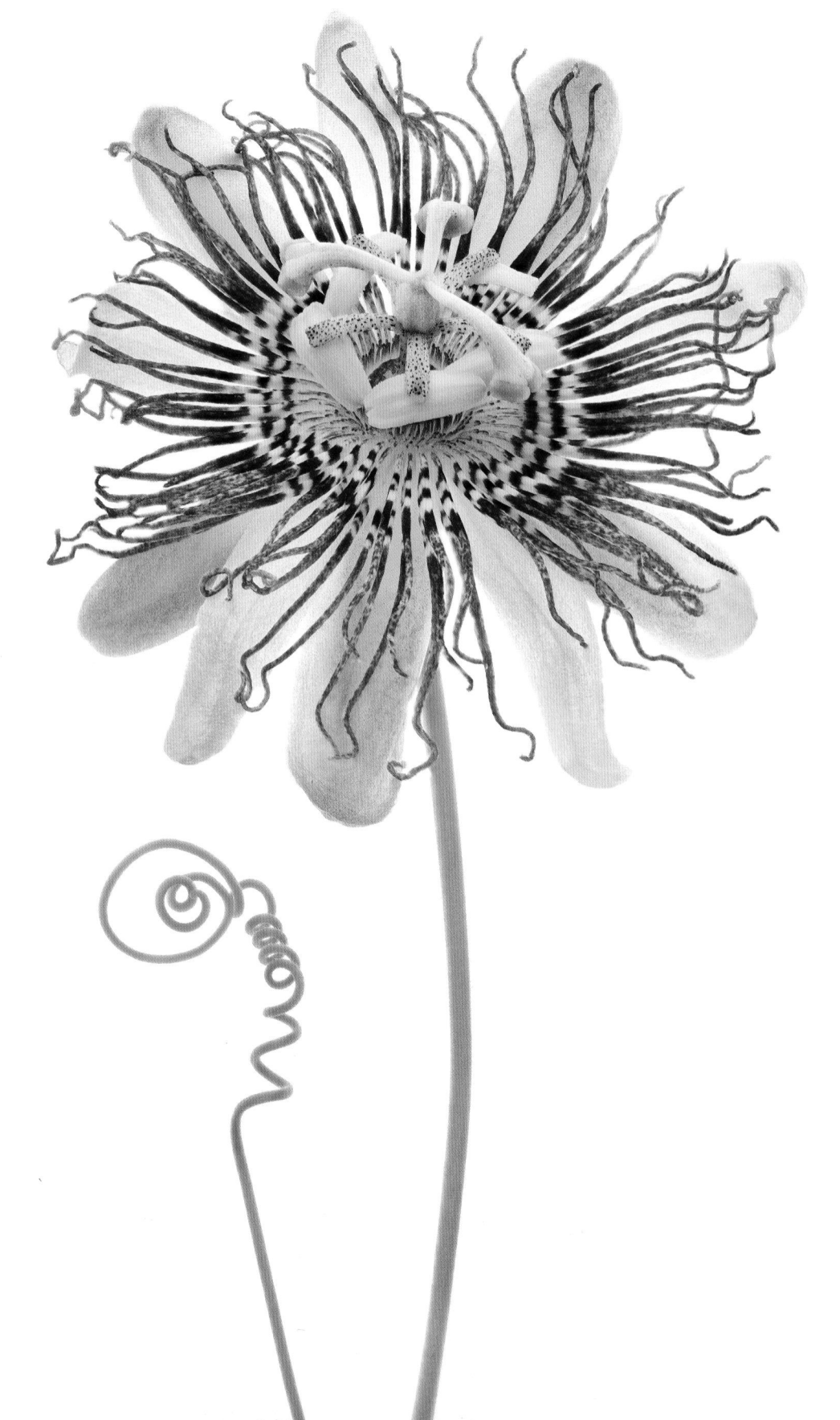

THE PASSIONFLOWER FAMILY

Passifloraceae

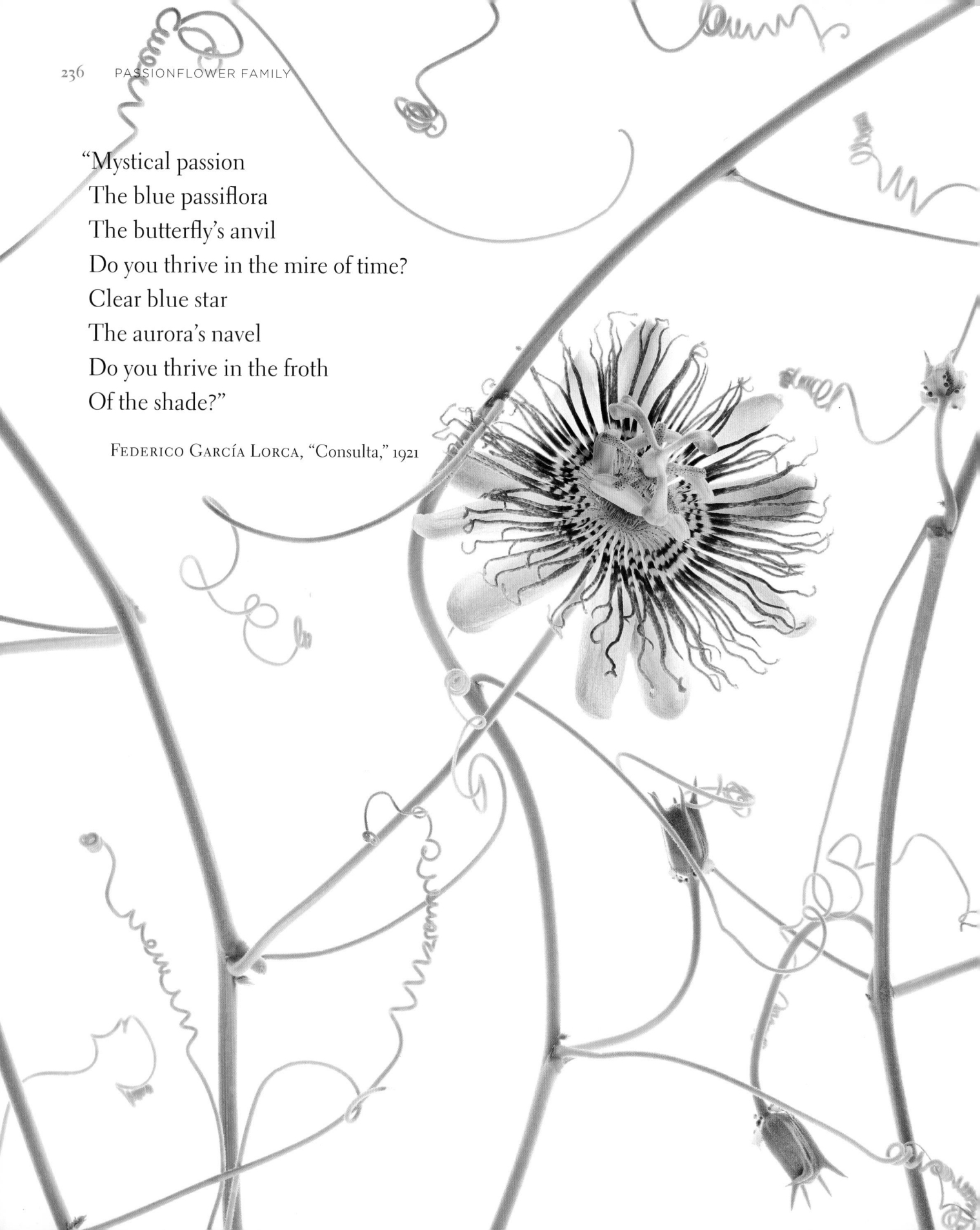

"Mystical passion
The blue passiflora
The butterfly's anvil
Do you thrive in the mire of time?
Clear blue star
The aurora's navel
Do you thrive in the froth
Of the shade?"

FEDERICO GARCÍA LORCA, "Consulta," 1921

The blooms of the passionflower vine, *Passiflora*, are so striking and ornate that throughout history admirers have appreciated them in detail. When early Spanish missionaries dispatched to Mexico first beheld these plants, they analyzed the plants in the context of Catholic stories of Jesus's final days. The pointy leaves suggested the piercing Roman spears (alternatively, the three lobes represent the Trinity); the five anthers, the five wounds; the tendrils on the vine, whips or cords. The column of the ovary represents the vertical cross. The three styles are nails, while the stamens are hammers. The fleshy, threadlike rays in the flower's center represent the crown of thorns, while the calyx is a halo. As for the petals and sepals, numbering ten, these evoke the ten loyal apostles (Judas and Peter were not counted). And the fact that individual flowers last only a single day was considered portentous. The common name became passionflower, and it is still used today.

Actually there are numerous related species, and Europeans quickly learned how to grow them, including under glass where they are not cold hardy. Blue passionflower, *Passiflora caerulea*; wing-stemmed passionflower, *P. alata*; and passion fruit, *P. edulis*, and their hybrids became fashionable. To this day, these twining, rambling vines remain popular subjects in warm greenhouses and are also enjoyed outdoors in mild climates, particularly in California and Florida gardens.

Pollinators vary from bees to wasps to hummingbirds to bats, depending on the species, variety, and setting. Some flowers are self-fertile, and some gardeners have had success with hand pollination. *Passiflora edulis* is widely cultivated in the Caribbean for its purple or orange-yellow, f. *flavicarpa*, fruits. It is also an ingredient in the popular juice blend Hawaiian Punch. Granadilla, *P. ligularis*, is also called passion fruit. All these species (and more) may be eaten fresh, made into syrups or preserves, or added to desserts and beverages, including cocktails.

To clarify, Passifloraceae contains only passionflowers, and there are many more than mentioned here — by some reckonings, up to 500 species. They come in other colors, notably red and pink, and some readily cross with one another. One might speculate that such a complex and exotic-looking flower, while it could and does vary somewhat in form, size, and color, would not diverge or develop true related genera easily. Evolution within this unique and uniquely beautiful family has gone deep but not wide.

LEFT AND PAGE 234 From the time early Spanish missionaries "discovered" it in Mexico, perhaps no other flower has so engaged public fancy. All parts of the passionflower, *Passiflora caerulea*, were assigned symbolic meaning relating to Jesus's final days: the five anthers signify the five wounds, the sepals and petals suggest the apostles, the center rays represent the crown of thorns, and so on.

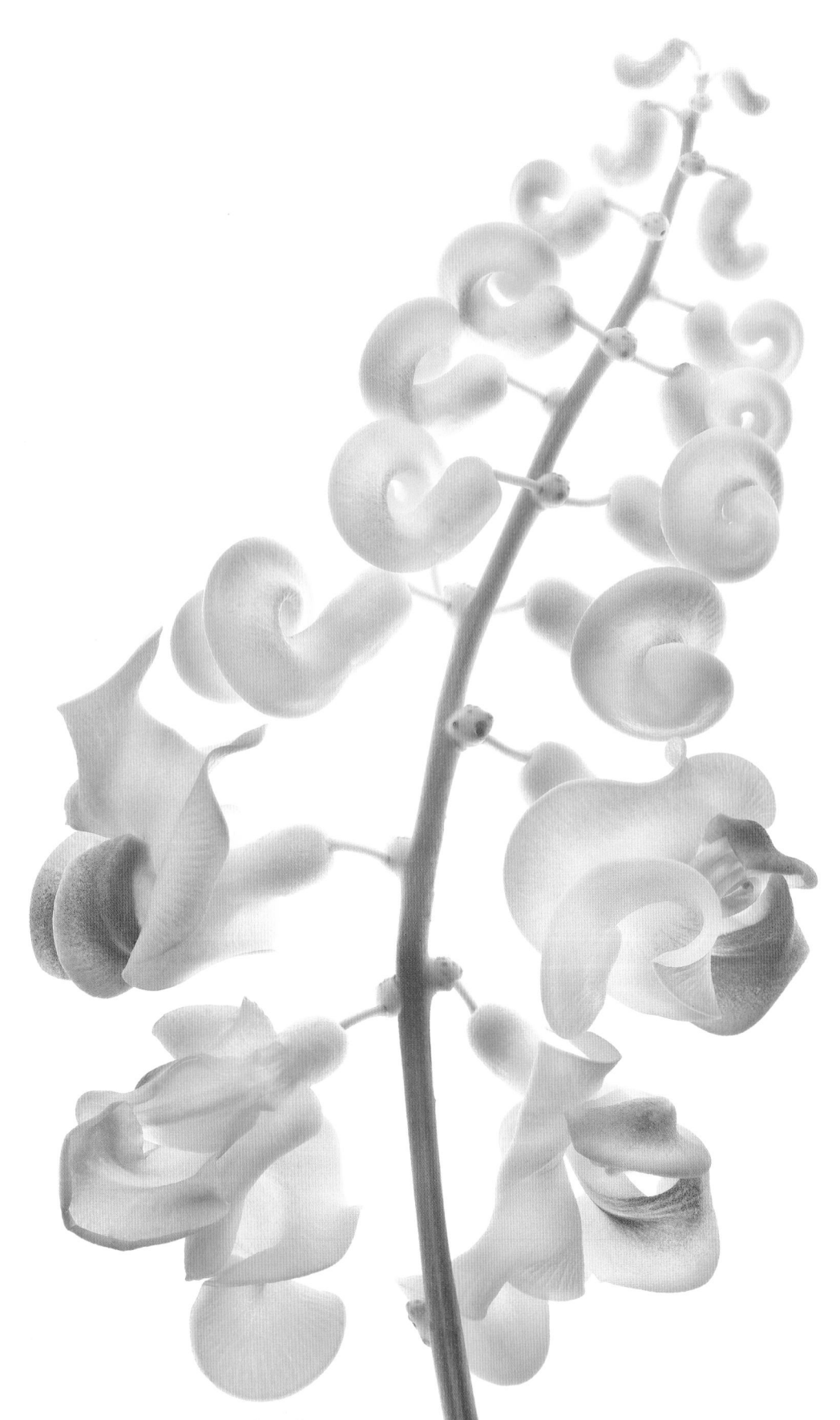

THE PEA FAMILY

Fabaceae (or Leguminosae)

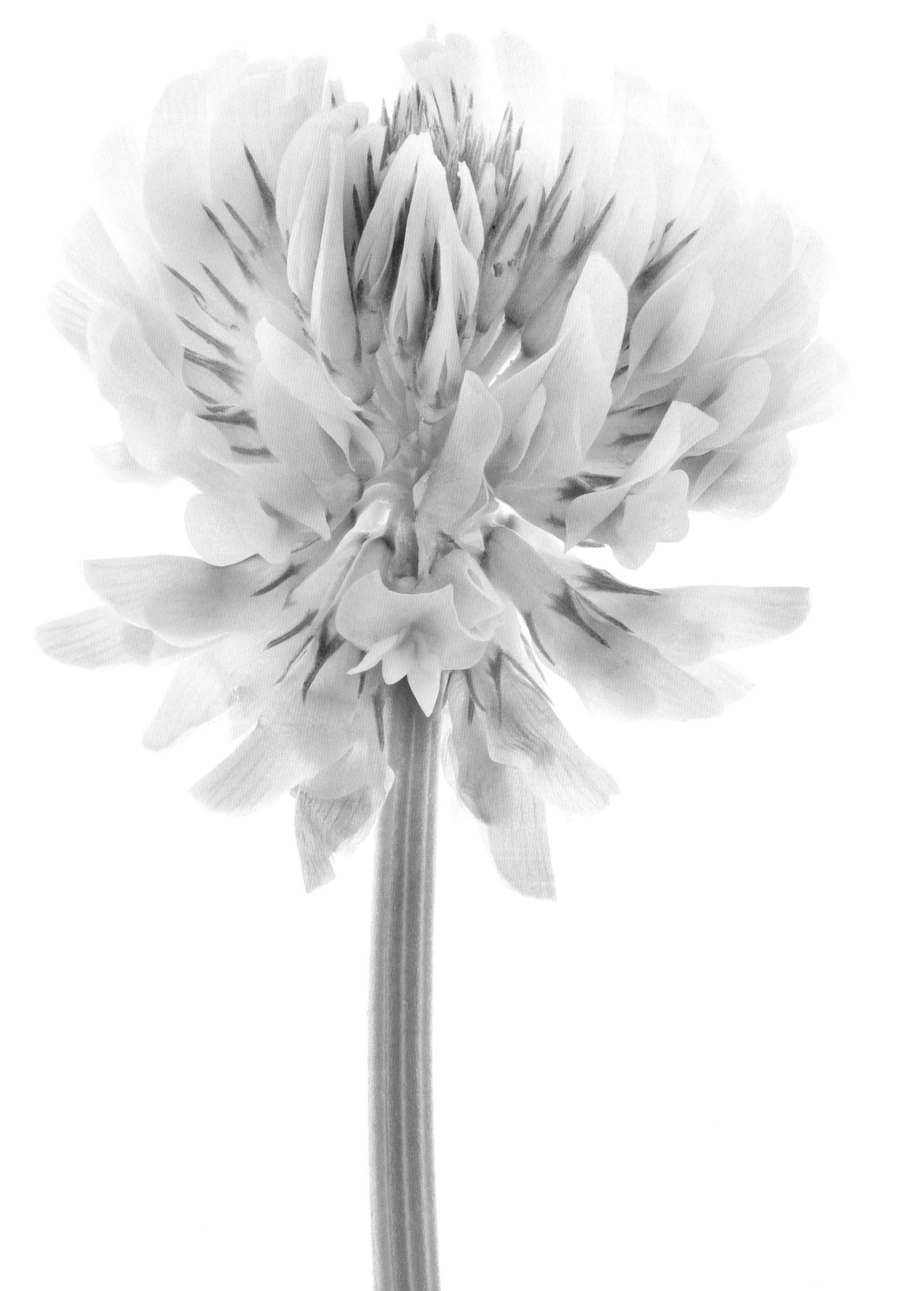

"To make a prairie it takes a clover and one bee,
   One clover, and a bee,
   And revery.
   The revery alone will do,
   If bees are few."

EMILY DICKINSON, "To make a prairie," 1924 (posthumous)

Dutch or white clover, *Trifolium repens*, often seen in lawns, may be considered an unwelcome pest. But it fixes nitrogen in the soil, and those browsing bees make delicious honey from the nectar. The blossoms are actually flowerheads consisting of as many as fifty small, five-petaled flowers.

PAGE 238 The weirdly pretty flowers of corkscrew vine, *Vigna caracalla*, are as curly as a seashell and as durable as a wisteria blossom. The fragrance has been compared to blooming magnolias. Evidently ants, not bees, navigate them to accomplish pollination.

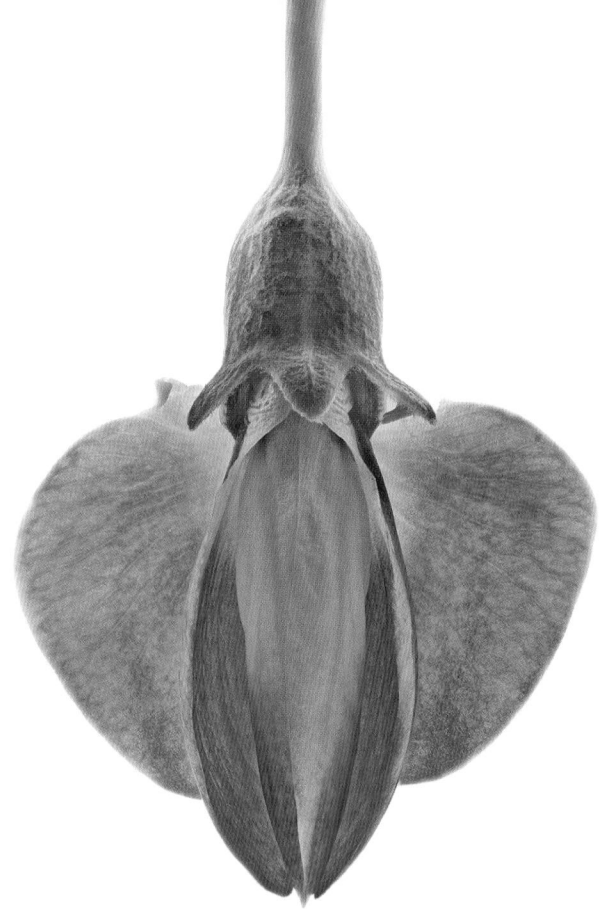

s there a gentler, more delicious perfume than that of blooming sweet peas swooning over a white picket fence on a summer day? Is there a more comforting feeling than lazing under a beach umbrella and opening your eyes to notice pretty beach pea flowers anchoring the sand dune nearby? Is there a sweeter sight than a toddler examining a corkscrew vine blossom with utter concentration and wonder? Is there a more photogenic, breathtaking view than a meadow of blooming lupines along the Maine coast or in California's Sierra Nevada foothills? And let's not forget the crisp, sugary crunch of a freshly picked sugar snap pea or the simple pleasure of salted peanuts to nibble on at a baseball game.

The pea family, Fabaceae or Leguminosae, is blessed with all of these fascinating, appealing species, and more, all over the globe and in a wide range of settings. Nearly 20,000 species are considered members, making this one of the largest plant families on earth. Indeed, pea family members span the gamut from low-growing weeds and groundcovers to herbaceous perennials, shrubs, vines, and even trees. And yet their deceptively simple-looking flowers are instantly recognizable, as are the pods that follow.

Undoubtedly you've observed upper and lower petals on those characteristic pea family flowers. But if you look more closely, you should always be able to count five petals, as well as five sepals, which are often fused.

ABOVE Chinese wisteria, *Wisteria sinensis*, is distinguished from the more widespread Japanese one, *W. floribunda*, in a few significant ways. Individual flowers are a bit bigger, carried in shorter racemes, and more lightly scented; the vines twine counterclockwise. Also, flowers on these racemes open together, instead of starting at the base.

RIGHT A close look confirms that Chinese wisteria is indeed in the pea family. The calyx clasps the inflated flower, which has two side wings, a banner, and a keel at the base. Pollen-dusted anthers and the stigma are concealed inside.

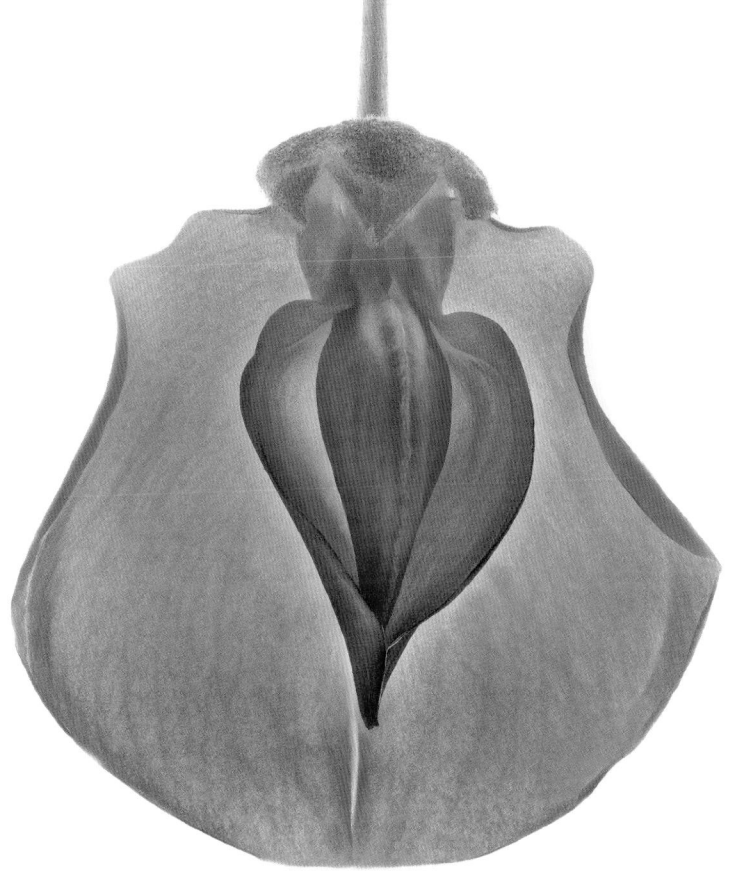

Stamens usually number ten, and there is a single elongated superior ovary with a curved style (pistil stalk). Bees and other pollinators relish these blooms, drawn by the irresistible combinations of fragrance, color, and a convenient or protected landing area, not to mention a good supply of pollen.

Because of their ability to self-pollinate, the "father of genetics," Gregor Johann Mendel, used these flowers to formulate his theories. Specifically, he studied pea plants, *Pisum sativum*, but his successors used the sweet pea, *Lathyrus odoratus*, in their research with equal success. Through the years, the plant's ease of growth and amenability to experimenting and tampering has also spawned lots of nifty variations for gardeners to enjoy—sweet peas come in a luscious array of colors and bicolors. The only hue that remains elusive is yellow.

There is, not surprisingly, variation in the general pea flower form. Three whimsically named parts of sweet peas have been identified in the subfamily to which they belong (the pea, or Faboideae): banner, wings, and keel. The larger upper petal, with two lobes, is the banner; the two wings are smaller and just below and embrace the still smaller keel, which is made up of two fused petals. Others with this flower form, in this subfamily, include wisteria vines (*Wisteria* species and cultivars), lupines (*Lupinus* species and cultivars), broom (*Cytisus*), blue false indigo (*Baptisia australis*), all the many clovers (*Trifolium*), and the plucky little yellow-flowered weed bird's-foot trefoil (*Lotus corniculatus*). Even the curious curly flowers of the corkscrew vine (*Vigna caracalla*) qualify.

Botanists have identified two other subfamilies. Caesalpinioideae is devoted mainly to trees. You might know a few. Redbud, *Cercis*, so radiant in rosy bloom, adorns springtime forests and is sometimes used in parks and campuses, and as a street tree. Honey locust, *Gleditsia triacanthos*, is a fairly winter-hardy and pollution-tolerant street tree, though wickedly thorny.

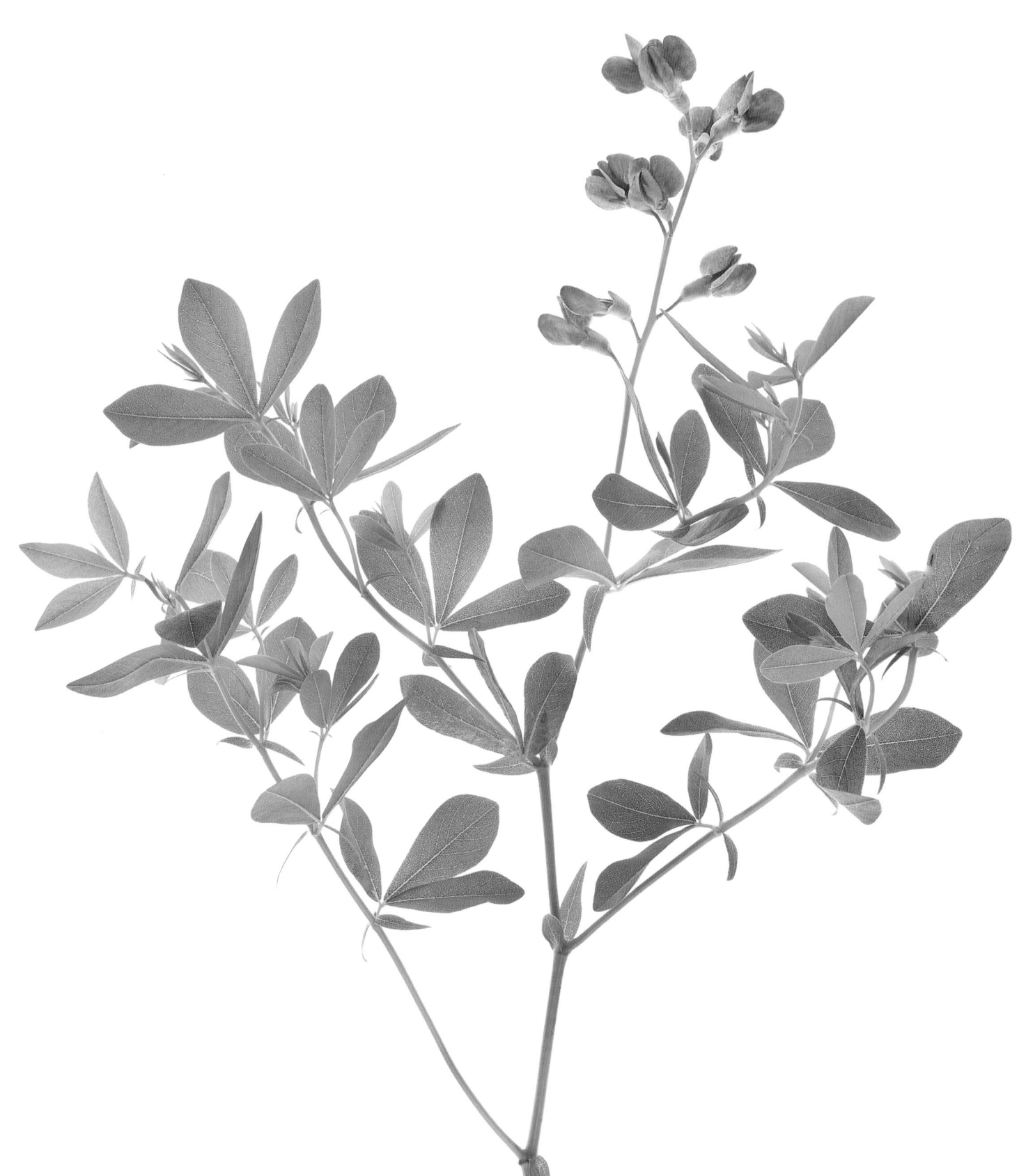

In mild-climate areas, you sometimes see the sensational red-flowered Royal Poinciana, *Delonix*. The flowers of each of these trees tend to be symmetrical, even if the petals are not all the same size, and they feature a prominent inner petal.

In the Mimosa subfamily, Mimisoideae, petals are quite small and the stamens steal the show; they are long, colored filaments generally arrayed in a globe shape. In addition to the mimosa, or pink silk tree, *Albizia julibrissin*, other members of this subgroup that you may recognize are *Acacia*; the fairy duster, *Calliandra*; and mesquite, *Prosopis*, all common mild-climate species. Another example is *Neptunia*, a sought-after aquatic plant that looks like a handful of emeralds cast upon the water surface and which, unfortunately, does not often bloom in cultivation.

The pea family is also widely valued for its seedpods—sometimes edible, often useful—and for the plants' ability to "fix" nitrogen in the soil in which they grow. But their flowers grab our attention first and foremost. Horticulturists have been able to select or develop a rainbow of colors from various members to enhance our gardens, vases, and public spaces, but the unique forms of the flowers remain unaltered and forever intriguing.

LEFT Spikes of deep blue join blue-green compound leaflets on the handsome, shrubby blue false indigo, *Baptisia australis*. They were used as a (poor) substitute dye for true indigo, *Indigofera*. Up close, individual flowers reveal the classic banner-wings-keel form of this family. A visiting bee's weight pulls down the keel, giving it access to the nectar while getting pollen dusted on his underside.

BELOW Rich yellow flowers adorn 'Carolina Moonlight', a robust hybrid between yellow wild indigo, *Baptisia sphaerocarpa,* and white false indigo, *B. alba*. It is perhaps the closest thing to a yellow lupine for those who are unable to grow lupines due to soil or climate. Like other baptisias, the beautiful inflated flowers eventually give way to inflated, decorative seedpods, but because it is a hybrid, resulting seeds will not come true.

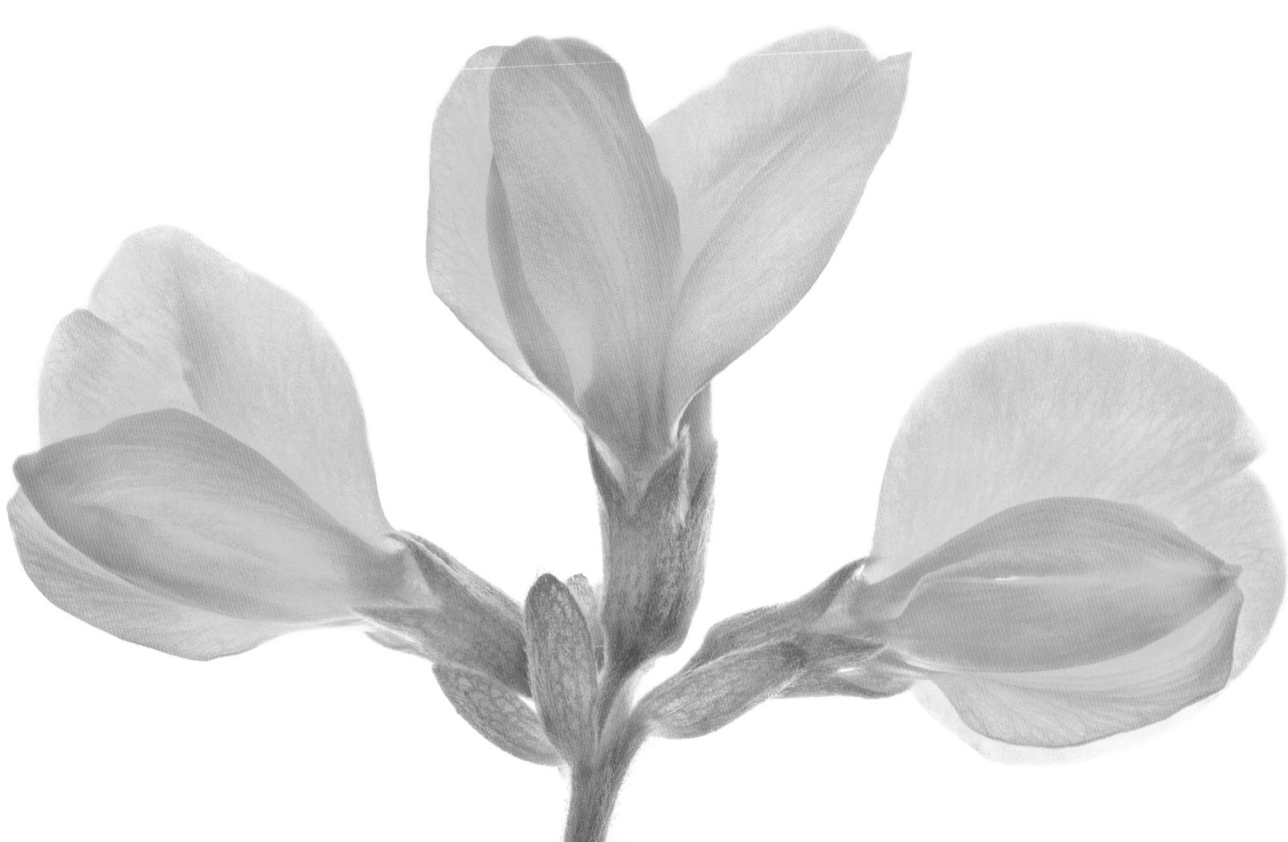

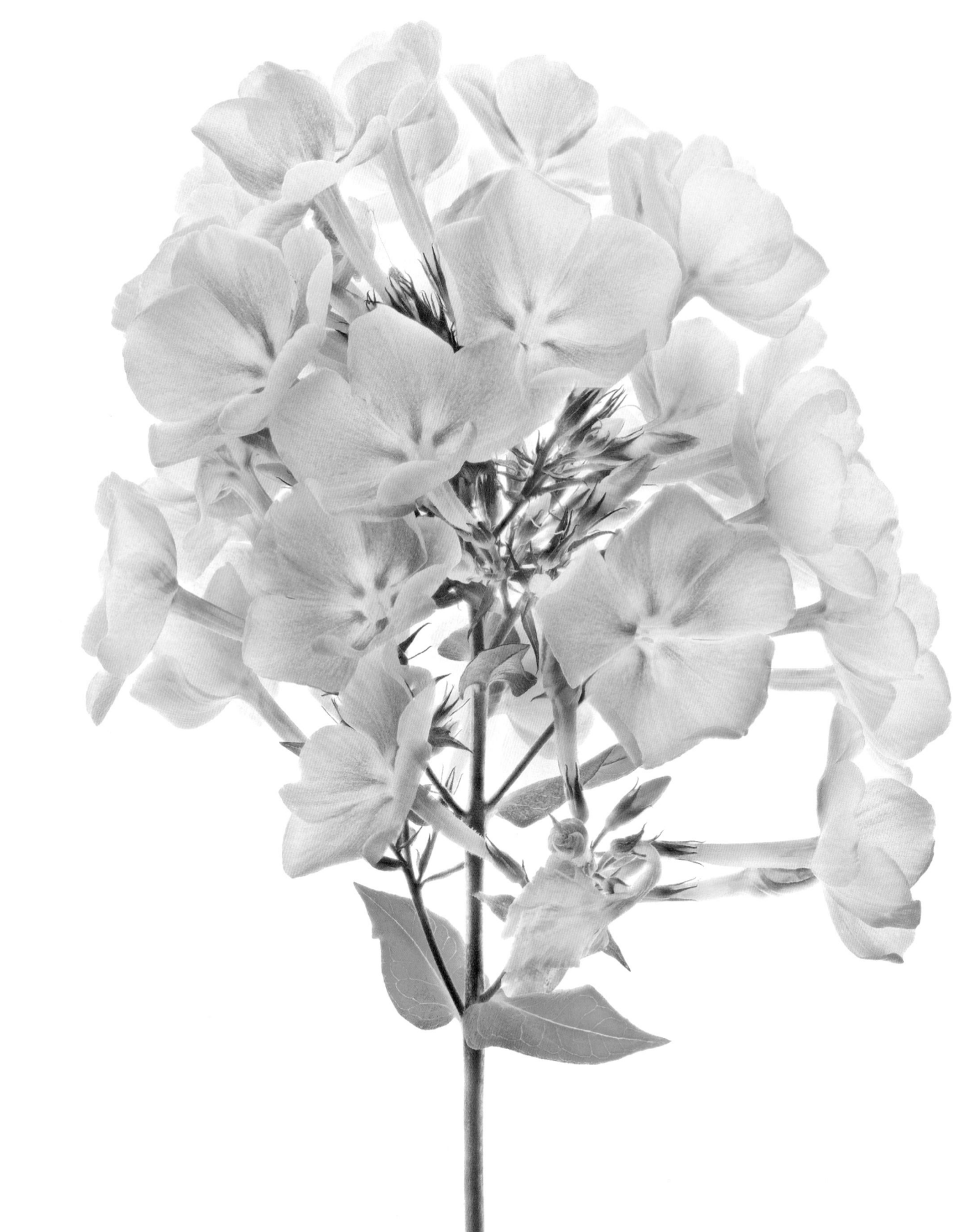

THE PHLOX FAMILY

Polemoniaceae

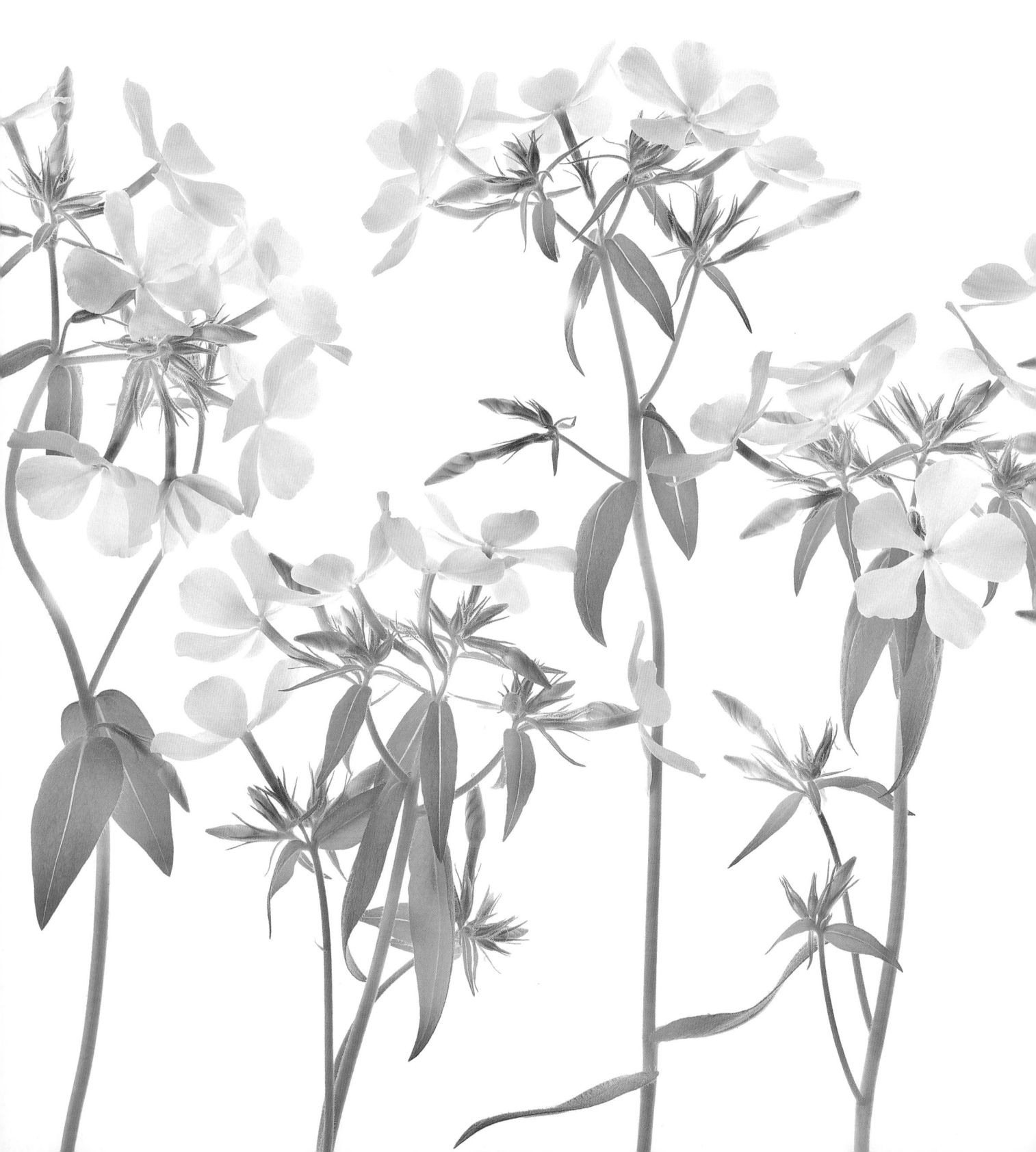

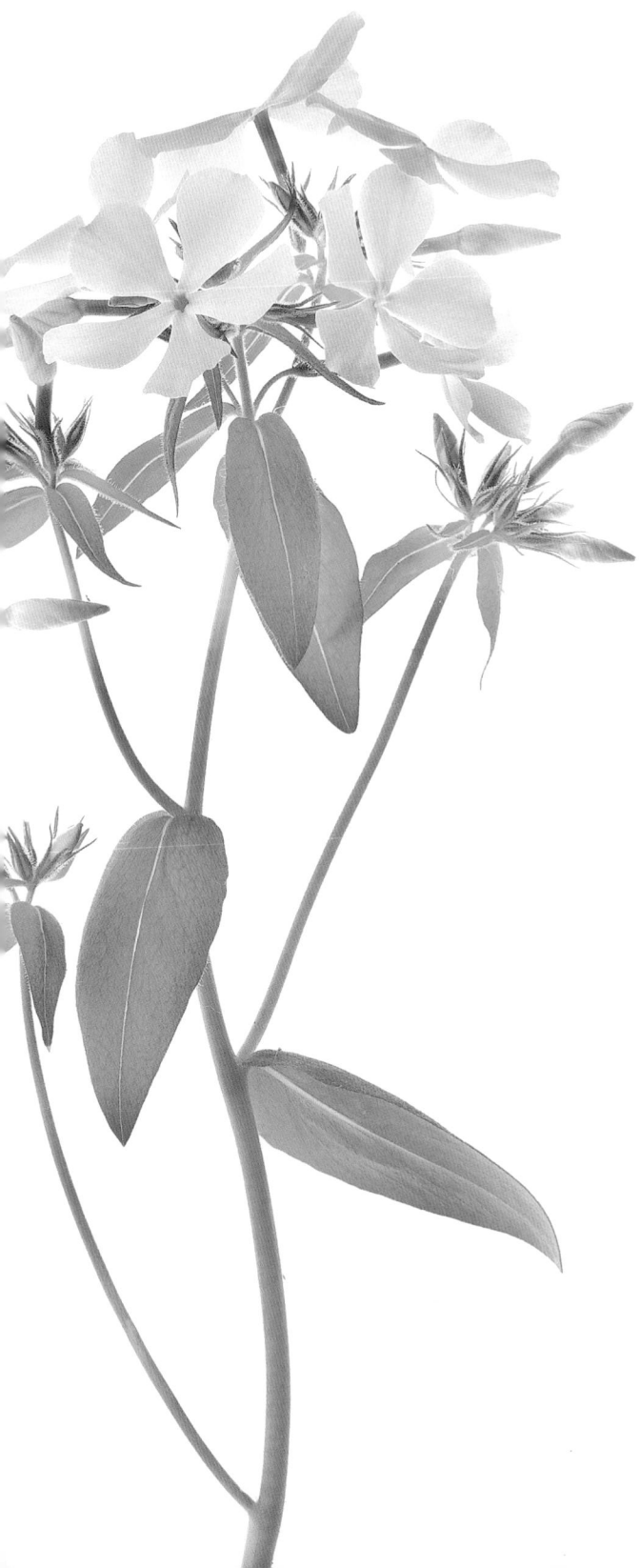

"A black cat among roses,
  phlox, lilac-misted under a quarter moon,
  the sweet smells of heliotrope and night-
    scented stock.
  The garden is very still.
  It is dazed with moonlight,
  contented with perfume."

AMY LOWELL, "The Garden by Moonlight," ca. 1918

Spicy perfume radiates from blooming woodland phlox, *Phlox divaricata*, in spring. Dainty five-petaled flowers are backed by five slim sepals. In their centers, you can detect one pistil and a three-branched style.

PAGE 246 The big, showy, sweet-scented phlox of gardens is *Phlox paniculata*, the queen of this appealing genus. Sturdy stems hold big, pyramidal clusters (technically panicles) aloft. These are composed of numerous 1- to 2-inch tubular flowers. A calyx has five lobes (sepals), a corolla has five lobes (petals), and there are five stamens.

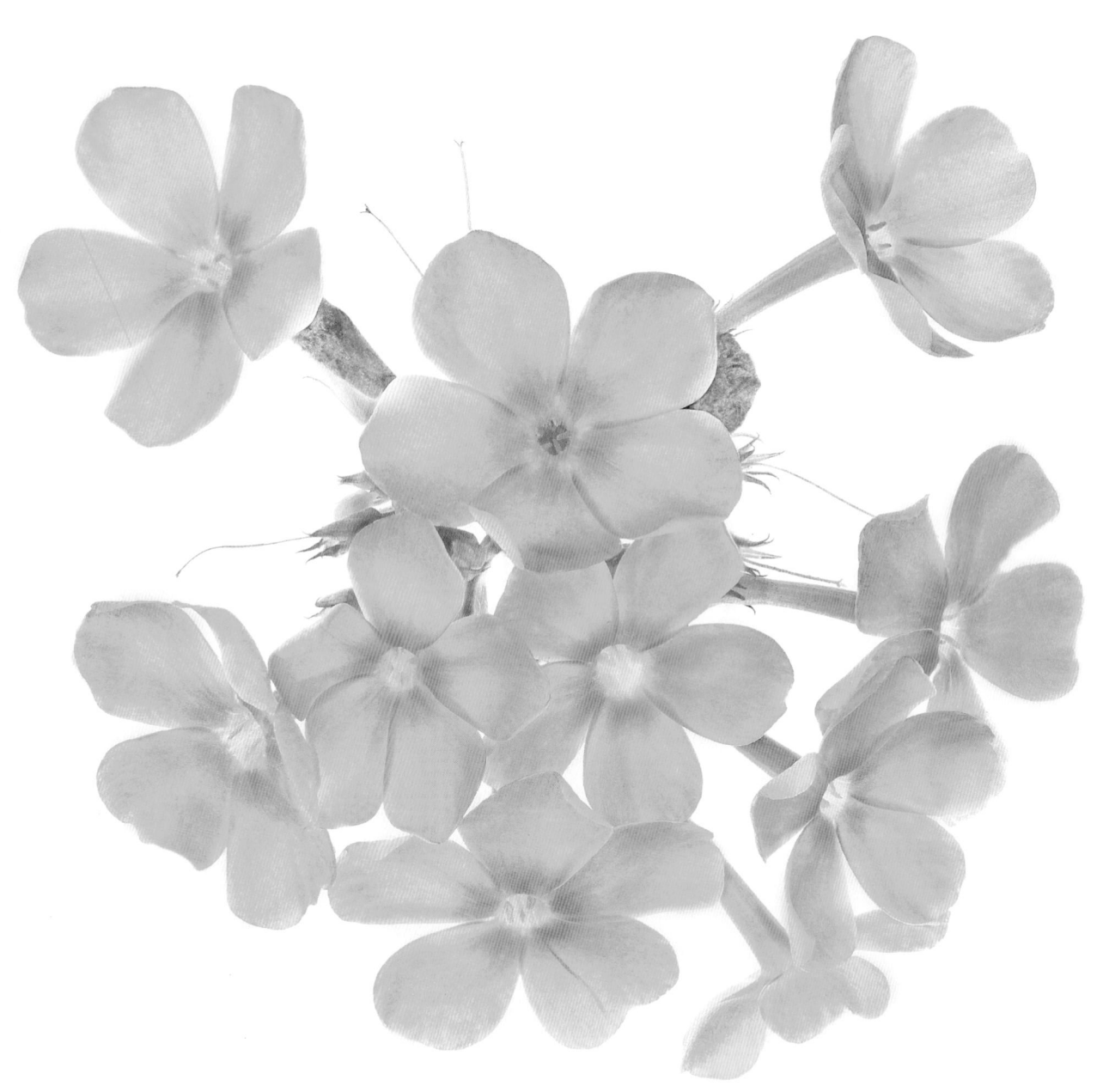

Although it is easy to understand that treasured garden flowers have wild origins, the phlox family, Polemoniaceae, appears to have made an especially colorful journey. Aside from phlox itself, there is a small handful of other species grouped with them.

The *Phlox* genus—with one exception, a little mat-former found in Siberia—is native to North America. There are about seventy species within it, many from the Mountain West area of the United States. Some are erect; others, low-growing cushions. They tend to be sweetly scented and to bloom in early to mid-spring. They come in a variety of colors, but mainly in the purple-pink-white range. Examples include *Phlox divaricata*, an erect woodland beauty with soft blue flowers; *P. subulata*, or moss pink, a mound-former studded with purple, pink, or creamy white blooms; and upright, spreading *P. carolina*, the Carolina phlox, valued for its light purple flowers. *Phlox drummondii*, or annual phlox, makes a low-spreading mound and comes in many pastel hues as well as red and yellow.

The annual one, *Phlox drummondii*, has an interesting story behind its species name, which commemorates the intrepid Scottish botanist James Drummond. On an 1835 expedition funded by the Glasgow Botanical Society, Drummond identified this species in the hills of Texas, braving blazing heat, cholera, and aggressive grizzly bears, which he scared off by beating on the large tin box he used to collect specimens. Seeds of this phlox made it safely back to Scotland. When it germinated and its beauty and productiveness were admired, it was promptly named in his honor.

The perennial species so popular in mixed borders and cutting gardens, erect-growing garden or summer phlox, *Phlox paniculata*, is originally an eastern North American native. First collected and sent back to France around 1730, hybridizers there (and, later, in England) admired and fooled with its big showy inflorescences, robust form, and lovely scent. Nowadays there is a bounty of gorgeous varieties for gardeners to choose from. Classic choices include bicolor white-and-pink 'Bright Eyes' and the mildew-resistant, pure white 'David'. These are not low-maintenance garden plants, however: They perform best in full sun in rich, fertile soil, need ample watering and protection from drying winds, and should be deadheaded to prevent self-seeding (which leads to mongrel offspring).

Phlox flowers are composed of five sepals and five fused petals, usually in either a disk or funnel form. Inside, five jaunty stamens alternate with the corolla lobes. These are almost always fused into the corolla tube, although you might have to use a magnifying glass to confirm. Ovaries, deep within the blossoms and later, after pollination, becoming seed capsules, always have three parts. Phlox flowers are usually carried in clusters, loose or dense, which adds to their impact. As for colors, among all the many wild and cultivated varieties you can find just about any color or bicolor (two toned or with pinwheel stripes) to beautify your flowerbeds, rock garden, and homegrown bouquets.

Phlox fragrance, form, and color conspire to draw many pollinators. Bees, moths, flies, butterflies, and hummingbirds are the primary agents. Alas, some phlox also attract caterpillars of moths, which nibble holes in the leaves. Groundhogs, rabbits, and deer relish eating entire plants.

It seems that within Polemoniaceae, variability in flower color and form is their great gift, whether we happen upon them along a roadside or on a hike, or invite them into our gardens. Neither grizzly bears nor powdery mildew should stop anyone from relishing their sweet perfume and captivating beauty.

Bumblebees and hummingbirds cannot resist sipping at the long-lasting, fragrant pink flowers of Carolina phlox, *Phlox carolina*. Contrasting, darker coloration in the center helps direct them to the waiting nectar.

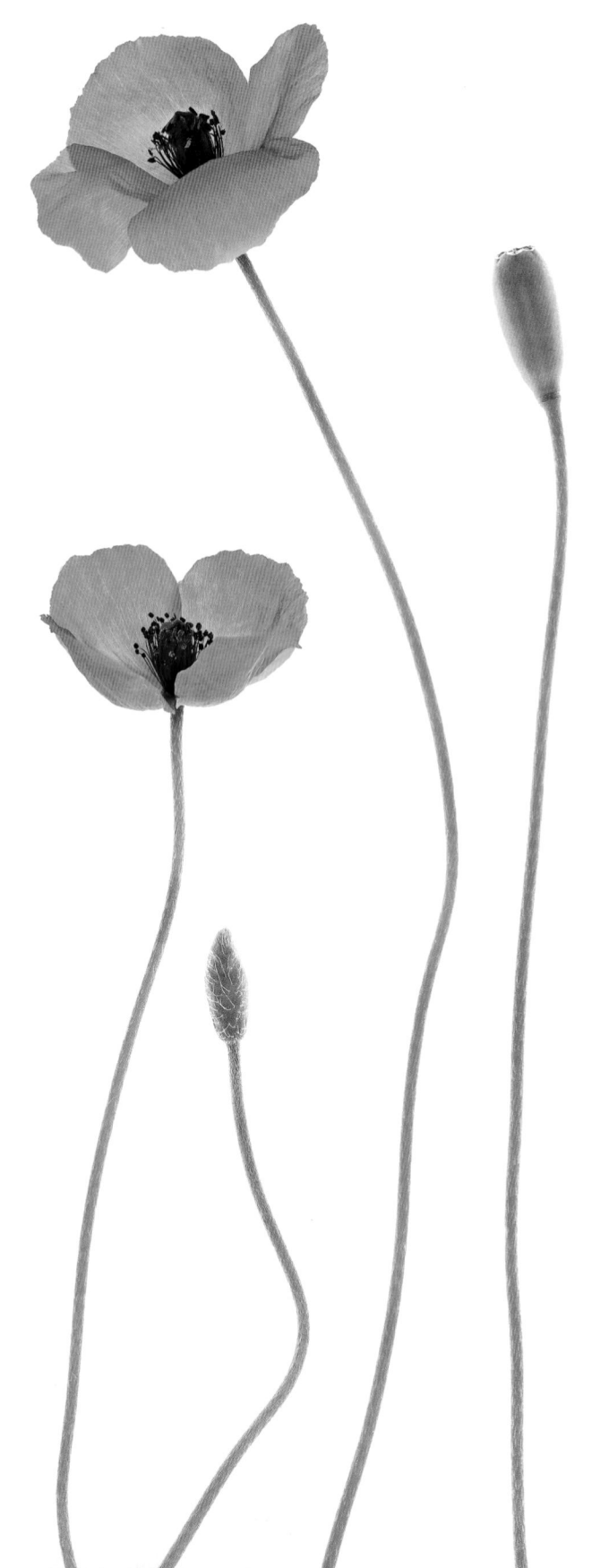

THE POPPY FAMILY

Papaveraceae

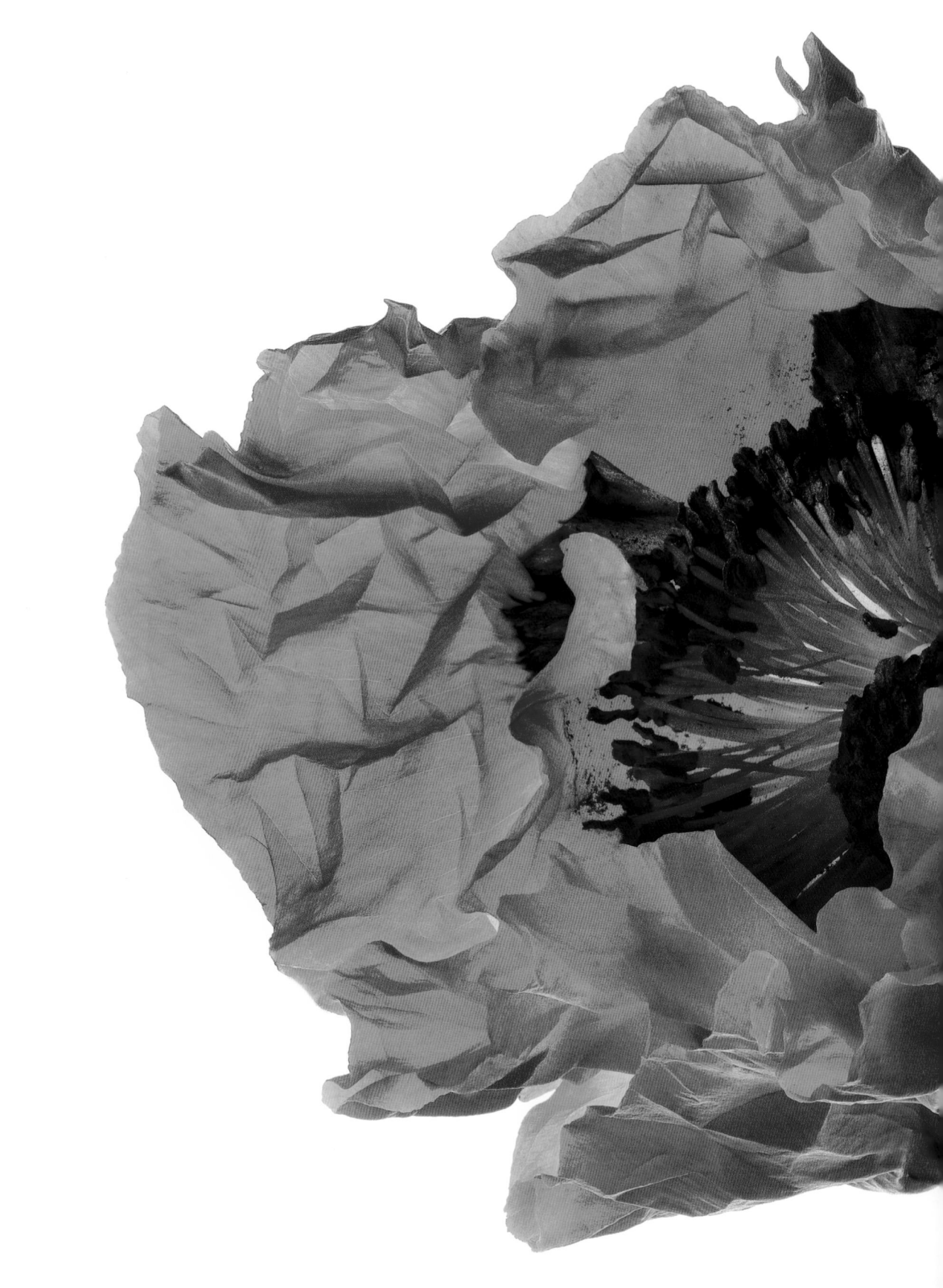

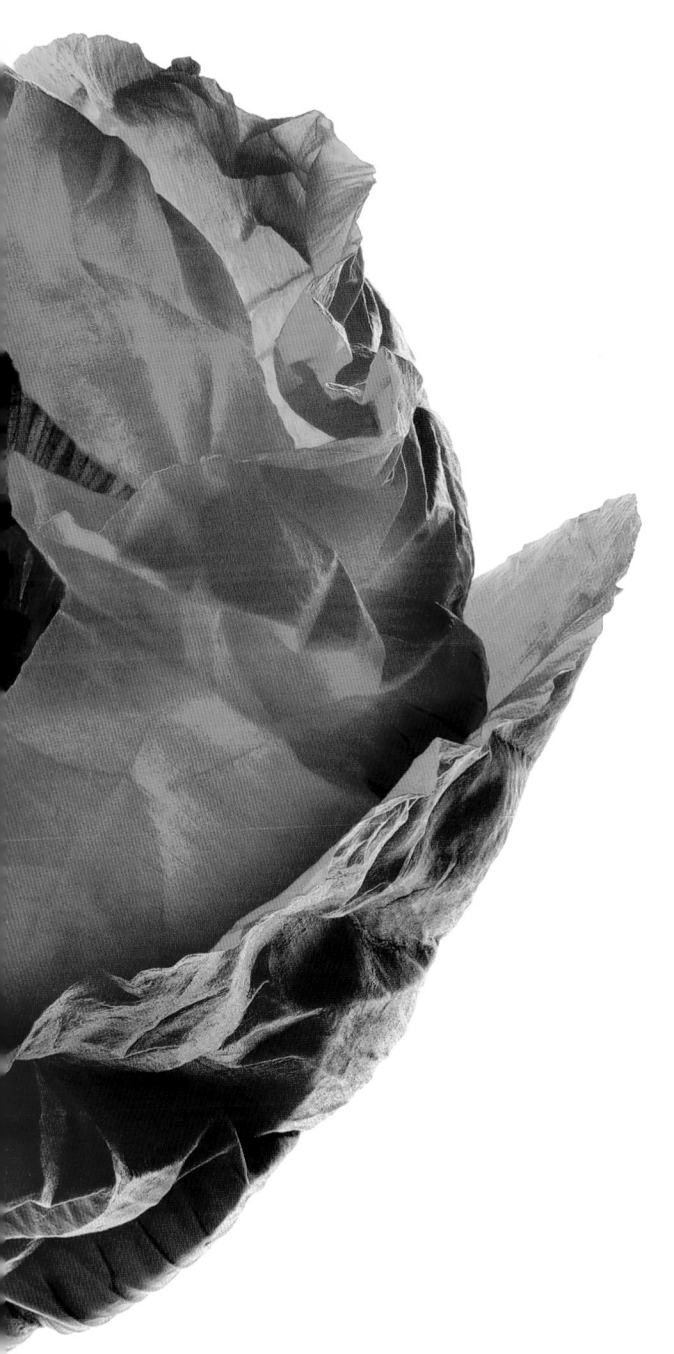

"Where there is space for them, poppies provide an air of unstudied fatness and richness that no other late-spring flower provides, not even the peony."

HENRY MITCHELL, *The Essential Earthman*, 1981

Possibly unrivaled in the plant world for pure, sexy beauty are the Oriental poppies, *Papaver orientale*. From their bountiful crepe-paper petals to the swollen-looking stigmatic disk in a flower's center, they exude a tactile, sensual appeal. The ruff surrounding the disk is a crowd of pollen-topped stamens.

PAGE 252 The sentimental favorite among poppies is arguably the corn poppy, *Papaver rhoeas*, also called red poppy or Flanders poppy. Four satiny petals are centered by a boss of dark stamens to form one of the world's most beautiful flowers.

From their colorful blossoms to their colorful history and uses, poppies (Papaveraceae) just might be the most fraught of all plant families. The color range is generally, well, hot. Whether radiant orange like California poppies (*Eschscholzia californica*), or bright red like field or corn poppies (*Papaver rhoeas*), or glowing yellow like celandine poppies (*Stylophorum diphyllum*), these flowers are always luscious. Nor does size seem to matter in their overall appeal. Big Oriental hybrids such as *P. orientale* 'Glowing Embers' dazzle, while wee ones, like the rock-garden darling *P. alpinum*, bring tall gardeners to their knees.

The sought-after Himalayan blue poppy, *Meconopsis betonicifolia* (*M. baileyi*), is something of a holy grail for some gardeners, not just for its exotic origins but also its unusual medium blue color. There are no cultivars of this beautiful species; it would be hard to improve upon such natural perfection.

Variations of hue aside, we must also hail poppy-petal texture, which only heightens the flowers' appeal. California poppy petals are especially smooth and silky, a seductive feature plant breeders have capitalized on, as in the popular, multihued 'Thai Silk Mix'. Oriental poppy petals also have a satiny feel but are frequently crinkled or crimped—and thus likened to crepe paper, although they are not as fragile. You long to touch them, to caress them.

The petals of some poppy species, especially the Orientals, include a contrasting blotch at the base. When blooms are freshly unfurled, this feature is tucked deep in the flower's interior and veiled from view (unless, of course, you draw near and peer in). The practical purpose of this mark is thought to direct insect attention

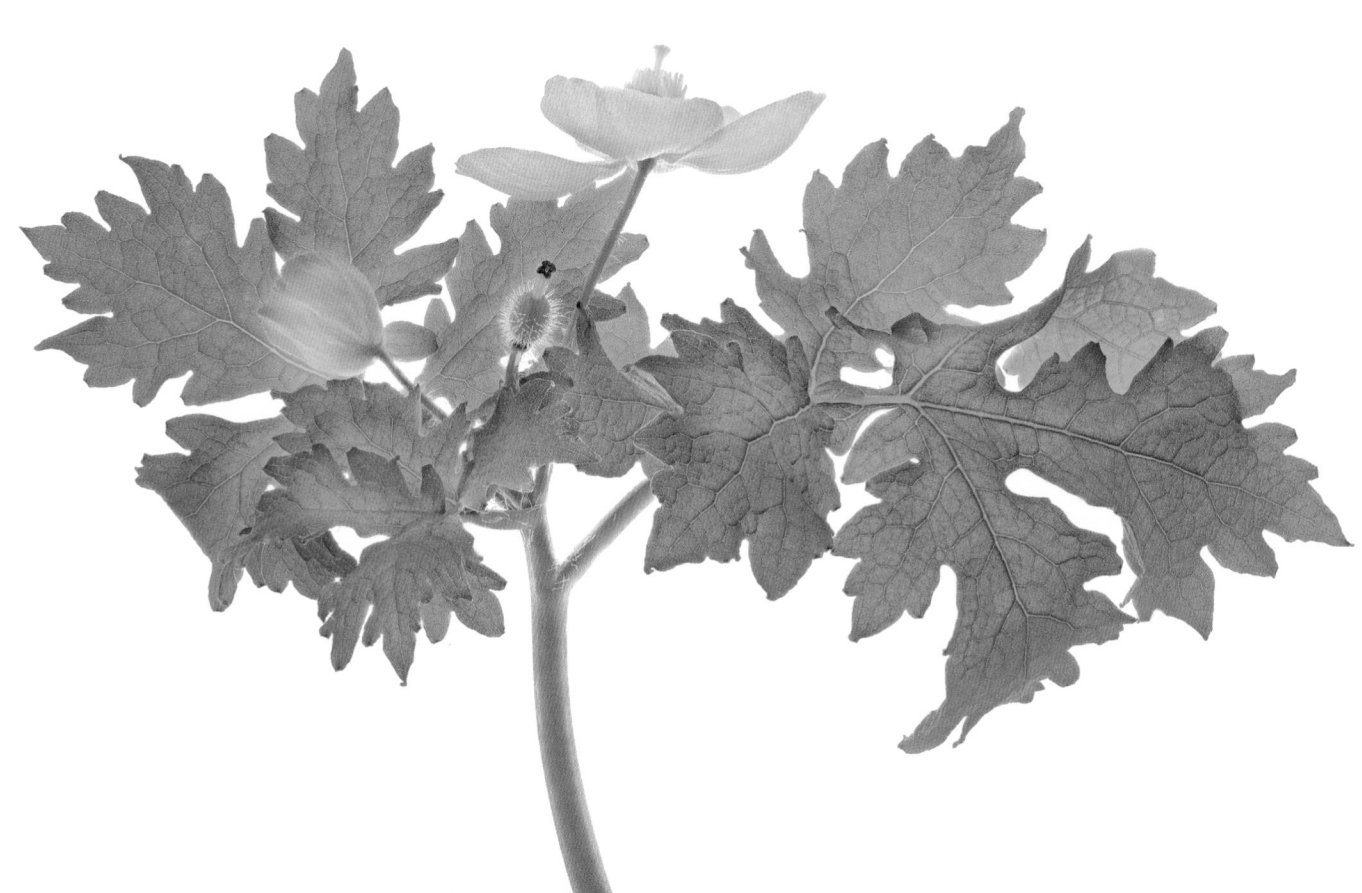

toward the center, where pollen awaits. The aesthetic benefit is smart contrast with the petal color.

The center of a poppy bloom is unique beyond the presence of spots. The large center structure (usually black or yellow) is the stigmatic disk, home to the female parts of the pistil where male pollen grains land and germination commences. Poppy stigmas unite in a radiating pattern to form this cushionlike disk—an uncommon sight in the plant world. As you might expect, the ovary (the repository of immature seeds awaiting fertilization) is down at the very base of the bloom. It is interesting that there is no neck or pathway down from the stigma to the ovary in poppies, as in many other flowers.

The stiff-looking things (usually black or yellow) surrounding the stigmatic disk are stamens, each one tipped with a minute, pollen-filled anther. There are usually dozens present in a thick ruff.

The various poppy blossoms are so spectacular that it may escape your notice that the form is very simple. Most species have only four—or, at best, eight—petals; double and more elaborate cultivars sport multiples of four. In any case, they tend to overlap, and are sometimes fringed along the top edges, furthering the illusion of bulk and complexity.

There is something show-offy about the way most poppy blossoms develop. Fuzzy, chubby little buds nod on flexible, fuzzy stems, looking almost frail and shy. As these fill and swell in size, they turn upward toward the sun and eventually spring open. The sequence is dramatic and decisive, like a ballerina leaping to life from a crouching position. Poppies really "pop."

These remarkable flowers loom large in human imagination and history. Where young men fought and died in the rough fields of Belgium, Flanders

LEFT You may recognize greater celandine, *Chelidonium majus,* because it is wending its weedy way into some damp area in your yard or neighborhood. It qualifies as a poppy family member: Four small petals spread out flat beneath the stamens.

BELOW While still green—that is, not yet ripe—a *Papaver orientale* seedpod subtly swells, and both petals and stamens eventually drop away. When it starts to change color, turning purplish or gray-brown, and holes develop between the top cap part and its base (think scuppers on a boat's deck), the tiny seeds within are ready.

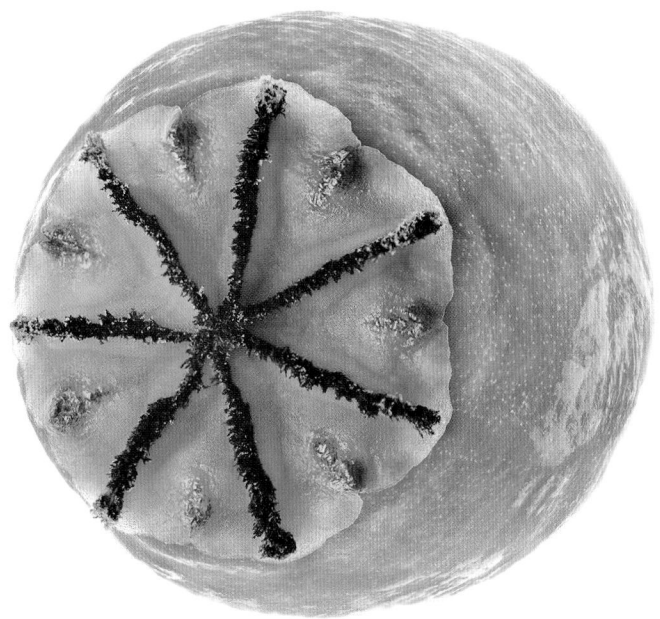

poppies (another name for field poppies) emerged. Blood-red beauty where blood was shed—a heart-rending image. Let's go directly to the famous poem, "In Flanders Fields," composed in 1915 by WWI soldier John McCrae:

> In Flanders fields the poppies blow
> Between the crosses, row on row,
> That mark our place; and in the sky
> The larks, still bravely singing, fly
> Scarce heard amid the guns below.
> We are the Dead. Short days ago
> We lived, felt dawn, saw sunset glow,
> Loved and were loved, and now we lie,
> In Flanders fields. . .

Veterans and those wishing to honor them pin small paper or cloth versions of these little red poppies to their lapels to commemorate the war dead on Memorial Day in the United States and Remembrance Day in Canada and Europe.

Then, of course, there is the opium poppy, *Papaver somniferum*, with its insinuating species name, which means "sleep inducer." This one has the long and notorious history of providing China, and in due course the rest of the world, with heroin and other opiates. The twenty-first-century conflict in Afghanistan, and indeed the dangerous and lucrative worldwide drug trade, reminds us that this particular poppy is still onstage.

All poppy parts are said to be poisonous, a designation herbalists deploy to warn you away, regardless of whether it is wholly or consistently true. In the case of

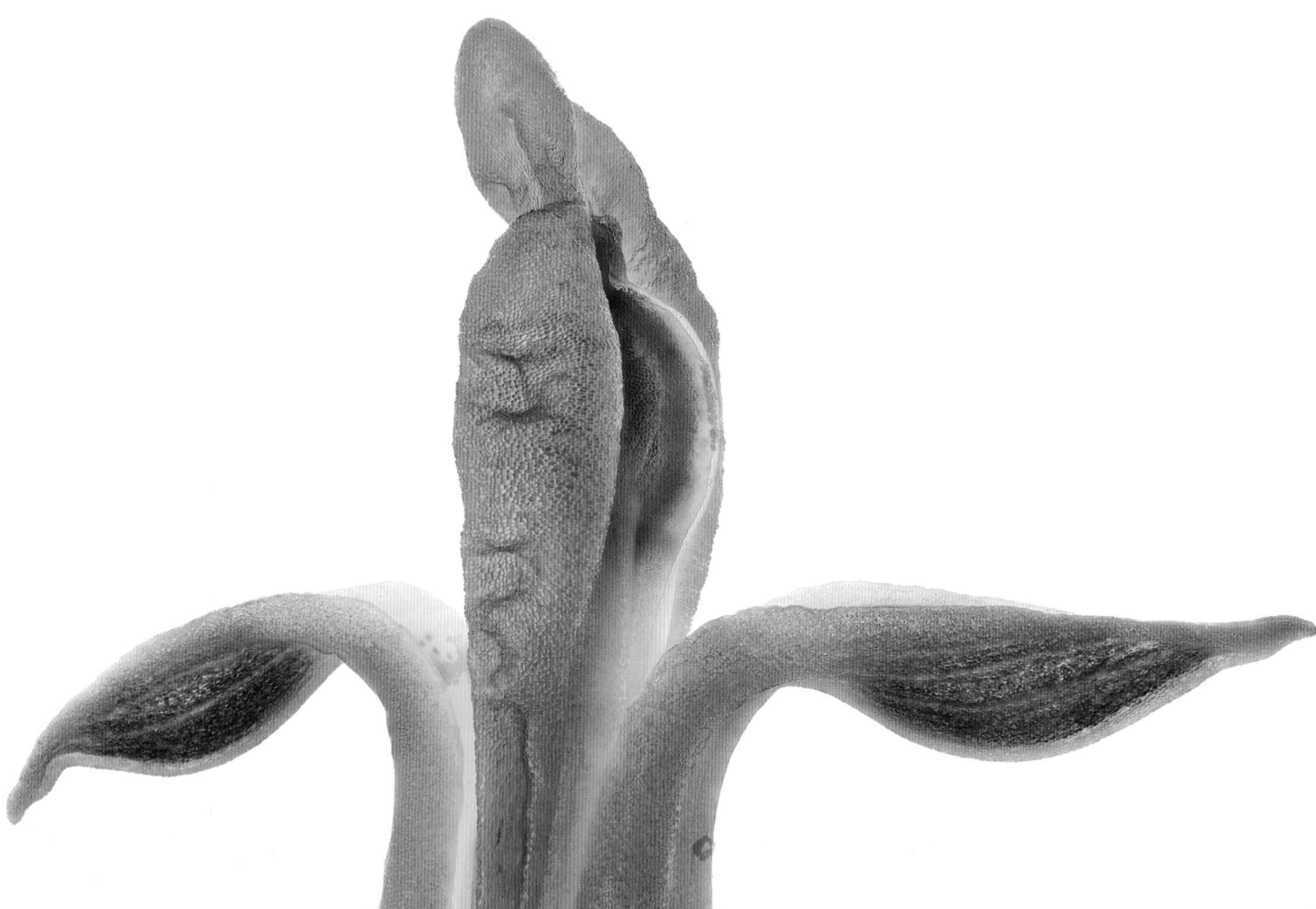

Papaver somniferum, the cautions are especially strident. According to science writer Michael Pollan, "Grown or purchased, fresh or dried, the seedpods contain significant quantities of morphine, codeine, and thebaine, the principal alkaloids found in opium." With such ingredients, however, this poppy has contributed its chemicals to legitimate, sanctioned medicinal uses, principally that of pain relief.

Powerful, addictive opium is derived not from the seeds, however, but from this plant's sap. To extract it, one scores an unripe (blue-green) seed capsule and harvests the bitter milky substance, latex, that emerges. Before you consider a career change from gardener to druglord, bear in mind that it would take a huge amount of sap to generate a significant amount of drug. Don't try this at home.

The sap is also made into a non-narcotic oil that is high in vitamin E. This oil has culinary and pharmaceutical applications, as well as long-established uses in making paints, varnishes, and soaps.

Speaking of edible poppy parts, those tiny, blue-black poppy seeds—the same kind you find in baked goods, including your favorite bagels—hail from this poppy. They add subtle peppery flavor and a satisfying crunch, but they will not get you high. That said (and none other than the iconoclastic investigative website snopes.com backs this up), ingesting a few poppy-seed bagels can cause you to flunk a drug test. In fact, foods containing the seeds are banned from federal prisons to avoid skewing random or routine tests on inmates.

Another surprise is that bleeding hearts, *Dicentra* species, are included with poppies. Shown here are *D. cucullaria* (NEXT SPREAD, LEFT), called Dutchman's breeches for its odd white flowers, *D. spectabilis* (NEXT SPREAD, RIGHT), the popular garden plant, and *D. formosa* (OPPOSITE), with its little pink flowers. Their generally locket-shaped flowers do have the requisite four petals—the outer ones are pouched, while the inner ones connect at the tip and clasp the stamens and pistil between them.

In addition to, or despite, all this dangerous baggage, opium poppies are no less beautiful than many of their cousins, and gardeners prize them. The good-size blossoms are plush with petals, leading to a noncontroversial alternative name, *Papaver paeoniflorum*, which reflects their similarity to plush peonies. Pollan has written extensively on their storied history in garden plots worldwide. You can probably enjoy their beauty in your garden if your local police have more important crimes to attend to.

No matter which sort of poppy you grow, the flowers do eventually jettison their petals (sooner than you'd wish) and leave behind distinctive seedpods. Those of the opium poppy are larger and bulbous, while California poppy pods are long and pointy, like a sultan's slipper. In any event, when poppy pods brown and dry, the seeds are ripe. Watch for the little "windows" around the crown to open. Snip them off and you can sprinkle the seeds just like with a pepper shaker.

Flower arrangers and crafters intervene before this happens, snipping off dried stems topped by dried pods and adding them to arrangements. Spray-painting them silver or red might strike you as tacky, but such sprigs can make an intriguing addition to a dried bouquet or wreath while sealing in the seeds and preserving the pod's form.

No doubt about it, poppies command attention. They are imperious. They are seductive. They are irresistible. If you fall under their spell, you'll forgive their relatively short bloom period (about two weeks or so in early summer for many), their short lives (annuals can self-sow or be sown again, while perennials just have to be replaced), and all the space they take up. You'll forgive their rather weedy, rangy foliage and the way they look so disreputable when they die back. You'll bow to their beauty every time they pop.

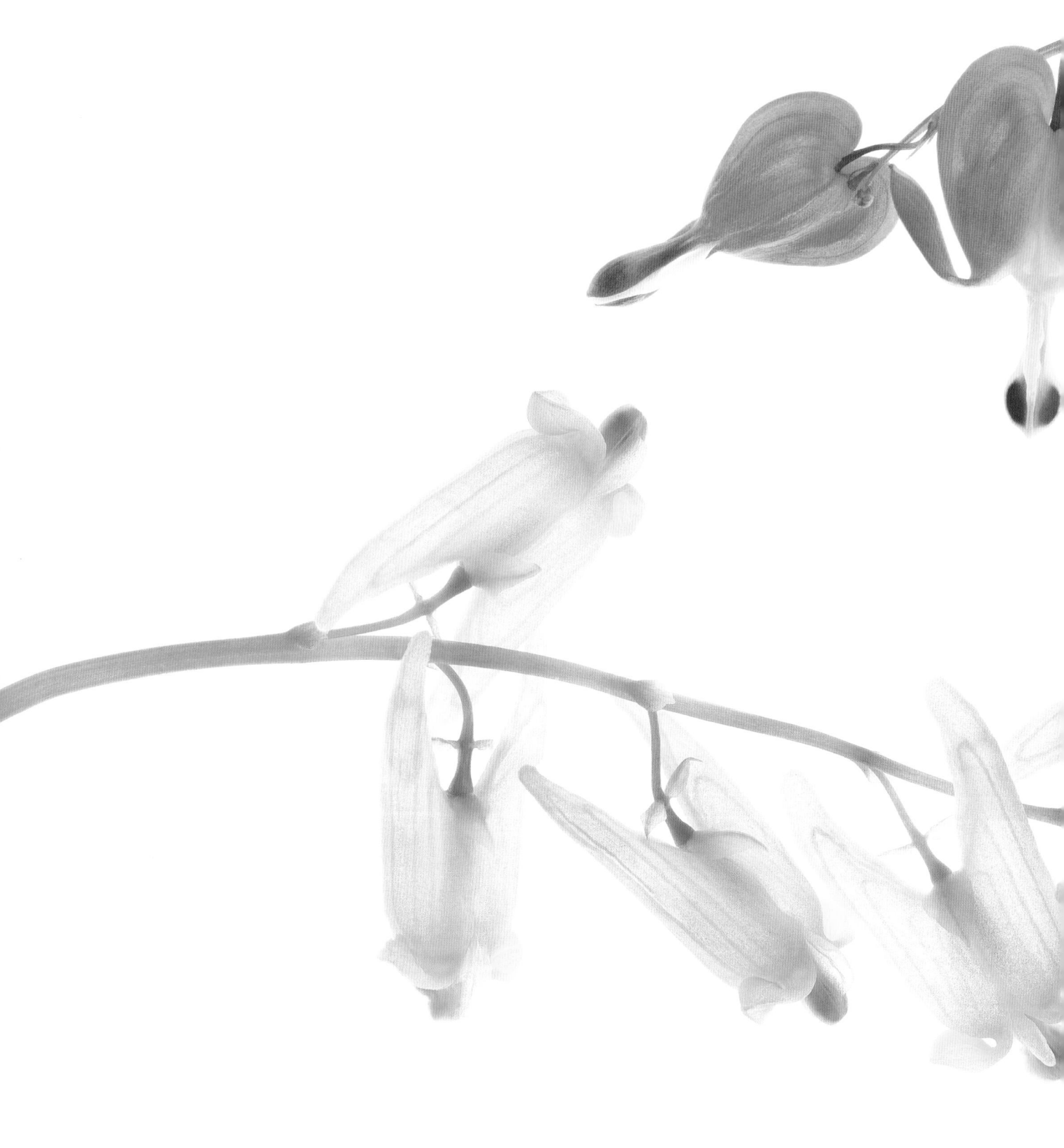

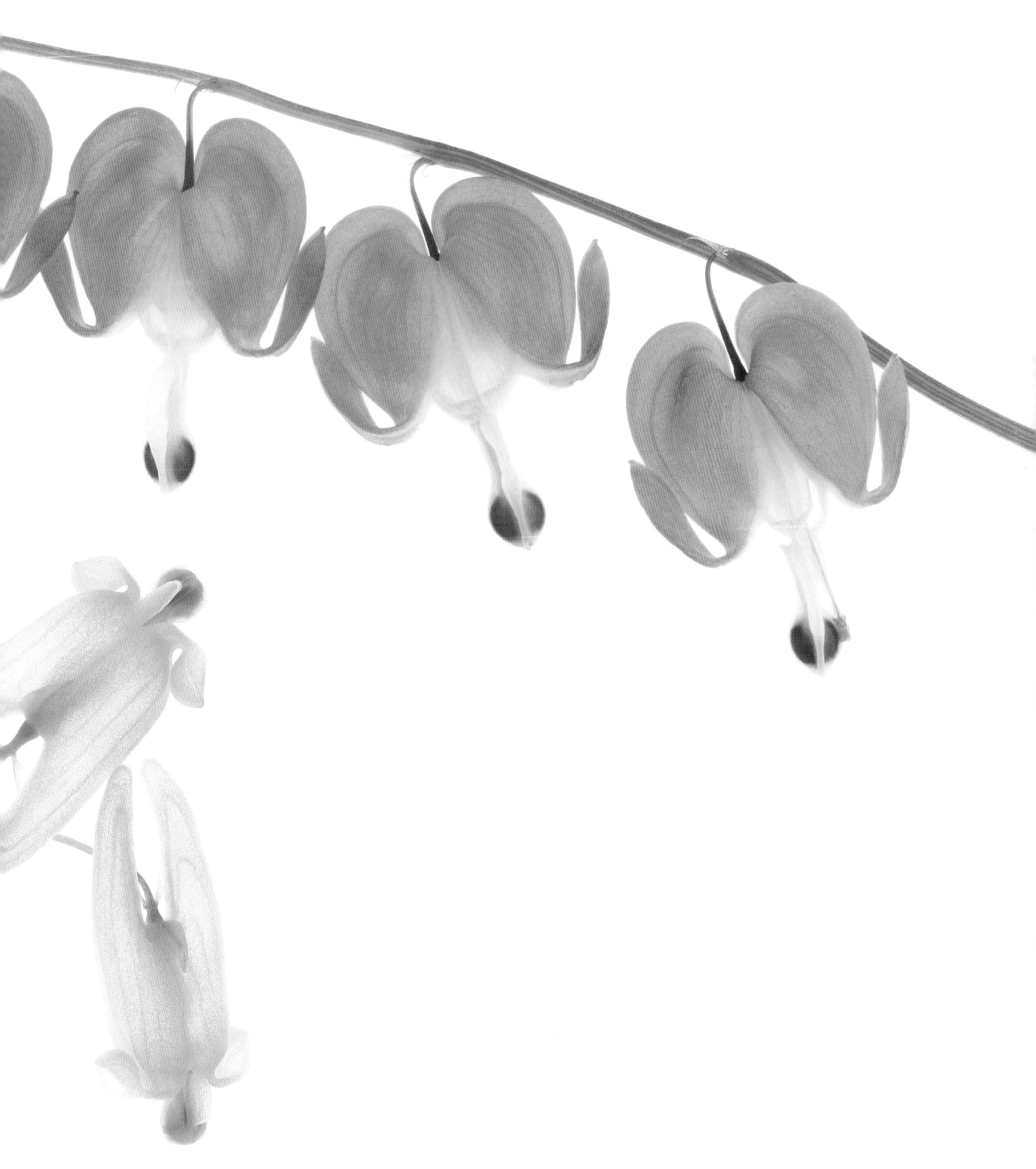

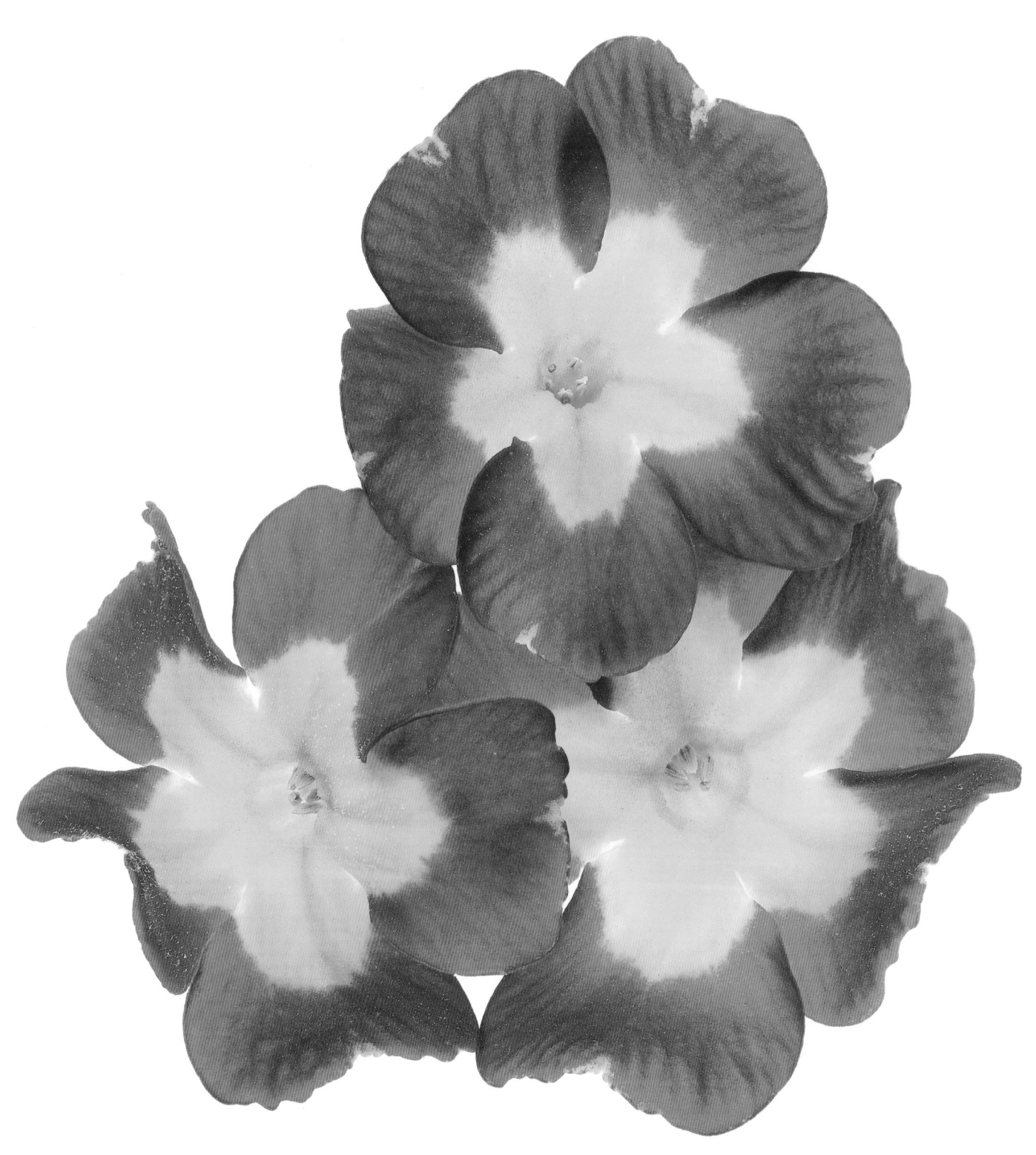

THE PRIMROSE FAMILY

Primulaceae

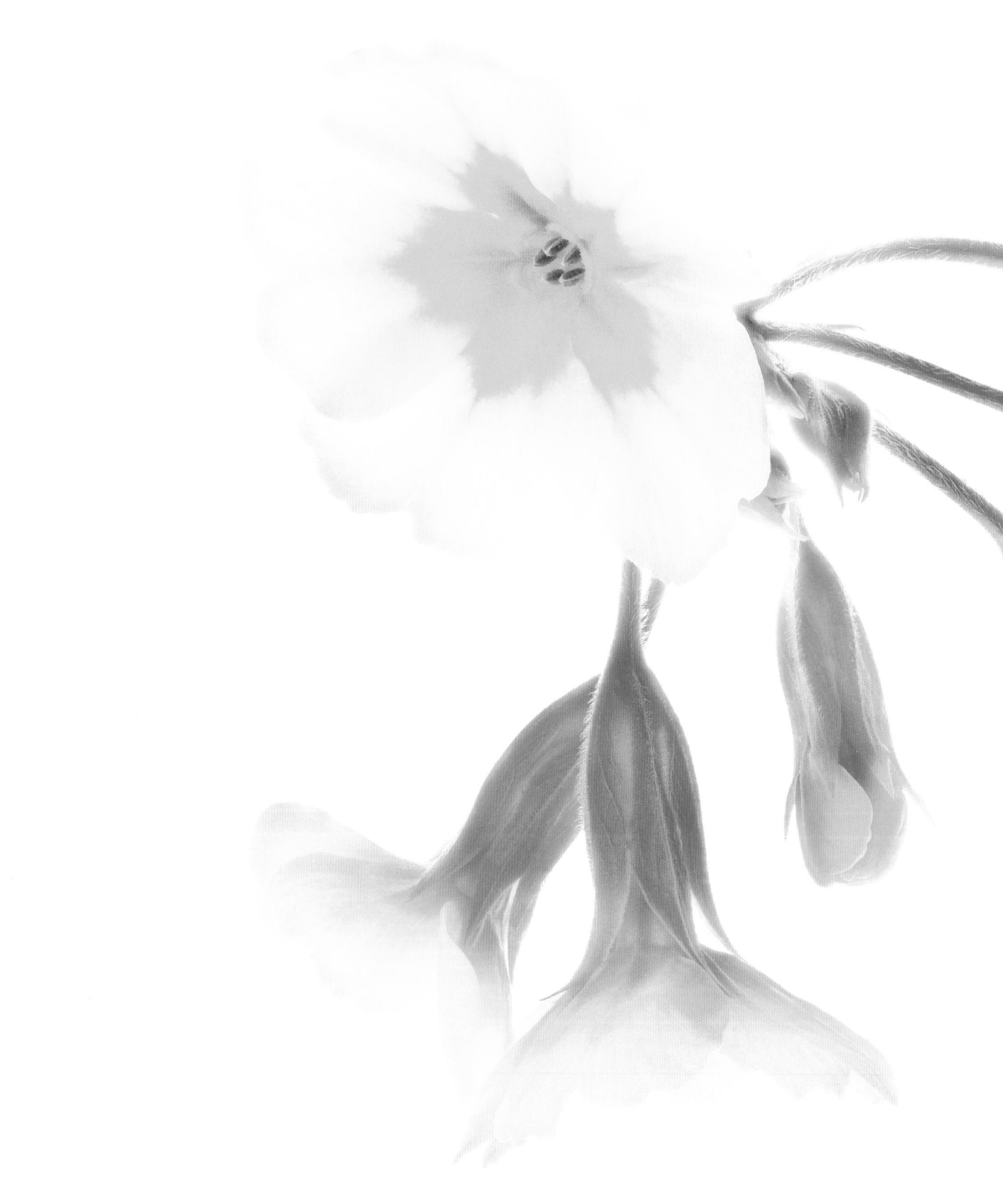

"O fairest flower,
No sooner blown but blasted,
Soft silken primrose
Fading timelessly"

JOHN MILTON, "On the Death of a Fair
Infant Dying of a Cough," ca. 1626

LEFT AND PAGE 262 Polyanthus primrose hybrids,
Primula vulgaris or *P. acaulis*, readily display the classic
features of their family: five petals, five sepals, and five
stamens. Less obvious are the subtle differences in pin
and thrum type flowers to allow pollination (details in the
text). Shown here are examples of "thrum" types.

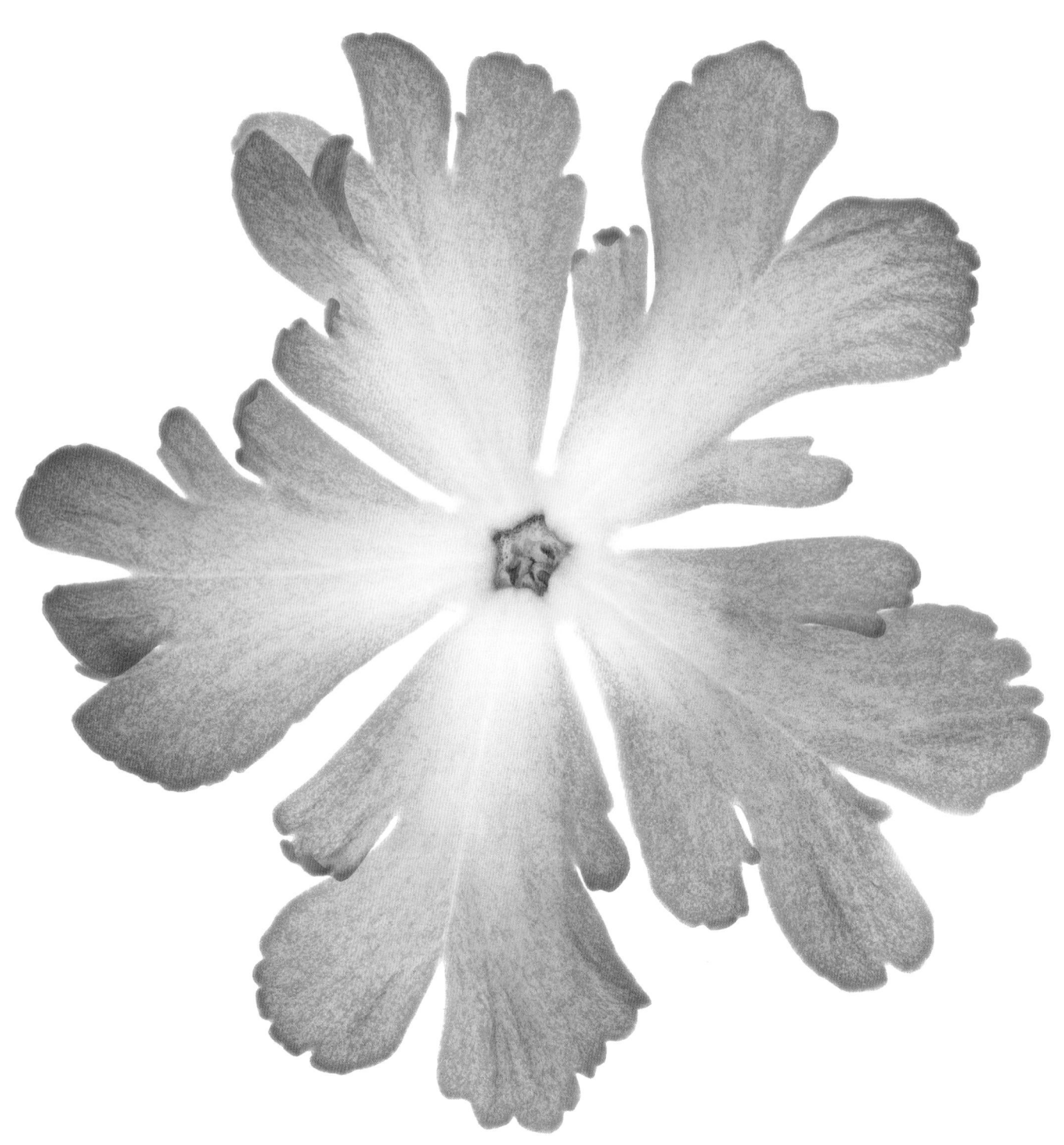

f you've ever been fortunate enough to see primroses naturalized in a woodland setting, you understand why they capture hearts. In the dappled shade of tall trees (their favorite growing conditions), perhaps with a stream dancing through (almost all of them love moisture), jaunty little blossoms raise their pretty heads above quilted basal rosettes in pastels, purple, pink, or creamy white.

The woodland ones that gardeners prize, hailing originally from China and Japan, are generally the Japanese or "candelabra" ones, *Primula japonica*; the so-called drumstick primroses, *P. denticulata*; and the Japanese star primroses, *P. sieboldii*. Any of these can reach about a foot high, and they display their bright blooms atop strong stalks. In *P. japonica*, they're carried in loose whorls; in *P. denticulata*, the inflorescence is indeed reminiscent of a plush drumstick; and the form in *P. sieboldii* is best described as crowded umbels.

English primroses or cowslips, *Primula vulgaris* (also referred to as *P. acaulis* or acaulis type), are lower growing and more compact in habit. When thriving, they usually present their flowers as a bouquet or umbel; the species is pale yellow. But both Asian and English primroses have been extensively hybridized, hence the broad color range as well as some double forms and some with contrasting-color eyes—the better for them to stand out, particularly in dimmer settings.

This is a large genus. Estimates vary from 400 to 500 species, with subdivisions, and their origins are diverse.

Some hail from the Himalayas, some the Mediterranean region, some the Arctic, and some California's High Sierras. Many of the alpine ones are the darlings of rock gardeners and are rarely found at nurseries. Collectors tend to raise them from seed, pamper them with humusy soil, and divide clumps every few years to maintain a patch's vitality.

If you want an easy primrose, your best bet is those enjoyed as annuals in the United States, the polyanthus primroses, *Primula polyantha*. They've been in cultivation a long time and are thought to have been derived from *P. vulgaris*, *P. elatior*, and *P. veris*. Their bold, cheerful colors are a welcome addition to spring gardens, including bulb displays, pots and window boxes, and as bedding plants within or at the edges of still-sleepy flowerbeds. Like their cousins, they prefer enriched soil and cooler temperatures.

Common to all primroses—rare or common, no matter what type of cluster they are carried in—is the form of the individual flowers. These are in five parts: five petals, often fused to form a five-parted corolla, with the calyx united at the base to form a tube (inside are anthers, style, and stigma); five stamens are attached opposite the petals; and the seed pod that follows germination is a five-valved capsule. The pistil alone is solo. The sweet nectar is a bit of a reach for many insects, so pollinators, particularly various butterflies and moths, usually have long tongues. Primrose fragrance, when present, is sweet and fresh, almost citrusy to some noses.

The frilly, playful-looking blooms of the Japanese star primrose, *Primula sieboldii*, are called salverform, which simply means that the tubular corolla ends up spreading out into flat lobes—five of them, as is typical of Primulaceae.

NEXT SPREAD Like perky butterflies, the five reflexed petals of *Cyclamen coum* are typical of the genus, but their tiny size (½ inch across, at most) and round shape is not usual. Darker blotches just above the white mouth make it easy for pollinators to find their way. Once seeds form, the flower stalks coil down to the soil surface to release them—hence the botanical name, derived from *kyklos*, the Greek word for circular.

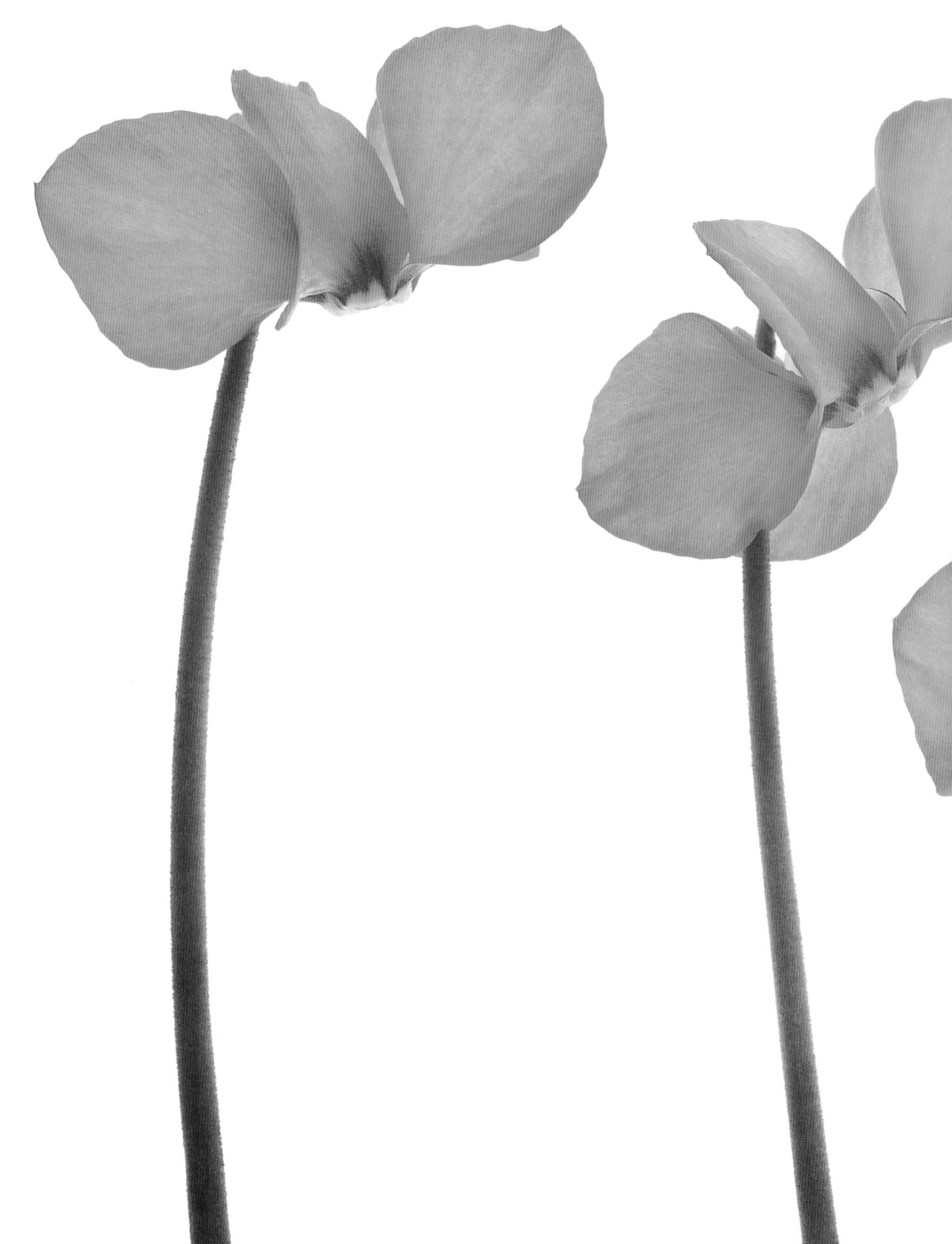

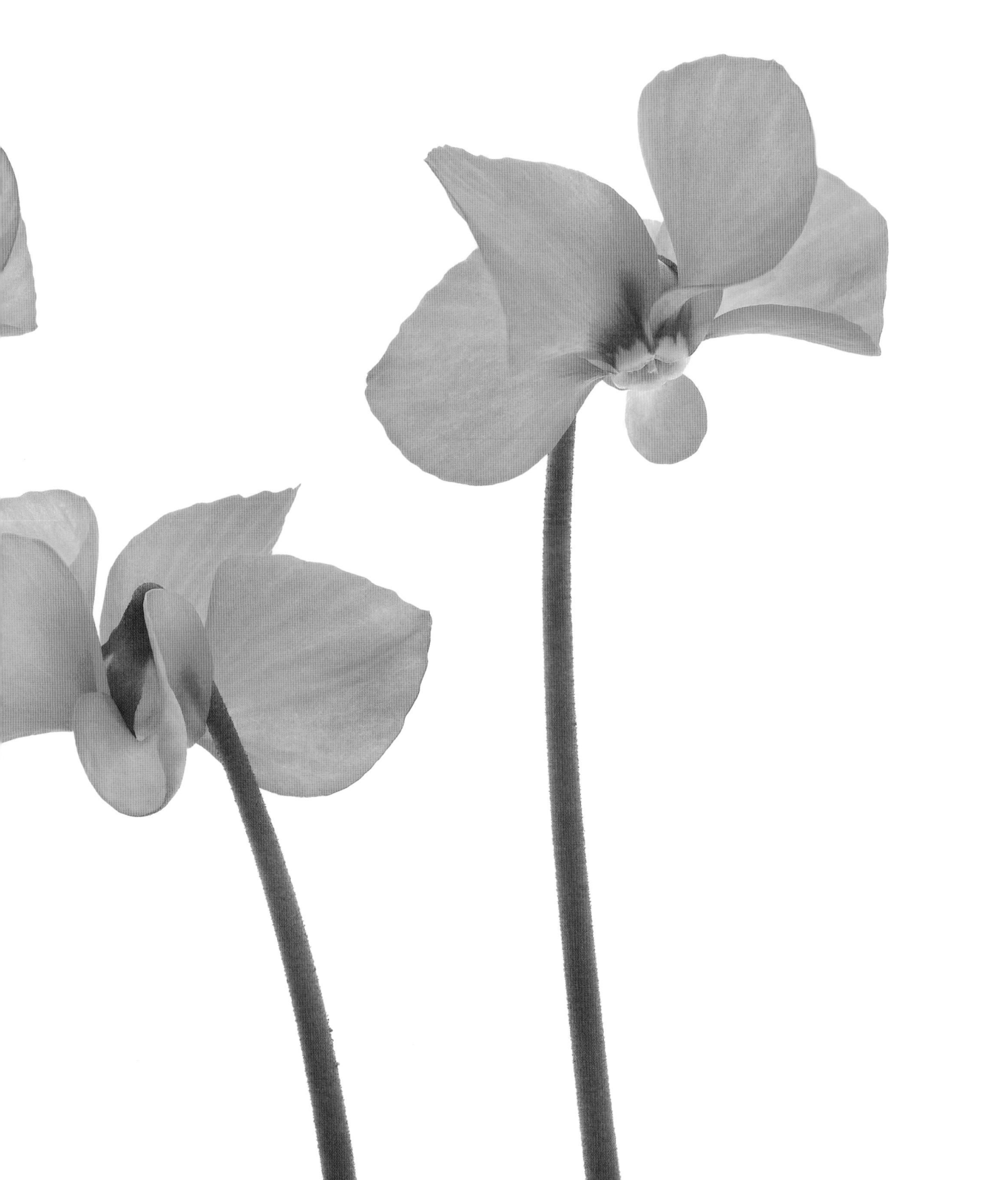

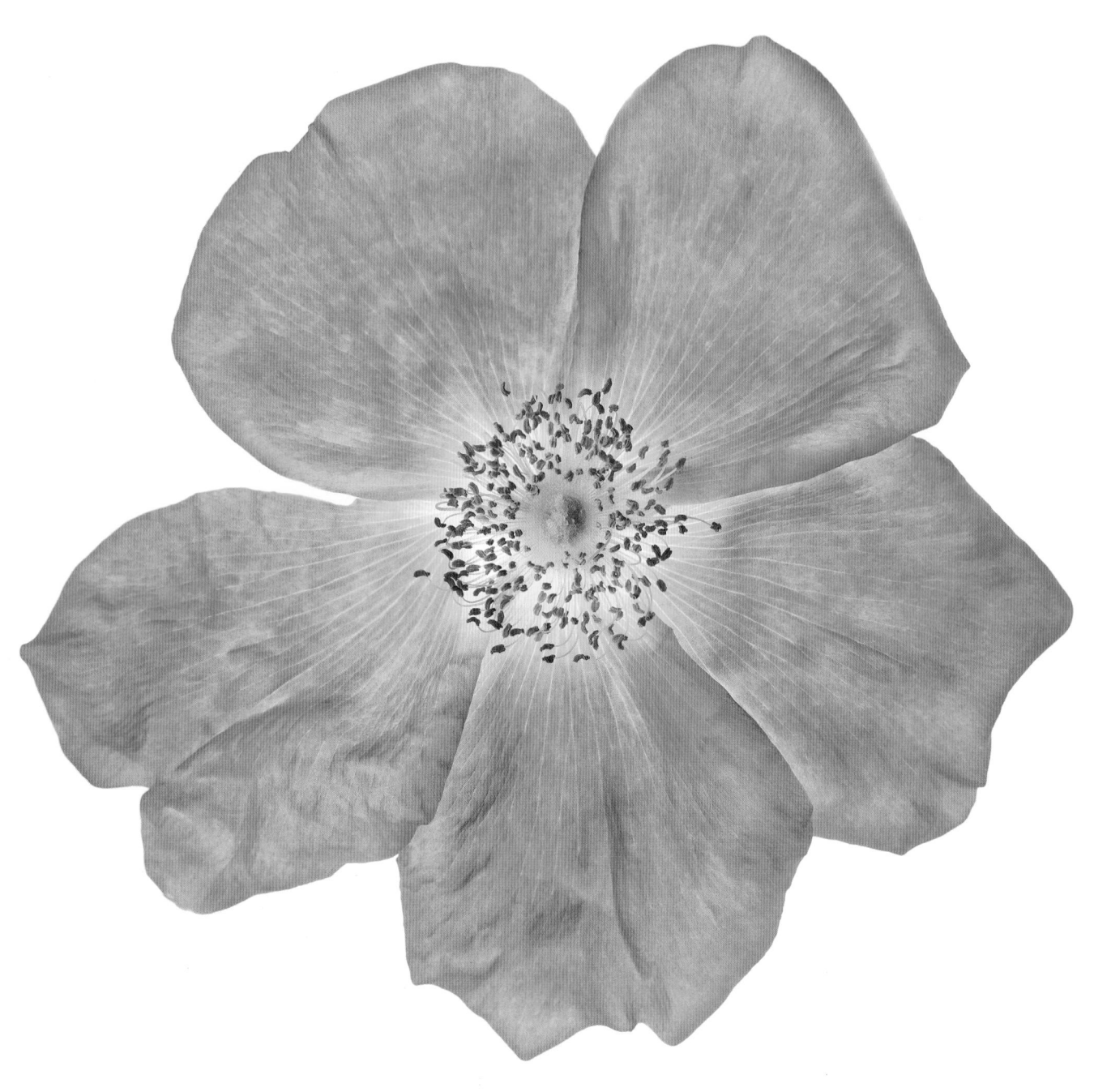

THE ROSE FAMILY

Rosaceae

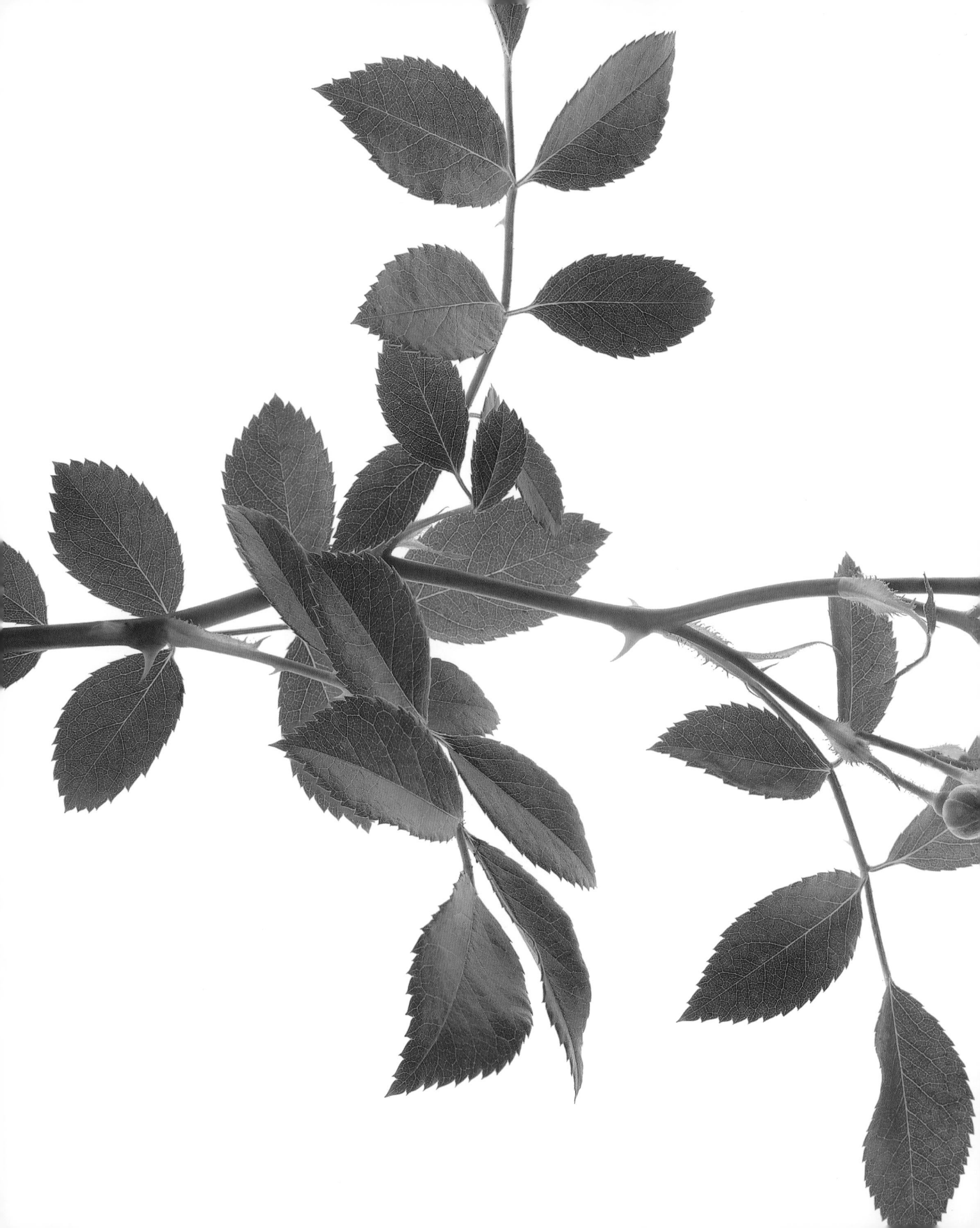

"The rose is a rose,
And was always a rose.
But now the theory goes
That the apple's a rose
And the pear is, and so's
The plum, I suppose.
The dear only knows
What will next prove a rose.
You, of course, are a rose—
But were always a rose."

ROBERT FROST, "The Rose Family," 1928

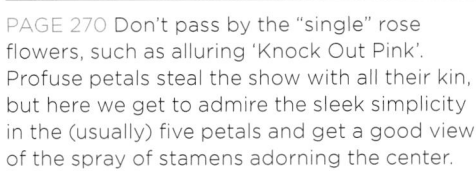

PAGE 270 Don't pass by the "single" rose flowers, such as alluring 'Knock Out Pink'. Profuse petals steal the show with all their kin, but here we get to admire the sleek simplicity in the (usually) five petals and get a good view of the spray of stamens adorning the center.

LEFT AND PAGE 275 Few flowers have been in cultivation, or been as beloved, as long as roses. This makes their heritage complicated and confusing, and there is a great deal of variation. But common to all, from the simplest five-petaled single-form blossoms to the plushest David Austin varieties, are those silken petals, the pincushion center of pistils and stamens, the characteristic foliage, (often) thorns, and the fall fruits or hips. Fragrance, where present, is generated by various aromatic chemicals and fatty acids embedded in the petals.

What's the first thing you do when you see a rose bloom? You stick your nose in it, close your eyes, and inhale deeply. No florist bouquet, no fancy bar of soap, no bottled perfume can quite compare with that delicious, sensual moment. If warm summer sun is on your back and has heated up and released the volatile oils in the petals, you'll linger or sniff twice. Humidity also helps contain and enrich the scent, because it slows or inhibits evaporation.

If you indulge in this pleasure often, or are lucky to visit a rose garden on a warm day, you will also notice that not all roses smell the same. Aside from the classic old-fashioned tea rose perfume, there are roses with distinct scents of citrus, sweet violet, sweet pea, freshly cut apples, raspberry, ripe currants, honey, fern, parsley, bay leaf, anise or sweet licorice, and spices like cloves and cinnamon. Meanwhile, flower form has exploded into a range of variations, from dainty little singles to plush blossoms packed with too many petals to count, along with a bounty of luscious colors and blends (save blue). Plant form, too, is variable: There are shrubs and climbers; cluster-flowered, lower-growing floribundas; ground-covering types; miniatures; and more.

If it sounds like this account is veering into jargon or connoisseur talk, realize that today's great array of roses is no accident. Roses have a long history. Their innate variability and amenable fertility have also promoted great variation.

Roses have been in cultivation so long that establishing their history is a matter of debate for archeologists examining the fossil record. Most agree, however, that the debut was approximately 5000 years ago in China. That said, native species thrive all over the world, except in extreme climates. However, it was not until late-eighteenth-century Europe that botanists began to study the group, enthusiastic gardeners began to tamper with and improve roses, and their popularity surged. Today finds the menu very full and the genetics, nomenclature, and classifications quite elaborate.

Rose flower form is a key reason the genus is so complex and varied. An open, round bloom with a stamen-studded center is an easy invitation to pollinators—chiefly bees but sometimes beelike flies—with an ample landing pad and plentiful, accessible pollen. Because the blossoms have both male and female parts, roses can and do self-pollinate. This means a foraging bee may take care of everything in a single visit, or buzz around creating inadvertent crosses.

Successful pollination leads to the development of hips, or the fruit of the rose, which spangle most plants by autumn and are full of seeds. Alas, if you were to raise these seeds, there is slim chance they would resemble the parent. Form is also diverse. Some are little reddish urns, while others look like plump orange berries. And it's true that rose hips are laden with vitamin C and safe to eat.

To create a new or improved rose variety, a hybridizer supersedes the bee's work and transfers male pollen onto the female parts of another rose at just the right moment (while keeping detailed records). But most roses already have complex and sometimes unknown backgrounds, so it's not always easy to guess the results. The big rose nurseries have huge greenhouses and fields to allow their breeders to grow and evaluate the results of their crosses. Sometimes seeds are frustratingly sterile. Sometimes not every desired quality is on one plant, which requires a careful shuffling of genes. This is why some modern roses are not fragrant, as sometimes developing a superior, florist-quality blossom means sacrificing scent. Or developing disease-resistant foliage equals staying with a high thorn count.

In any event, perhaps only one seedling in thousands meets the breeder's goals and standards. All this diversity and experimentation is a good thing, though. Lodged in this expansive gene pool are resistance to rose

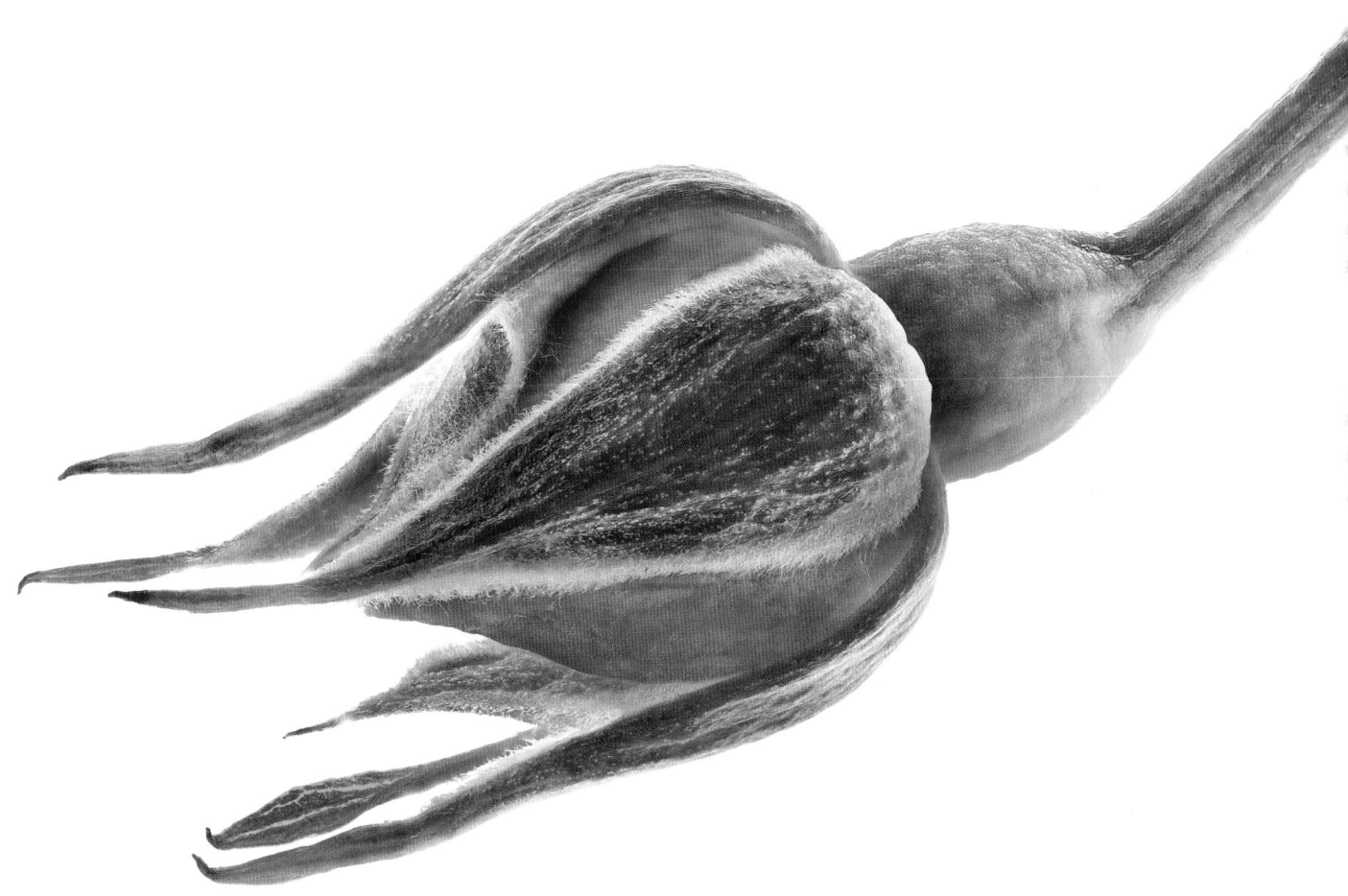

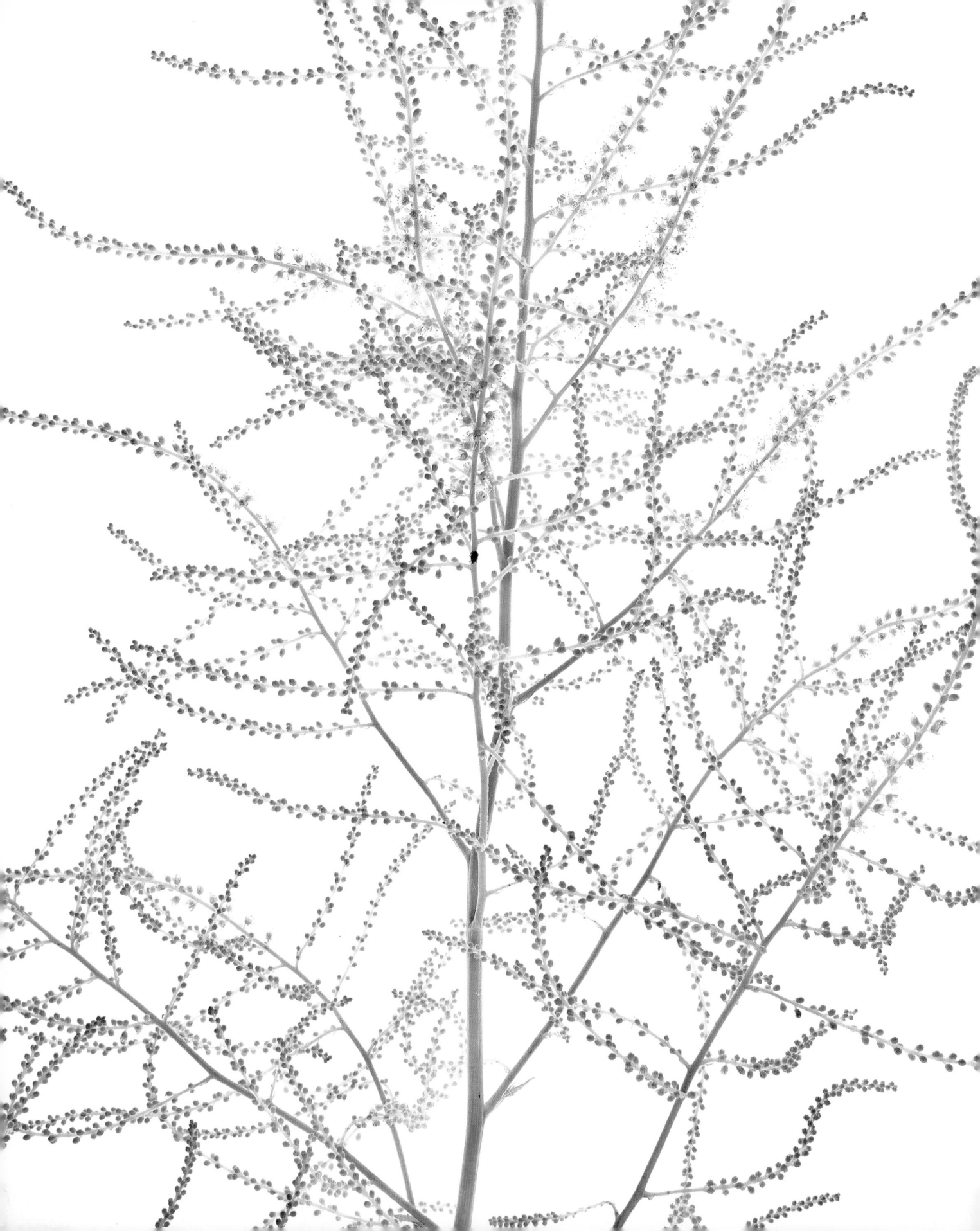

diseases and pests, the ability to bloom at different times or over longer periods, and plentiful options in scent, color, flower form, blooming habits, hip appearance, and plant growth habit.

Roses dominate this family by sheer numbers, as well as the big place they have held in public attention and affection for so long. But there are plenty of other species. These days, botanists place nearly 3000 besides *Rosa* in Rosaceae. Many members, to look at their flowers or fruits, are obviously relatives, such as cinquefoil, *Potentilla*, apples and crabapples (*Malus* genus), berry bushes (*Rubus* species, such as raspberries and black-berries), strawberries (*Fragaria*), hawthorn trees (*Crataegus*), and mountain ashes or rowan trees (*Sorbus*). Some of these have benefited from breeding and selection, not just to produce garden-friendly plants (smaller, more compact, and, where applicable, less thorny) but to capitalize on disease resistance and improve flower color or fruit appearance and quality. As with roses, the pollination process is not tricky.

In the flowers of these relatives, there is the same form of the simplest species roses: five sepals, five petals, a flat-disk form of ample petals when fully open, and (usually) both male and female floral structures on the

LEFT The teeny-tiny, creamy white, five-petaled flowers of goatsbeard, *Aruncus dioicus* (*A. vulgaris*), are borne on dense plumes above dark green foliage. Male and female ones are on separate plants; the males are showier, thanks to plentiful stamens.

BELOW What is a rose thorn, really? These sickle-shaped hooks are technically prickles, outgrowths from the stem's bark. You can push them off with your thumb, as any florist or flower arranger will attest, whereas with a true thorn, a modified stem, you cannot.

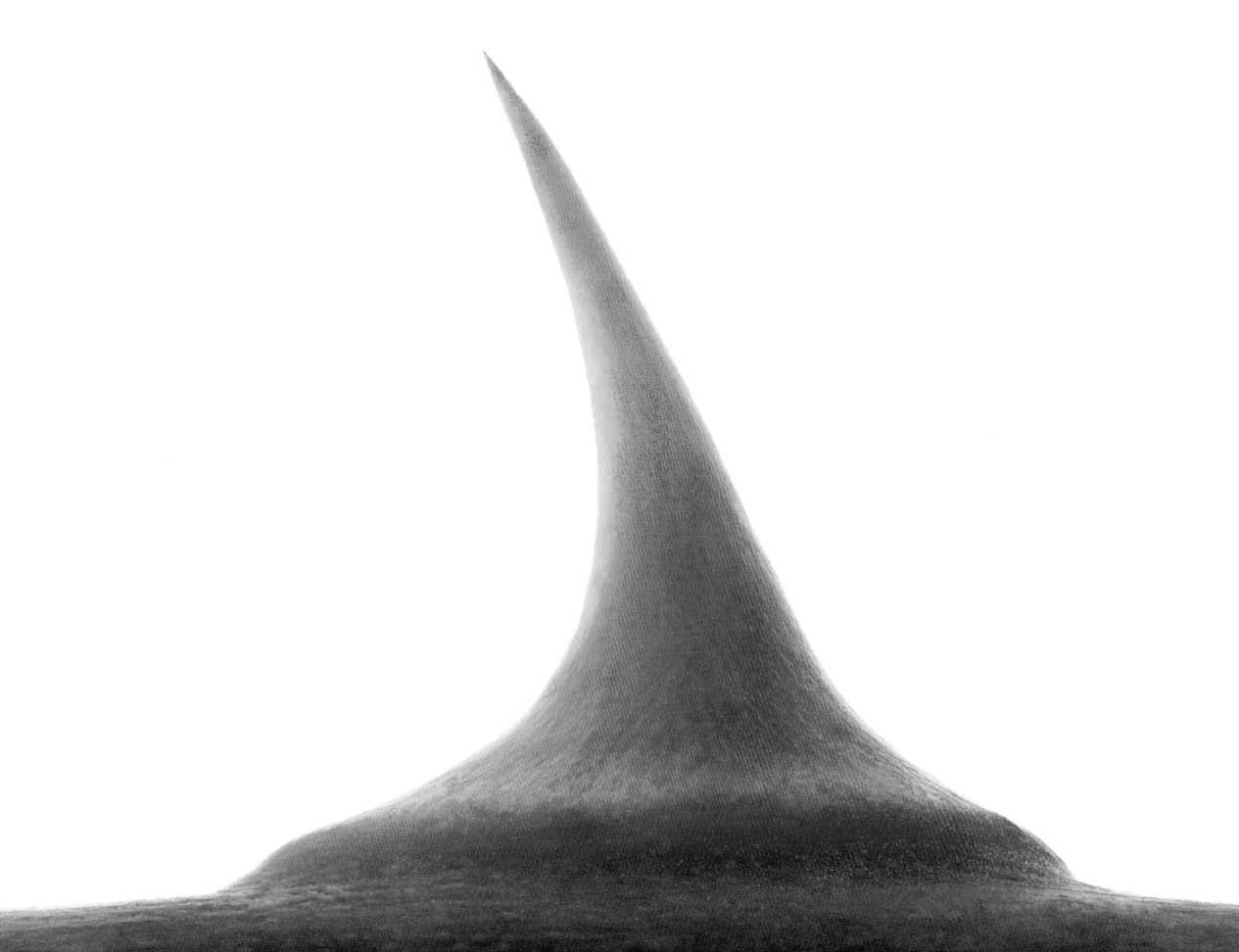

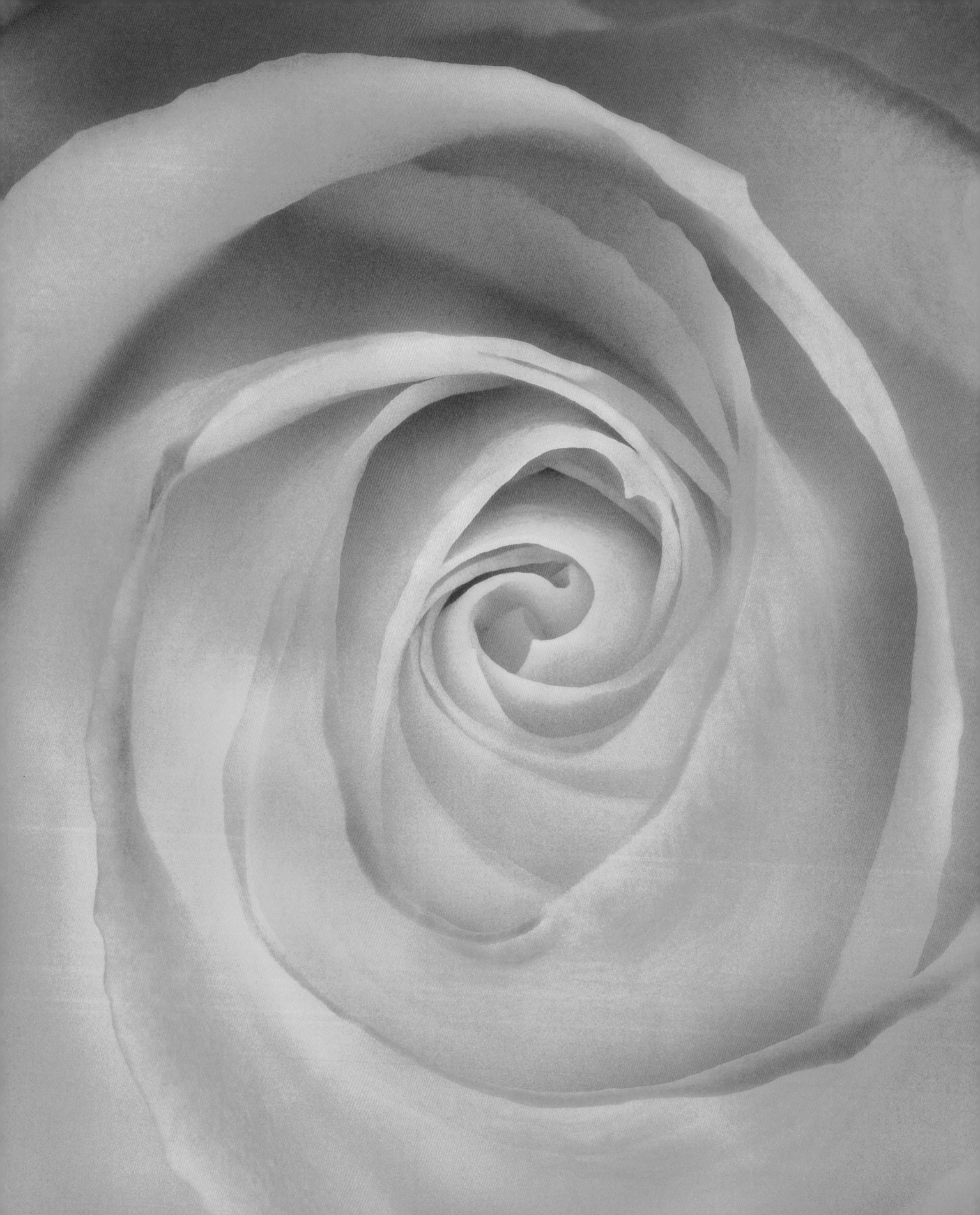

same flower. Another thing to look for in this family is the epicalyx, green parts below and alternating with the sepals—an extra layer of support or protection for flower and fruit, if you will. Other distinguishing characteristics of this family, like details of the carpels and where the ovaries are positioned, are rather technical. You'll be certain enough with this information.

Some members are not as readily identified as such. Examples include lady's mantle, *Alchemilla*, with its

Many modern roses, like this gorgeous hybrid tea, are dense with petals. Warm weather and passing days coax a blossom into unfurling, but if there are many petals, the pistils and stamens tucked within may never be fully exposed.

velvety scalloped leaves and wands of chartreuse flowers, and goatsbeard, *Aruncus*, an herbaceous plant of shrublike proportions with handsome creamy white flower plumes. Lady's mantle flowers lack true petals, and those of goatsbeard are actually dioecious (slightly different-looking male and female flowers on separate plants). Yet, sure enough, very close inspection still reveals that characteristic five-part form. If you grow either of these in your yard along with some roses, you might try combining their flowers in a vase. The florist principle of contrasting larger blooms with sprays of smaller ones always looks lovely, plus now you know that such a bouquet keeps it all in the family.

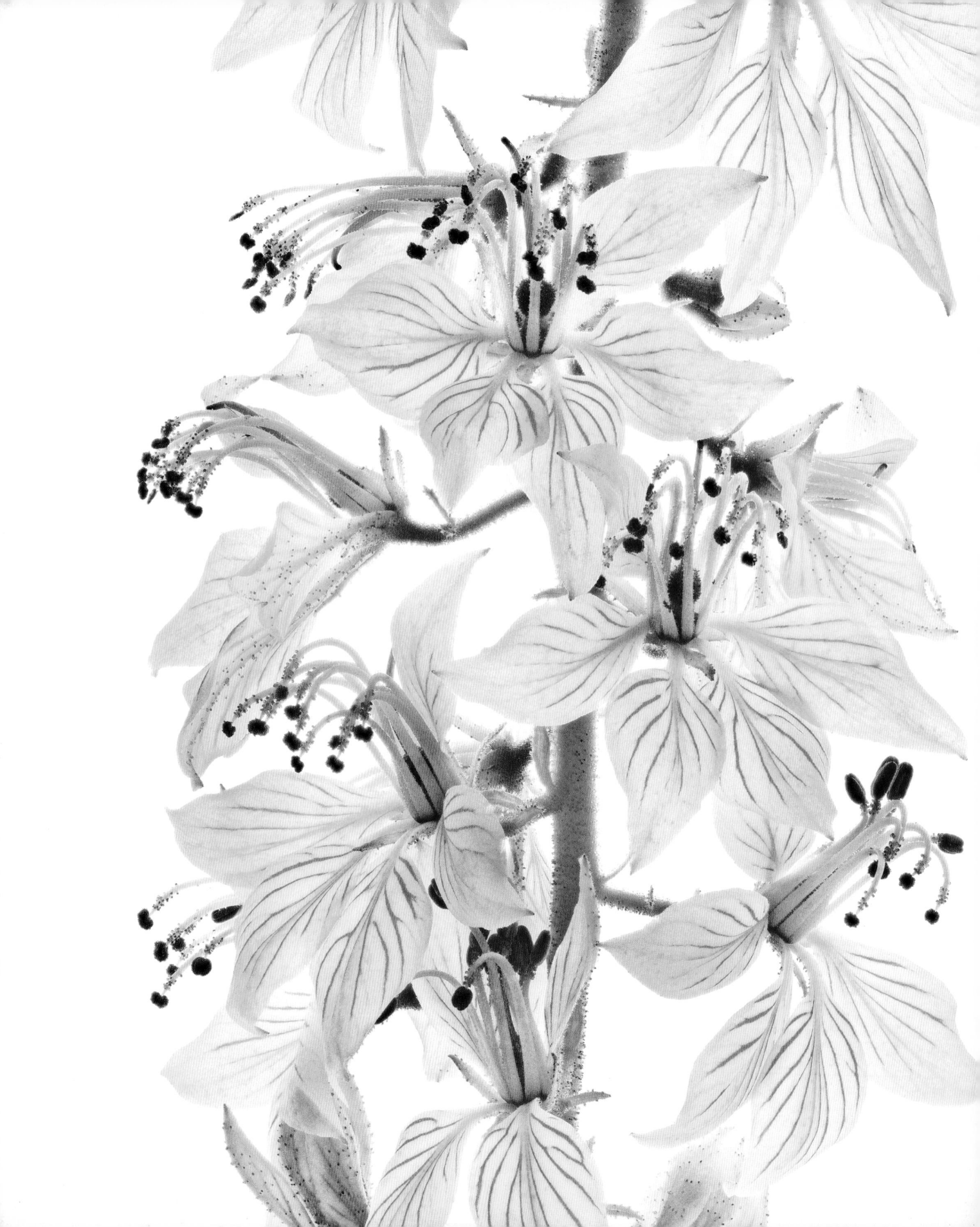

THE RUE FAMILY

Rutaceae

"We opened two halves of a miracle,
 Congealed acid trickled from the hemispheres of a star,
 The most intense liquor of nature,
 Unique, vivid, concentrated,
 Born of the cool, fresh lemon,
 Of its fragrant house,
 Its acid, secret symmetry. . ."

PABLO NERUDA, "Ode to a Lemon," 1957

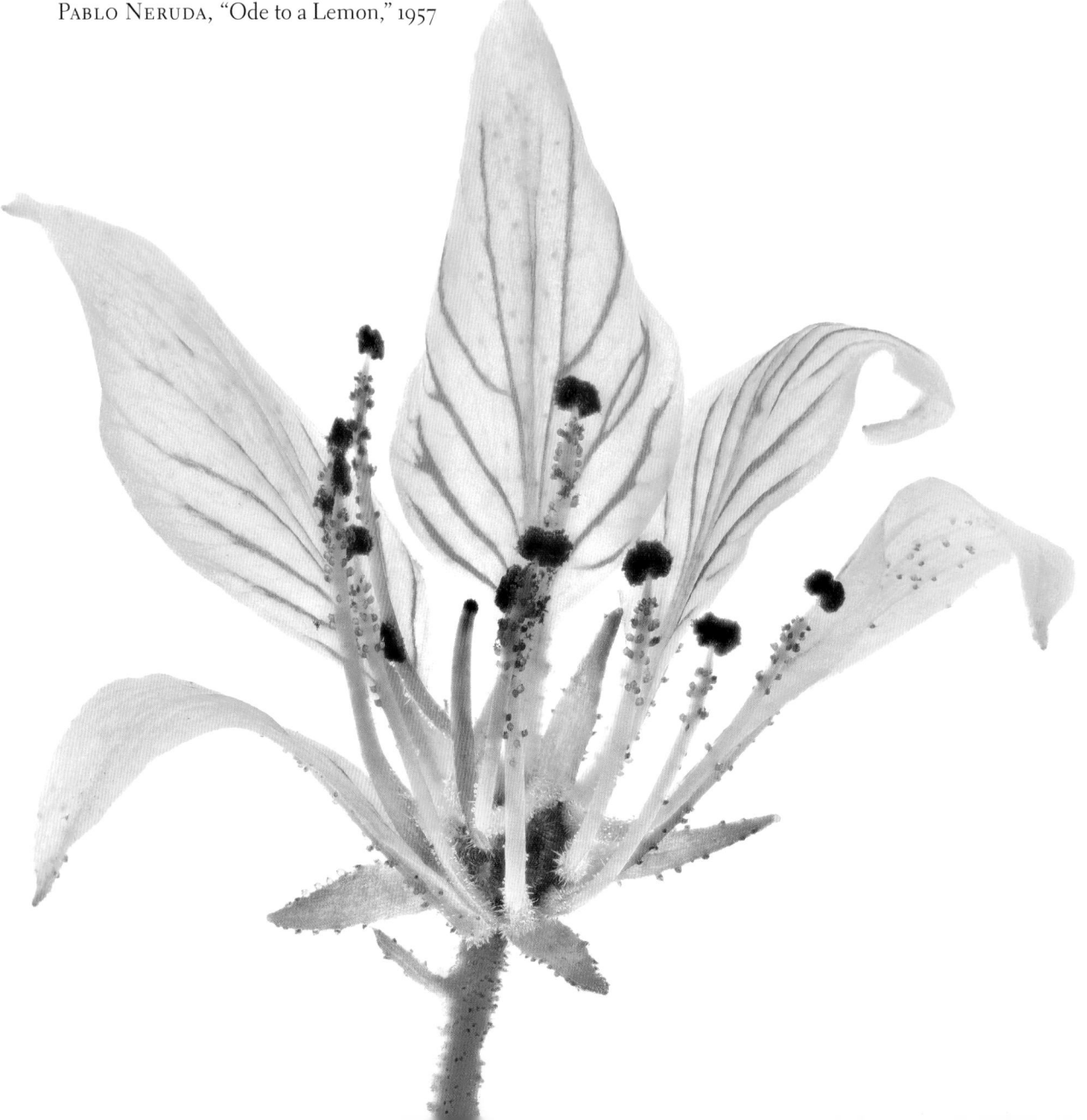

Among the many strange and wondrous plants in the world, members of the family Rutaceae are truly curious. You might know rue, *Ruta graveolens*, the ferny, bluish-leaved herb that provides attractive color in an herb or a flower garden. Or you might have avoided it, having heard—correctly—that handling it can cause the hands and other exposed skin to break out in unpleasant, uncomfortable blotches or water blisters after exposure to sunlight. Evidently, volatile oils in glands found all over the plant undergo a chemical reaction when exposed to light. These oils also account for rue's strong, acrid scent.

One of its close relatives, *Dictamnus albus*, is odder still. It too has aromatic foliage that contains volatile oils; all plant parts are considered poisonous. Among its common names are gas plant and burning bush—clues to an odd quality. The short-lived but showy flower racemes, which last for only about two weeks, give way to spikes of small, star-shaped seedpods, which actually have a pleasant, citrusy fragrance. Reportedly a lit

match held to the base of a one of these spikes causes a dramatic burst of flame as the oil ignites, even as the rest of the plant is spared. Or you can squeeze a pod and quickly light a match for a quick little burst of flame. Alternatively, bring the pods inside and leave them on a table or other open area to dry; they will explode and shoot their seeds around the room, while the pod's lining will make crackling noises. All potentially fun parlor tricks for an adventurous gardener.

And thus we come to the most surprising inclusions in this family: the citrus fruits. Yes, oranges, grapefruits, lemons, limes, and more are members of Rutaceae. Their scent is much more pleasant than their eccentric relatives and, of course, they are not poisonous to ingest, nor particularly medicinal. Also incorporated are some familiar ornamental shrubs, elegant Japanese skimmia, *Skimmia japonica*, and the spiny-stemmed hardy orange, *Poncirus trifoliata* (*Citrus trifoliata*).

Given such diversity in foliage and growth habit, we look to the flowers to find something that unites all these plants. All have five, rarely four, petals. All are fragrant—some sweet, some not—and are almost always carried in groups or clusters, both qualities that entice pollinating bees and flies. Essential oil is present in all members, whether it is appealing, useful, or an irritant. Thus this quirky, interesting family makes its unique place in the plant world.

LEFT AND PAGE 280 The gas plant, *Dictamnus albus*, is something of a novelty because the oil within the small seedpods is supposedly flammable. The species has white flowers; shown here is ravishing var. *purpureus*, whose brightly veined petals and matching anthers argue for growing it for the blooms alone.

THE VIOLA FAMILY

Violaceae

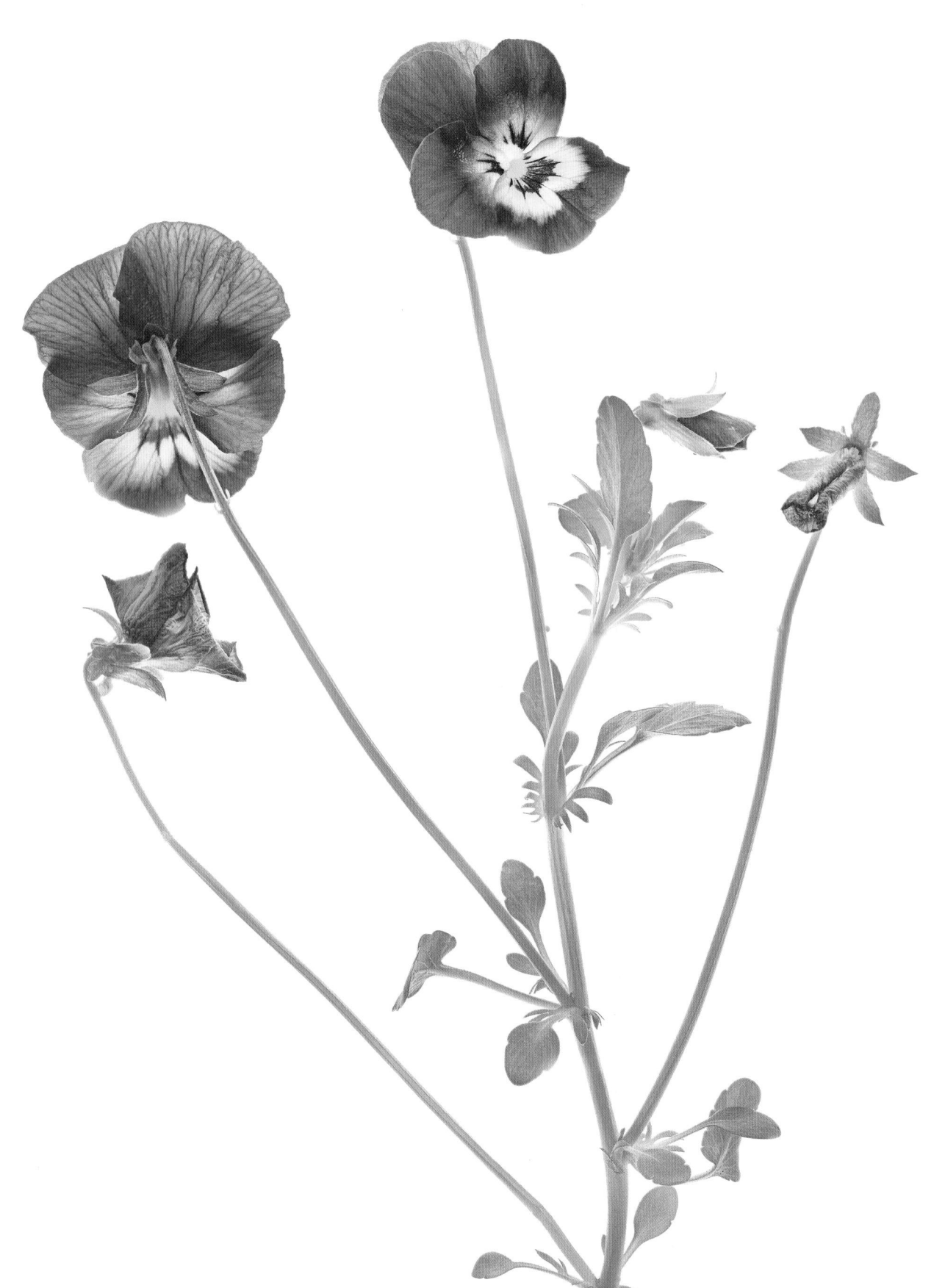

"O wind, where have you been,
That you blow so sweet?
Among the violets
Which blossom at your feet.

The honeysuckle waits
For Summer and for heat
But violets in the chilly Spring
Make the turf so sweet."

CHRISTINA ROSSETTI, *Sing-Song, a Nursery Rhyme Book*, 1872

Pretty as can be, but sly like a fox—that's violets. In cultivation since ancient times, members of the violet family, Violaceae, are found in temperate climates worldwide. They are among the early bloomers of spring in cool woodlands and gardens, where they spread perky color and often waft a sweet fragrance. But don't let charm and small flower size deceive you. If they find the growing conditions they prefer—decent soil, sufficient moisture, enough sunshine—violets spread rampantly, almost scandalously. They have a few tricks up their sleeves, so to speak.

Violet blossoms harbor some complex, even weird, qualities. For starters, there are actually two kinds of violet flowers, spring and summer. You know the spring ones, of course, for they are much showier. Borne in the leaf axils, each has five petals: a spurred lower petal, two lateral petals, and two upward-facing upper petals.

BELOW AND PAGES 284, 286, AND 290 Violets, violas, pansies—they all are closely related, look alike, and charm with their perky personalities, bright colors, and (in many) sweet scents. Petal markings are meant to guide pollinators to the interior store of pollen; in some, they are whimsically but accurately called whiskers, as if the flower were a little face.

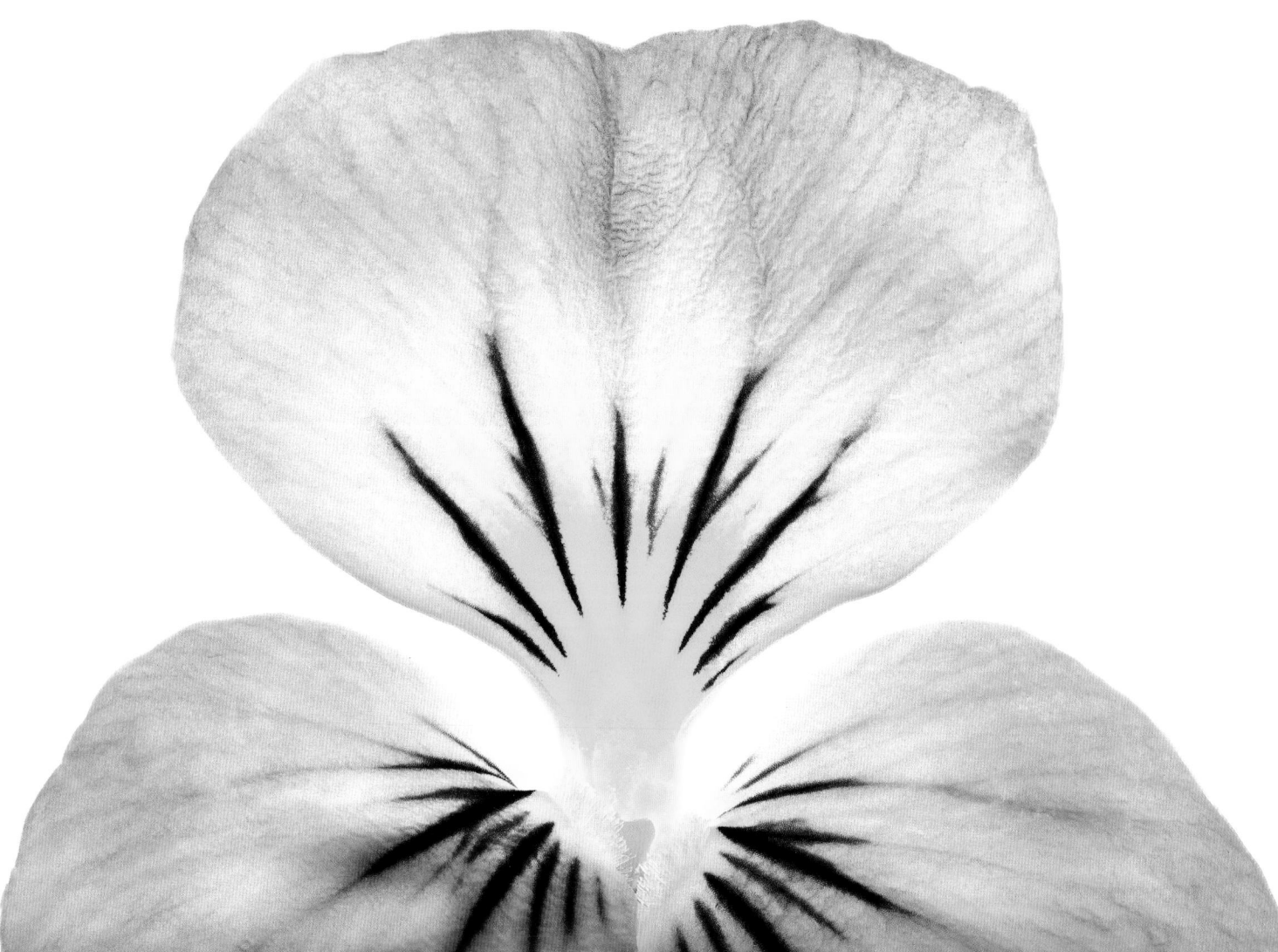

This configuration makes it easy for pollinating insects—mainly bees and butterflies—to land and dine on nectar. Contrasting color in the center or throat of the flower, as well as lines (whiskers) on the petals, help direct traffic to the sweet spot.

If you watch very closely when bees stop by, it's fun to witness their technique. They tend to land on the bottom petal, flip completely over, grasp the top petal, and only then stick their head into the center of the flower.

Peer under the bottom of the lowest petal and you'll spy a tiny sac where the nectar is stored. Curiously, the five green sepals are attached not at their base but in their middle; they project forward around the petals and backward around the stem's base. The backward-facing feature has a practical purpose: It shields that little nectar spur or sac, thwarting any insects that might try to access it from behind and therefore fail to help pollinate the flower.

But wait, there's more. Once a flower is successfully pollinated, its stalk bends down, bringing the ripening fruit closer to the ground. It takes time, but over a period of weeks the fruit elongates into a capsule with three parts, turning light brown as it slowly dries out. Eventually all three parts open and, looking in, you'll see tiny green seeds. They do not just fall to the ground when ripe, though. Instead, as their valve of the capsule continues to dry out, it contracts, compressing the seeds and ultimately shooting them out in all directions. Violet seeds can land several feet away.

Summer flowers, on the other hand, develop weeks after those have faded away. These stay undercover, underneath the leaves that remain, close to the ground.

They never actually unfurl or try to entice pollinators. Instead, they self-fertilize and then develop the shooting capsules described above, further propagating even more violets. If you have just a few violet plants, beware: It's only a matter of time before you have them in abundance.

Within this family are many species violets, which can be hard to tell apart. Most have the signature heart-shaped leaves and are either perennials or tender perennials grown as annuals. The common blue violet (its spring flowers are more accurately purple-blue) is *Viola papilionacea*. The adorable little Johnny-jump-up is *V. tricolor*, which is appropriate given that its flowers are purple, yellow, and white. Bird's foot violet, *V. pedata*, has light purple-blue blossoms and gets its name from its deeply segmented leaves. The sweet violet, purple-flowered *V. odorata*, is the one most often used for perfumes, syrups, and extracts. The common pansy is a hybrid that goes by the overall name of *V. ×wittrockiana*. Pansies have larger flowers (up to 4 inches across) and have enjoyed plentiful tinkering and selection, so gardeners can choose from everything from solid colors (pure orange 'Padparadja' is just gorgeous) to contrasting or freckled bicolors and mixes. The tribe also encompasses some shrubs and even trees, mainly found in the tropics and rarely seen in cultivation.

So don't be fooled by the delicate, diminutive look of violets. Be prepared to have more each year, or to spend time plucking out unwanted offspring. Consider it an opportunity to come eye to eye with their intriguing flowers and seed capsules.

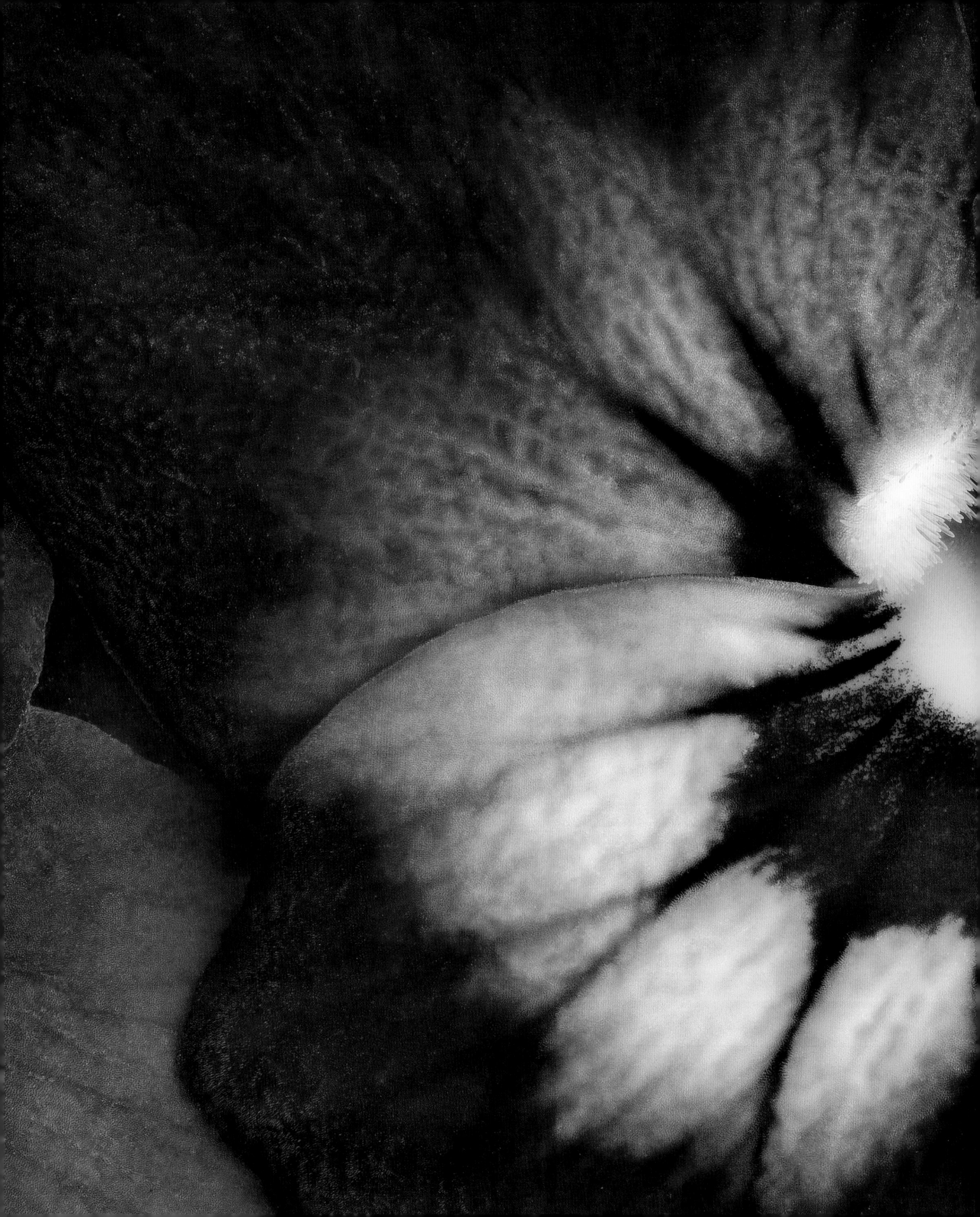

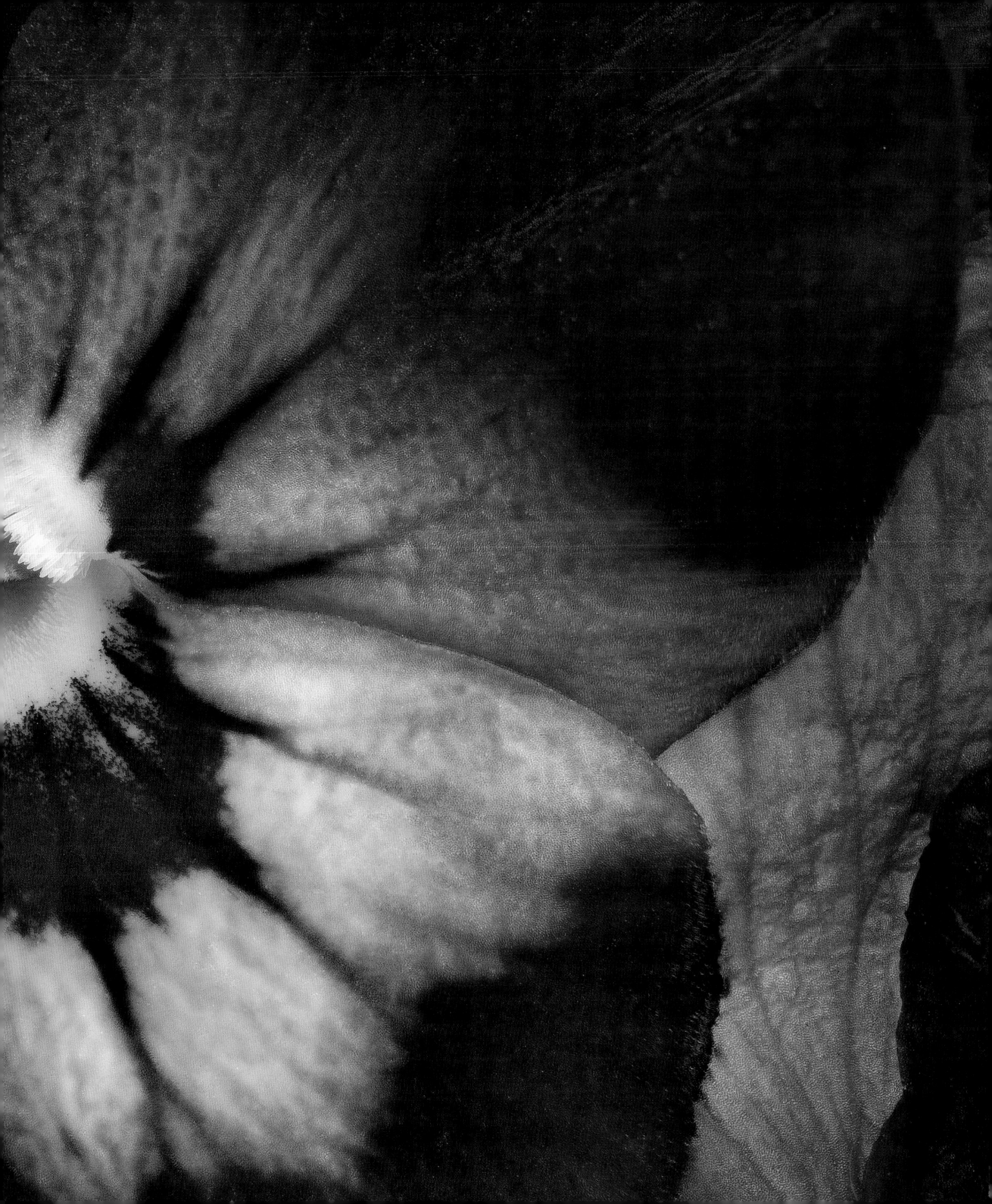

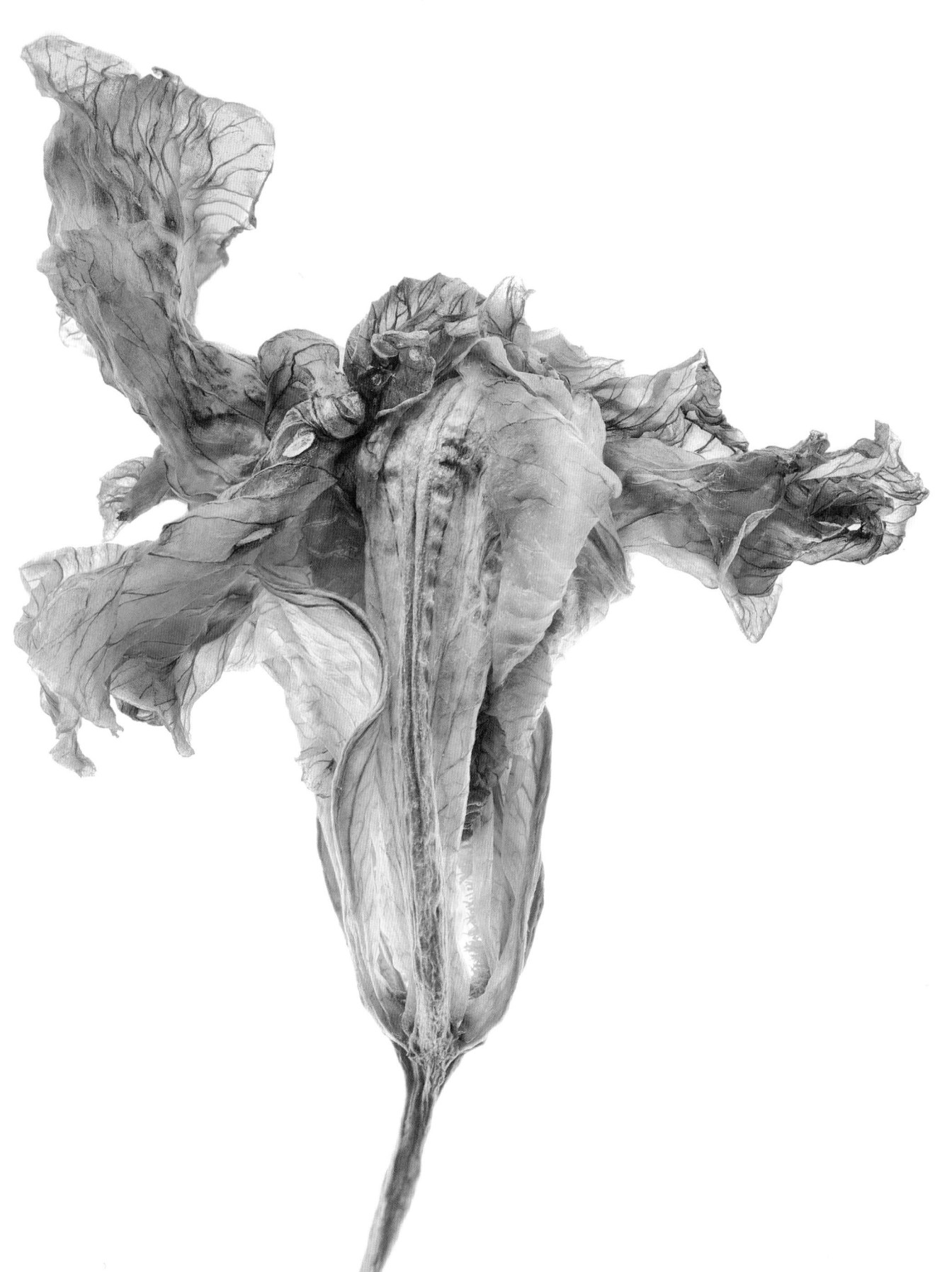

While I was composing the captions for Bob's magnificent photos, the last part of putting this book together, I thought I'd come a long way in seeing flowers. Flower form intricacies and pollination's intimate secrets were clearer to me. All the blooms in this book had gained new piquancy and, in my view, glory—a journey I hope you have been able to take with us.

When I got to the morning glory family, I understood that the funnel-form flowers are technically known as corollas because of several fused petals. I wondered exactly how many fused petals. A quick search of my scientific references was inconclusive; this particular detail did not seem to be of interest to the botanists. Peering at Bob's photographs, I decided the correct answer was five.

Then, it being a fine, blue-sky autumn day, I took a break and went outside. I find I now look at my flower garden differently, more searchingly. Who is pollinating the caryopteris? Which family includes nasturtium? Draped over the fence were my morning-glory vines, bejeweled with that day's flowers. I walked over, plucked a single blossom, and brought it back indoors. Yes: five parts to that corolla, clearly defined with contrasting color.

What astonished me was what that lone morning-glory blossom did later, indoors. Cut off from the sustenance provided by the vine, it slowly faded. That

Some flowers don't just fade and crumple, but rather contort themselves into artful—to our eyes, anyway—shape and go out in style. Like this iris.

evening, I expected to encounter a crumpled mess—the texture is more frail than tissue paper. Instead, I found it had rolled up very neatly and evenly, like a petite striped umbrella. And at the very tip, there was a tidy, elaborate twist, like the finishing touch on a fancy job of gift-wrapping. It completely amazed me that the flower had the energy, wherewithal, and instinct to finish its life in this fashion.

Witnessing this behavior also humbled me. I remember being a touch skeptical when I first read the lofty words of naturalist John Burroughs:

> To the scientist Nature is a storehouse of facts, laws, processes, [but] to the artist she is a storehouse of pictures; to the poet she is a storehouse of images, fancies, a source of inspiration; to the moralist she is a storehouse of precepts and parables.

And then this fleeting little flower, in its final hours, made me stop and consider. Do other flowers do something similar? And, will I, in my own final moments one day, honor and retain the form I was born to and cultivated all my life, and do my best to finish in beauty? Burroughs may not have been reading too much into all this, after all. To the botanist, gardener, or anyone who enjoys plants, nature is intricate, marvelous, inspirational. To the photographer—if I can dare to speak for Bob—nature is intricate, marvelous, inspirational. May anyone who sees flowers come and go in wonder and delight.

BIBLIOGRAPHY

Brickell, Christopher, and H. Marc Cathey, eds. 2004. *The American Horticultural Society A–Z Encyclopedia of Garden Plants*. Rev. U.S. ed. New York: DK Publishing.

Clausen, Ruth Rogers. 1989. *Perennials for American Gardens*. New York: Random House.

Cole, Rebecca. 2002. *Flower Power: Fresh, Fabulous Arrangements*. New York: Clarkson-Potter.

Cruso, Thalassa. 1975. *Making Things Grow, A Practical Guide for the Indoor Gardener*. New York: Alfred A. Knopf.

Dirr, Michael A. 2009. *Manual of Woody Landscape Plants*. 6th ed. Champaign, Illinois: Stipes Publishing Company.

Dunn, Teri. 2003. *Beautiful Roses Made Easy, Mid-Atlantic and New England Edition*. Nashville, Tennessee: Cool Springs Press.

Durant, Mary, 1976. *Who Named the Daisy? Who Named the Rose? A Roving Dictionary of North American Wildflowers*. New York: Congdon & Weed.

Eiseley, Loren. 1959. *The Immense Journey*. New York: Vintage Books.

Ellis, Barbara W. 1999. *Taylor's Guide to Annuals*. Boston: Houghton-Mifflin Company.

Ellis, Barbara W. 2001. *Taylor's Guide to Bulbs*. Boston: Houghton-Mifflin Company.

Glattstein, Judy. 1998. *Flowering Bulbs for Dummies*. New York: IDG Books Worldwide, Inc.

Halpin, Anne. 1990. *The Naming of Flowers*. New York: Harper & Row Publishers.

Hugo, Nancy Ross. 2011. *Seeing Trees*. Portland, Oregon: Timber Press.

Kowalchik, Claire, and William H. Hylton. 1998. *Rodale's Illustrated Encyclopedia of Herbs*. Emmaus, Pennsylvania: Rodale Press, Inc.

Mitchell, Henry. 2003. *The Essential Earthman: Henry Mitchell on Gardening*. Bloomington, Indiana: Indiana University Press.

Norris, Kelly D. 2012. *A Guide to Bearded Irises: Cultivating the Rainbow for Beginners and Enthusiasts*. Portland, Oregon: Timber Press, Inc.

Pollan, Michael. "Opium, Made Easy," *Harper's*, 1 April 1997.

Roberts, June Carver. 1976. *Born in the Spring: A Collection of Spring Wildflowers*. Athens, Ohio: Ohio University Press.

Seguin-Fortes, Marthe. 2001. *The Language of Flowers*. New York: Sterling Publishing Co., Inc.

Sheehan, Thomas J., and Robert J. Black. 2007. *Orchids to Know and Grow*. Gainesville, Florida: University Press of Florida.

Spellenberg, Richard. 2001. *The Audubon Society Field Guide to North American Wildflowers—Western Region*. Rev. ed. New York: Alfred A. Knopf.

Stuckey, Maggie. 1999. *The Houseplant Encyclopedia*. New York: Guild America Books.

Thieret, John W., and William A. Niering. 2001. *The Audubon Society Field Guide to North American Wildflowers—Eastern Region*. Rev. ed. New York: Alfred A. Knopf.

Thomas, Graham Stuart. 2004. *Perennial Garden Plants: Or the Modern Florilegium*. Portland, Oregon: Sagapress, Inc./Timber Press, Inc.

As a member of the Buttercup family (Ranunculaceae), the Christmas rose, *Helleborus orientalis*, has the requisite five-parted form, including five colorful petals that are actually sepals.

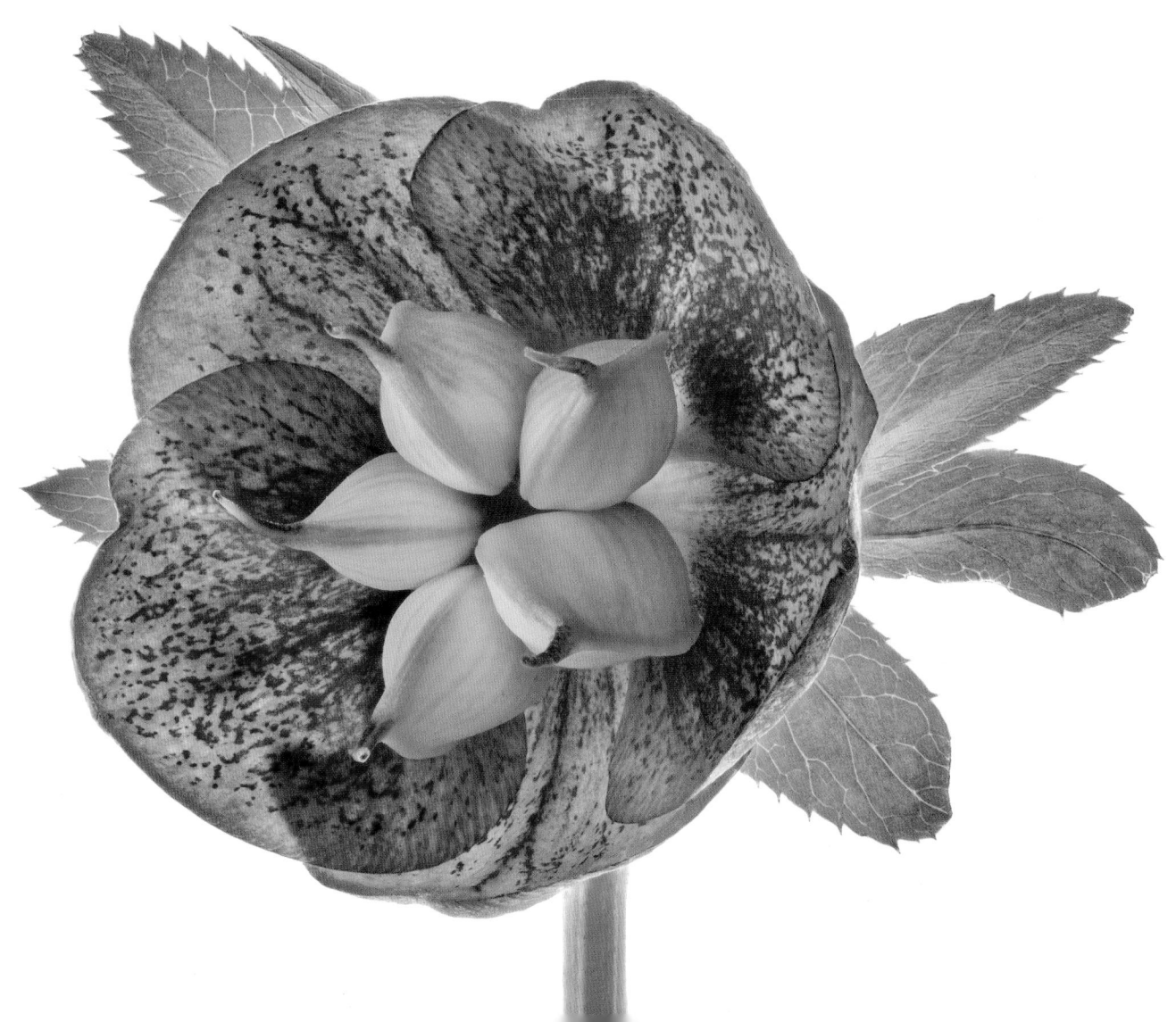

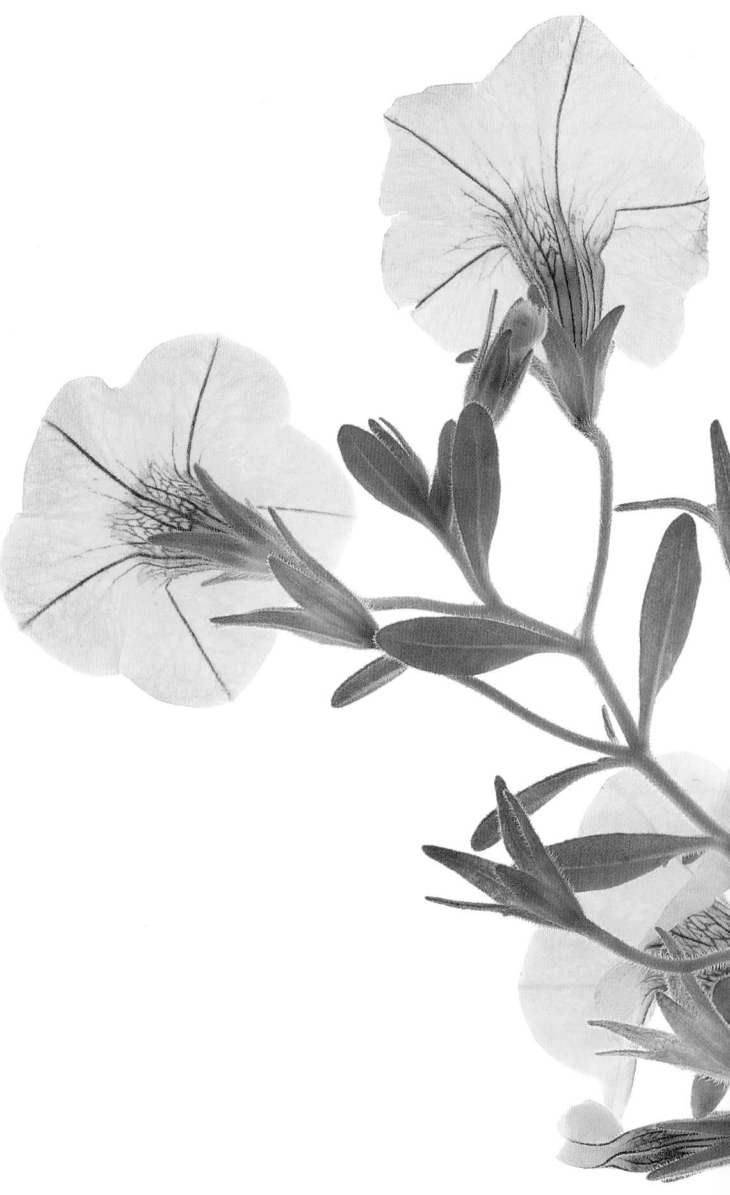

Calibrachoa 'Saffron' looks uncannily like a miniature petunia. They are indeed closely related; both are members of the nightshade family.

PAGE 304 Million Bells, *Calibrachoa*, like all members of the nightshade family, has flowers in parts of five. Their five petals are fused into a bell or trumpet shape called a corolla. In the center, five stamens surround a single pistil.

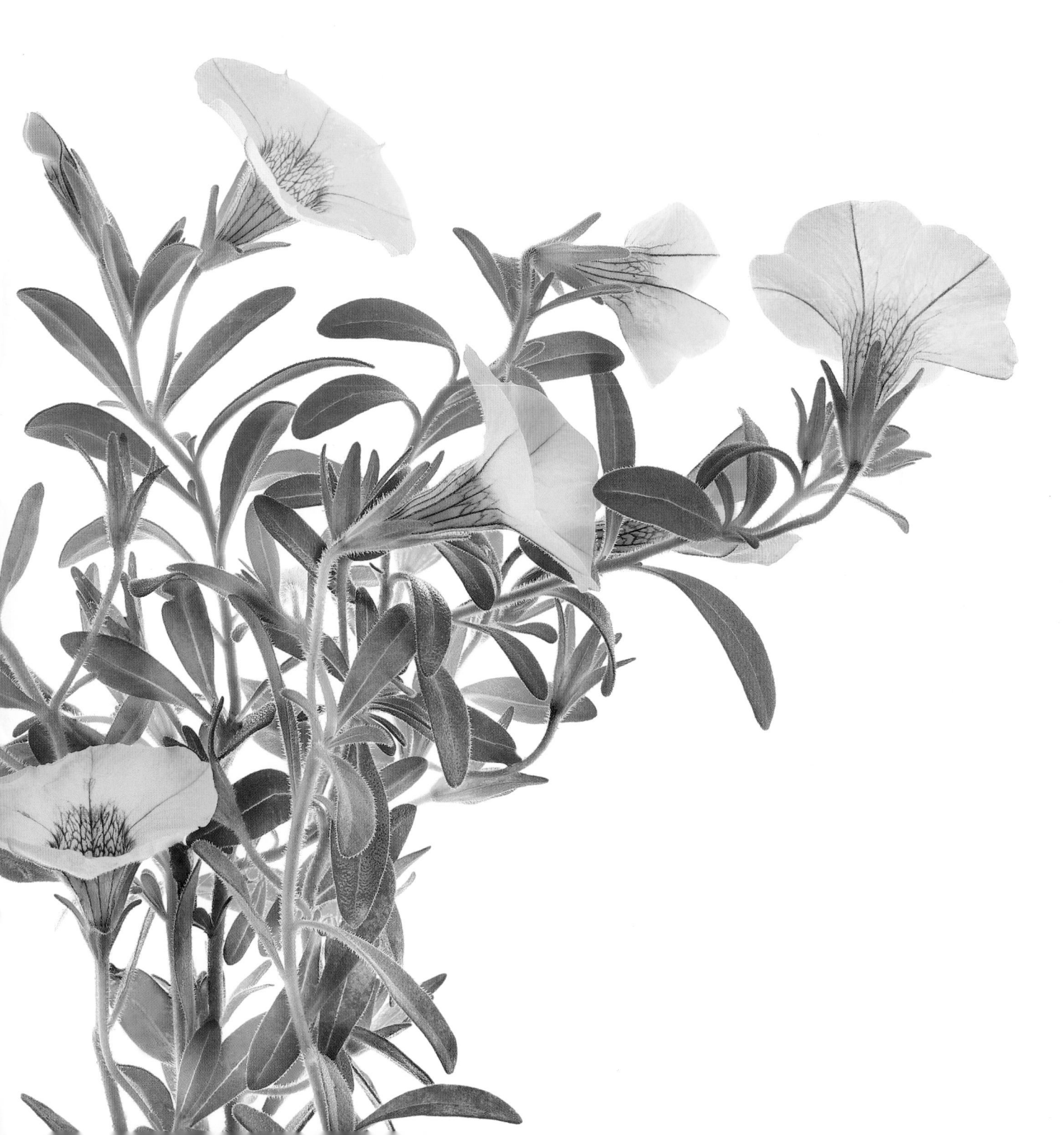

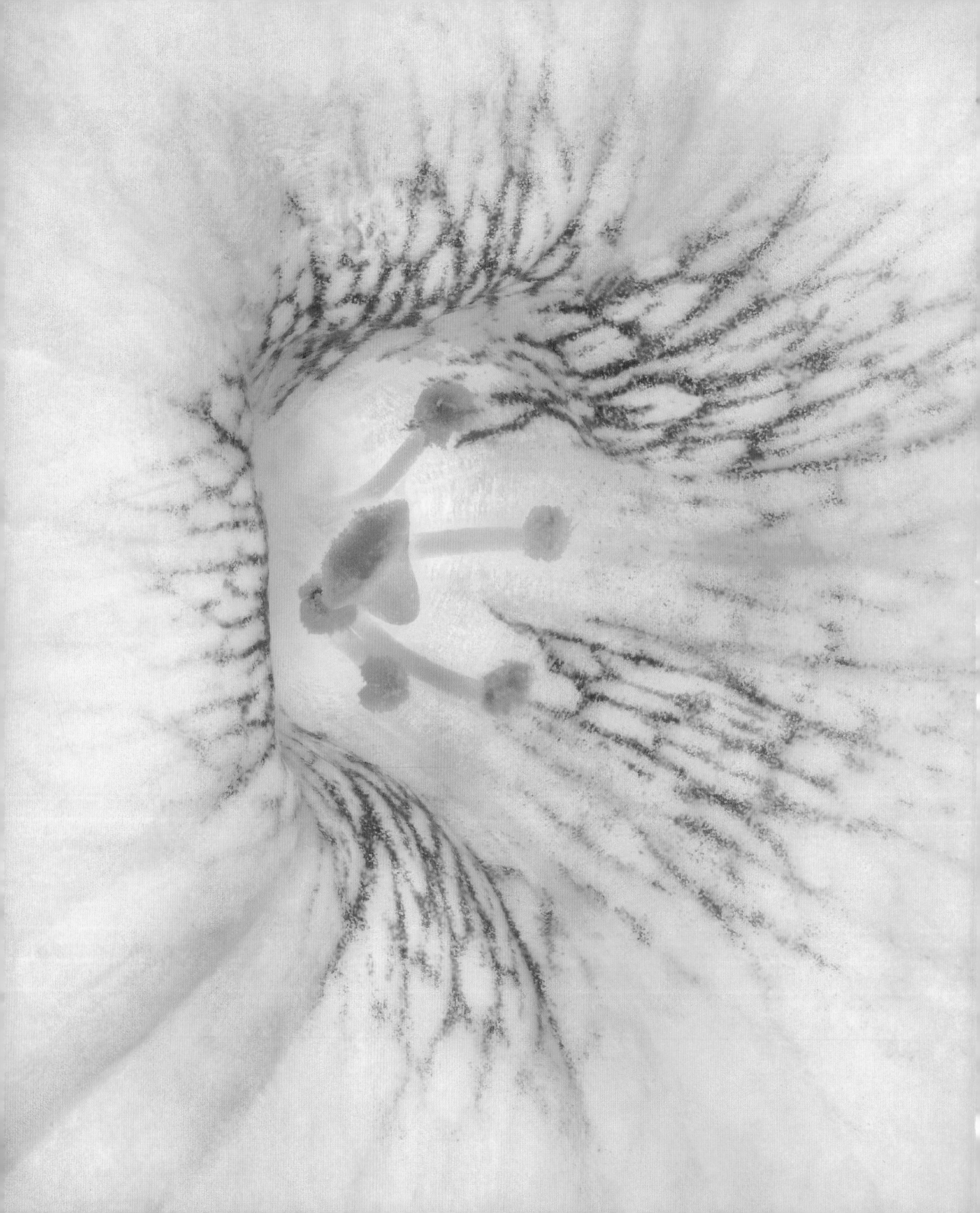